VISUALIZATION
OF
RECEPTORS

Methods in Light and Electron Microscopy

Edited by

Gérard Morel

CRC Press
Boca Raton   New York

Publisher:	Robert B. Stern
Editorial Assistant:	Carol Messing
Project Editor:	Debbie Didier
Marketing Manager:	Susie Carlisle
Direct Marketing Manager:	Becky McEldowney
Cover design:	Dawn Boyd
PrePress:	Kevin Luong

Library of Congress Cataloging-in-Publication Data

Visualization of receptors : methods in light and electron microscopy
/ edited by Gérard Morel.
    p. cm.
    Includes bibliographical references and index.
    ISBN 0-8493-2644-3
    1. Cell receptors—Structure.   2. Nuclear receptors
(Biochemistry)—Structure.   3. Microscopy.   4. Electron microscopy.
I. Morel, Gérard.
QH603.C43V57   1997
572—dc21

96-54597
CIP

THE EDITOR

Gérard Morel, Ph.D., Dr. es Sc., is a Research Director at the National Center of Scientific Research (CNRS), Institut Pasteur, Lyon, France.

Dr. Morel obtained his M.S. and Ph.D. degrees in 1973 and 1976, respectively, from the Department of Physiology of Claude Bernard University — Lyon. He was appointed as Assistant of Histology at the same university in 1974 and became Doctor es Science in 1980. He was appointed by CNRS in 1981 and became Research Director in 1989.

Dr. Morel is a member of the American Endocrine Society, International Society of Neuroendocrinology, Société Française des Microscopies, Société de Biologie Cellulaire de France, Société de Neuroendocrinologie Expérimentale de France, Society of Neuroscience, and Society of Cell Biology.

Dr. Morel has been the recipient of research grants from the European Community, INSERM (National Institute of Health and Medical Research), La Ligue Contre le Cancer, l'Association de Recherche Contre le Cancer (ARC), and private industry.

Dr. Morel is the author of about 150 papers, 15 chapters, and 4 books. His current major research interests include the internalization and cellular trafficking of ligand and receptor molecules (in particular, nuclear receptors for peptides), the regulation of gene expression, and paracrine interactions.

CONTRIBUTORS

Andrés Beiras
Department of Morphological Sciences
School of Medicine and Hospital General de Galicia
University of Santiago
Santiago de Compostela, Spain

Christine Brisson
Institut Pasteur de Lyon
Lyon, France

Annie Cavalier
Biologie Cellulaire et Reproduction
Université de Rennes I
Campus Beaulieu
Rennes, France

Valérie Coronas
Laboratoire de Physiologie Neurosensorielle
Université Claude Bernard
Villeurbanne, France

Michèle Crumeyrolle-Arias
Laboratory of Pharmacology, Neuro-immunology, and Endocrinology
Institut Pasteur de Paris
Paris, France

Didier Decimo
Unite de Virologie Humaine
Ecole Normale Superieure
Lyon, France

Máximo Fraga
Department of Pathology and Forensic Sciences
School of Medicine and Hospital General de Galicia
University of Santiago
Santiago de Compostela, Spain

Marie José Freund-Mercier
Laboratoire de Physiologie Générale
Université Louis Pasteur
Strasbourg, France

Rosalía Gallego
Department of Morphological Sciences
School of Medicine and Hospital General de Galicia
University of Santiago
Santiago de Compostela, Spain

Tomás García-Caballero
Department of Morphological Sciences
School of Medicine and Hospital General de Galicia
University of Santiago
Santiago de Compostela, Spain

Béatrice Grandclément
Pathologie des Fibroses
Institut Pasteur de Lyon
Lyon, France

Marie Jeanne Klein
Laboratoire de Physiologie Générale
Université Louis Pasteur
Strasbourg, France

Peter E. Lobie
Institute of Molecular and Cellular Biology and Defence Medical
  Research Institute
National University of Singapore
Singapore

Hugues Lortat-Jacob
Institut Pasteur de Lyon
Lyon, France

Pierre-Marie Martin
Laboratoire de Cancérologie Expérimentale
Faculté de Médecine Nord
Marseille, France

Hichem C. Mertani
Institut Pasteur de Lyon
Lyon, France

Gérard Morel
Institut Pasteur de Lyon
Lyon, France

Emmanuel Moyse
Laboratoire de Physiologie Neurosensorielle
Université Claude Bernard
Villeurbanne, France

Allal Ouhtit
Unit of Multistage Carcinogenesis
International Agency for Research on Cancer
World Health Organization
Lyon, France

Pierre Poulain
INSERM
Lille, France

Sylvie Ricard-Blum
Institut Pasteur de Lyon
Lyon, France

Brice Ronsin
Institut Pasteur de Lyon
Lyon, France

Daniel Seigneurin
Laboratoire de Cytologie
CHU
Grenoble, France

Masahisa Shimada
Department of Anatomy
Osaka Medical College
Osaka, Japan

Lalit K. Srivastava
Department of Psychiatry
Douglas Hospital Reseach Center
McGill University
Montreal, Quebec, Canada

Marie-Elisabeth Stoeckel
Laboratoire de Physiologie Générale
Université Louis Pasteur
Strasbourg, France

France Wallet
Laboratoire de Cancérologie Expérimentale
Faculté de Médecine Nord
Marseille, France

Mark R. Walter
Department of Pharmacology and Center for Macromolecular
 Crystallography
University of Alabama at Birmingham
Birmingham, Alabama
U.S.A.

Elisabeth Waltisperger
Laboratoire de Physiologie Générale
Université Louis Pasteur
Strasbourg, France

Maryvonne Warembourg
INSERM
Lille, France

Masahito Watanabe
Department of Anatomy
Osaka Medical College
Osaka, Japan

Graham K. Wood
Departments of Neurology and Neurosurgery
Douglas Hospital Research Center
McGill University
Montreal, Quebec, Canada

CONTENTS

DEDICATION

To my wife, Françoise
To my boys, Marc-Franck and Guerric
To Christelle

VISUALIZATION
OF
RECEPTORS

Methods in Light and Electron Microscopy

Chapter 1

BIOCHEMICAL DIVERSITY OF RECEPTORS

Lalit K. Srivastava
Graham K. Wood

CONTENTS

0-8493-2644-3/97/$0.00+$.50
© 1997 by CRC Press LLC

LIST OF ABBREVIATIONS

AA	arachidonic acid
AMPA	α-amino-3-hydroxy-5-methyl-4-isoxazole propionic acid
CAM	cell adhesion molecules
cAMP	cyclic adenosine 5′-monophosphate
CAR	cell adhesion receptors
CEA	carcinoembryonic antigen
CNTF	ciliary neurotrophic factor
CRE	cAMP response element
CREB	cAMP response element binding protein
DAG	diacylglycerol
DCC	delete in colon carcinoma
ECM	extracellular matrix
EFG	epidermal growth factor
FGFR	fibroblast growth factor receptor
FNIII	fibronectin type III
GABA	γ-amino-butyric acid
GDP	guanosine 5′-diphosphate
GM-CSF	granulocyte macrophage colony stimulating factor
GR	glucocorticoid receptor
GRE	glucocorticoid response element
GTP	guanosine 5′-triphosphate
HSP	heat shock protein
5-HT	5-hydroxytryptophane
ICAM	intracellular adhesion molecule
IFN	interferon
Ig	immunoglobulin
IGF	insulin-like growth factor
IL	interleukin
IP_3	inositol triphosphate
JAK	janus kinase
LAMP	limbic system membrane protein
LeuCAM	leukocyte cell adhesion molecules
LIF	leukemia inhibitory factor
LIF-R	LIF receptor
LTP	long-term potentiation
MAdCAM	mucosal addressin cell adhesion molecules
MAG	myelin-associated glycoprotein
NCAM	neural cell adhesion molecules
NMDA	N-methyl-D-aspartate
OBCAM	opioid binding cell adhesion molecules
PDGF	platelet-derived growth factor
PKC	protein kinase C
POMC	pro-opiomelanocortin
RE	response element

SH2	src homology 2
TGFα	tumor growth factor α
VCAM	vascular cell adhesion molecule

I. INTRODUCTION

Receptors play a key role in communication between cells. Secreted ligands can act on receptors to communicate information over long distances, such as hormones which are released by localized endocrine cells and act throughout the body.[1] Receptors can also recognize ligands secreted within a restricted environment, as in the case of prostaglandins, and allow communication by secreted ligands between juxtaposed cells or on the cell itself, such as in the case of neurotransmitters released at synapses. Finally, cell surface proteins involved in cell-cell interaction act as receptor-counter-receptor and translate adhesion signals into altered behaviors of cells. In the chain of events of cellular communication, receptors convert extracellular signals, which may be in the form of secreted ligands or membrane-bound proteins, into intracellular signals. This seemingly simple step initiates complex intracellular reactions that differentially respond to and are selective to given ligands. Secreted ligands can act only on selective subsets of cells that express the complementary receptor. A hormone released into the bloodstream can potentially affect every cell in the organism, but localized cellular expression of receptors will limit its action to certain areas. It is not uncommon to have multiple receptors for the same ligand. Thus, different receptor subtypes for a certain ligand allow a diversity of responses from cells expressing the receptor subtypes. For example, a ligand can cause the opening of an ion channel through one subtype of receptor and cause changes in the cellular metabolism through another subtype. Thus, cell-cell communication through receptors allows a wide range of coordinated processes to occur throughout the organism.

Binding of the ligand to its receptor initiates a cascade of events that can be as simple as causing a flow of ions or as complex as modulating transcription of specific genes. Initiation of the cascade generally occurs by a conformational change in the receptor caused by ligand binding, which can directly initiate the cascade or cause it by activating an effector molecule. The cascade results in physiological responses that work to regulate activities such as growth and development, and to maintain homeostasis.

Visualization of receptors, at the level of the cells and tissues is an important step toward elucidating their function. A knowledge of the anatomical localization of receptors is necessary to determine which system they function within. Ultrastructural localization offers a glimpse at the fate of the receptor or the receptor-ligand complex within cells. The culmination of the study of receptor localization can offer novel methods of intervention for treating clinical disorders.

A. STRUCTURAL AND FUNCTIONAL DIVERSITY

There is a wide array of structural diversity in secreted ligands, ranging from amino acids or amino acid derivatives to glycoproteins. However, receptors can be grouped into six major types: tyrosine kinase receptors, cytokine receptors, G-protein coupled receptors, ionotropic receptors, DNA binding receptors, and adhesion receptors. The G-protein coupled receptor superfamily is notable in that its members can recognize the full structural array of ligands and each receptor converts them into a signal common to the superfamily (activated G-protein).[2] The remaining receptors recognize a restricted variety of ligands; however, again each superfamily uses a common mechanism of signal transduction. There are additional receptors that do not fall into these six families. For the sake of offering a better explanation of receptors that do fall into the six superfamilies, other receptors will not be reviewed. However, the principles that will be presented for the six receptor superfamilies can offer a guideline to receptors not specifically mentioned that fall within the superfamilies and to the receptors that can not be included in any of the known superfamilies.

The receptor superfamilies are separated by structure and function. Due to the difficulties of crystallizing proteins, the bulk of structural information concerning receptors is derived from pharmacological, biochemical, and genetic approaches.[2] Within each superfamily there are structural differences, but there are also certain functional domains that share common structures (homologous domains). Functional differences between superfamilies reside in how the receptors initiate the signal transduction cascade. Once the receptors have initiated the signal cascade, there are overlaps in the resulting mechanisms between superfamilies. There are similarities in protein phosphorylation, ion channel opening, and selective activation of gene expression downstream of receptor initiation of the signal cascade. Therefore, when comparing superfamilies it must be kept in mind that despite wide ranges of structure within families, and functional similarities between families, the defining feature of each superfamily is that the functional initiation of signal transduction and the corresponding structure are generally similar.

B. ENDOGENOUS AND EXOGENOUS LIGANDS

As previously mentioned, endogenous ligands (ligands that are physiologically active in cells) are diverse in structure and size. The receptor recognition of those ligands is remarkably selective. At physiological concentrations, endogenous ligands bind only to their complementary receptor(s). Exogenous ligands (ligands synthesized chemically that mimic or antagonize endogenous ligand activity), however, are not as selective. In several of the following sections an attempt will be made to list the classes of the receptors and their endogenous and exogenous ligands. However, due to the ever-expanding number of ligands these lists are not exhaustive.

II. G-PROTEIN COUPLED RECEPTORS

The G-protein coupled receptor superfamily is the largest and most diverse family of receptors (for reviews see References 2 and 3). Since the cloning of the receptor for rhodopsin there have been approximately 300 or more G-protein receptors cloned,[2] and the total number is expected to exceed 1000.[3] The importance of the family is exemplified by the fact that nearly 60% of clinically relevant drugs act on G-protein coupled receptors.[3]

G-protein coupled receptors recognize a wide array of ligands and they all initiate transduction by activating G-proteins. The physiological effects of G-protein coupled receptors are varied. Examples include serotonin receptor subtypes which, apart from 5-HT$_3$, are coupled to G-proteins. Serotonin receptors are expressed by neurons throughout the brain in regions such as the cerebral cortex, nucleus accumbens, cerebellum, amygdala, hippocampus, hypothalamus, and thalamus, which are each innervated by serotonin releasing neurons of the raphe nucleus. The serotonin system has been implicated in depression since agents that decrease the uptake of serotonin are an effective treatment of the disorder. Another example is offered by dopamine receptors, of which there are five subtypes that are all coupled to G-proteins. They are expressed by neurons of the neocortex, limbic forebrain, striatum, and pituitary stalk. The importance of dopamine in delusion, hallucinations, and locomotion can be demonstrated by the effectiveness of antipsychotics, which are dopamine receptor blockers, in treating schizophrenia. Also, the involvement of dopamine in locomotion is demonstrated by Parkinson's disease, a disorder that is characterized by locomotor blunting, which involves disrupted dopaminergic transmission to the striatum. A final example is offered by opioid receptors, which differ in that they are expressed throughout the central nervous system, specifically in the periaqueductal gray region and the superficial dorsal horn of the spinal cord, and are not restricted to the brain. The physiological action of opioids is to modulate nociception of primary sensory neurons. These few examples of G-protein coupled receptors offer a glimpse of the varied range of ligands they bind and their physiological potential.

A. STRUCTURE AND FUNCTION

The G-protein coupled receptor superfamily has a basic structure that does not differ extensively among individual receptors. The importance of the structure is exemplified by its conservation from *E. coli* to humans. Each receptor contains seven hydrophobic domains separated by intervening hydrophilic domains.[3] The distribution and amino acid sequence of the hydrophobic regions, each of which contains about 20 to 25 amino acids, is consistent with α-helical domains that span the membrane. Accordingly, the intervening hydrophilic domains correspond to alternating extracellular and intracellular loops.[3] The N-terminal sequence is exposed to the extracellular space and the C-terminal is exposed intracellularly. The overall structure, as depicted in Figure 1, was derived from that of bacteriorhodopsin and has yet to be viewed with high-resolution techniques, so the details of the association of the domains are unknown.[4]

When the primary sequences of individual receptors are compared, homology is predominantly in the hydrophobic transmembrane sequences while the hydrophilic sequences of connecting loops tend to diverge.[3] The sequence divergence between receptors in the hydrophilic domains suggests their involvement in the molecular specificity for ligand and G-protein coupling. However, this is not necessarily the case for all ligands. As mentioned, G-protein coupled receptors bind a large range of ligands; in doing so there must be an equal range of structures between receptors that bind a biogenic amine compared to, for example, those that bind glycoproteins.

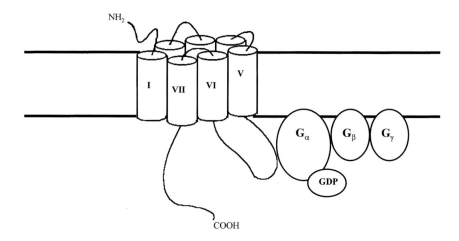

FIGURE 1 Overall structure of a G-protein coupled receptor in a lipid bilayer associated with the three components of a G-protein. I, V, VI, and VII are the sequential transmembrane domains from the N-terminal to the C-terminal.

Using site-directed mutagenesis and chimeric receptors, it has been demonstrated that biogenic amines bind within the transmembrane sequences of receptors. The majority of this work has been performed with the α- and β-adrenergic receptors. Replacing a single aspartate residue found in transmembrane domain 3, which is highly conserved among biogenic amine receptors, with asparagine results in reduced specific agonist binding in the mutant adrenergic receptors.[5] Also, a section of transmembrane region 7 was determined to be important for biogenic amine binding by constructing chimeric α_2/β_2-adrenergic receptors.[3] These results, and others, suggest a view of biogenic amine binding in which the ligand binds within a hydrophilic pocket consisting of transmembrane regions 3, 5, 6, and 7.[3]

G-protein coupled receptors recognizing peptides as ligands share some of the binding principles as the receptors recognizing biogenic amines, with variations to accommodate the larger size. Studies of the neurokinin 1 receptor, which binds the peptide neurokinin, demonstrated that the peptide binds the receptor within transmembrane regions 2 and 7. However, potentially due to the size of the peptide, it has also been shown to bind the first two extracellular loops.[2]

Protein ligand binding again differs from that of the smaller peptides. Studies on the glycoprotein hormone receptors have revealed that the ligand is bound almost entirely by the N-terminal extracellular domain. The N-terminal functions as a true domain, retaining high affinity for the ligand when expressed without the transmembrane domain. However, the transmembrane region is not without function in ligand binding. Glycoprotein hormones also interact with residues in the transmembrane domain. Due to each class of ligand associating with the transmembrane sequence despite size and structure differences, it has been postulated that regions within the transmembrane domain are responsible for receptor activation.[3] In the case of the larger ligands, the extracellular domain would then function to line up the ligand so that a smaller portion can interact with the transmembrane domain.

The activation of G-protein coupled receptors by ligand binding leads to the formation of a high-affinity ternary complex consisting of the ligand, receptor, and G-protein. The association of either ligand or G-protein is believed to initiate a conformational change in the receptor.[6] The ternary complex model states that the binding of the G-protein to the receptor feeds information back to the ligand binding domain of the receptor.[6] Therefore, G-proteins association can modify the affinity of G-protein coupled receptors for their complementary ligand. This modification is used in many experimental protocols to deplete endogenous ligands through reduction of the affinity of the ligand for the receptor by uncoupling the G-protein using guanine nucleotides.

The so-called switch in the receptor that leads to activation is thought to be a conserved sequence of asparagine-arginine-tyrosine in the second cytoplasmic loop.[3] The arginine residue is the most highly conserved of the three and has attracted the most attention. Site-directed mutation of the arginine to asparagine in β-adrenergic receptors results in the destruction of guanine nucleotide-sensitive agonist binding.[7] In muscarinic M_1 receptors the same mutation results in low-affinity agonist binding and eliminates agonist-induced enhancement of guanine nucleotides.[8] These results have led to the following model: when the arginine residue is near the binding pocket of the receptor it is assumed to be in the off position; it is in the on position when it moves to a position near the cytosolic portion of the receptor where it can interact with associated G-proteins.[3] Movement of the arginine residue is thought to be caused by rearrangement of the transmembrane sequences.[3] The conserved arginine residue is required for G-protein activation; however, a G-protein binding capability is also required for G-protein receptors to be functional.

The protein-protein contact site of G-protein coupled receptors for G-proteins has been localized to transmembrane regions 5 and 6 and the intervening cytoplasmic loop in muscarinic, angiotensin II, pituitary adenylyl cyclase activating peptide, and thyroid stimulating hormone receptors.[3] Through site-directed mutagenesis the adrenergic receptor G-protein coupling site was further localized to cytoplasmic regions adjacent to transmembrane sections 5, 6, and 7.[3] Accordingly, organisms will alter the sequence between transmembrane sections 5 and 6 and use it as a method of regulating G-protein coupling. For example, dopamine D_2 receptors are expressed in multiple forms by alternate splicing. The long isoform has a 29-amino-acid insertion within the intervening sequence of transmembrane sections 5 and 6, presumably causing distinct coupling to G-proteins, thus providing a mechanism of regulating dopamine receptor activation.[9,10]

As mentioned, G-protein coupled receptors initiate signal transduction by activating G-proteins (for review see Reference 11). G-proteins are composed of three subunits, α, β, and γ, each having different subtypes; however, different G-proteins are best characterized according to the nature of the α subunit. The α subunit contains the guanine nucleotide binding site and GTPase activity. G-proteins associate with receptors as heterotrimers, and upon activation (in response to receptor ligand interaction) the α subunit releases GDP, binds GTP, and dissociates from the β and γ subunits. As long as the ligand remains bound to the G-protein coupled receptor it will bind and activate G-proteins, leading to amplification of the signal. The α subunit remains active until its GTPase activity

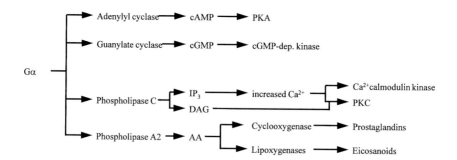

FIGURE 2   Signal transduction events produced by different Gα subtypes following stimulation.

cleaves a phosphate from GTP, forming GDP. Activated α subunits act on effector systems such as adenylyl cyclase, calcium channels, and phospholipase C-β (Figure 2). The effector activated depends on the subfamily of Gα, and currently four subfamilies of Gα are recognized: G_s, G_i, G_q, and G_{12}.[11] The effector systems activated by G-proteins lead to an entire cascade of events that can affect anything from gene expression to ion channel activity.

G-protein receptor activity can be regulated by changes in G-proteins or in the effector molecules that G-proteins act on. Any change in the level of G-proteins or effector molecules will result in a concomitant change of G-protein coupled receptor activity. G-protein coupled receptors themselves can also be directly regulated through protein trafficking, degradation, and phosphorylation.[12] An example of phosphorylational control is offered by desensitization following repeated stimulation. A number of G-protein coupled receptors have been shown to be regulated by phosphorylation by a family of kinases, aptly named G-protein coupled receptor kinases.[12] When a ligand is bound to the receptor the kinases can bind to and phosphorylate the receptor at selected serine or threonine sites located within the cytoplasmic loop 3 or the C-terminal section of the receptor. Once the receptor is phosphorylated an arrestin protein binds to the phosphorylated region, which will block receptor G-protein coupling. Therefore, G-protein coupled receptor kinases function to uncouple G-protein coupled receptors, but only when a ligand is bound to the receptor, acting to desensitize the receptor.[12]

B. ENDOGENOUS AND EXOGENOUS LIGANDS

The endogenous ligands for G-protein coupled receptors can be grouped into eight distinct types: biogenic amines, amino acids and derivatives, peptides and proteins, nucleotides or nucleosides, fatty acid derivatives, phospholipid derivatives, multistructural odorants, and retinal.[3] The ligands for odorant receptors are unique in that they are not endogenous and the receptors are not used for intercellular communication. Retinal is also an exception in that it remains bound to the receptor but will only cause activation when it is excited by a photon. The remaining groups act in the classical fashion with many endogenous ligands and constructed exogenous ligands (see Table 1).

TABLE 1   G-Protein Coupled Receptors Grouped by Type of Ligand[a]

Receptor	Sub-type	Endogenous Ligand	Exogenous Agonist	Exogenous Antagonist
		Biogenic Amines		
Dopamine	D_1	Dopamine		*SCH 23390**
			SKF 38393	SCH 39166
			Fenoldopam	SKF 83566
	D_2	Dopamine	N0437	(–)-Sulpiride
			Bromocriptine	*YM 09151-2**
			*Quinpirole**	Domperidone
				Remoxipride
	D_3	Dopamine	*Quinpirole**	S 14297
			PD 128907	S 11566
			7-OH-DPAT*	U 99191
				*YM 09151-2**
	D_4	Dopamine	*Quinpirole**	*YM 09151-2**
				Clozapine*
	D_5	Dopamine	*SKF 38393*	*SCH-23390**
Adrenergic α_1	α_{1a}	Norepinephrine	*Cirazoline*	Phentolamine
			Methoxamine	*WB 4101*
			Phenylephrine	*5-Methylurapidil*
				*Prazosin**
	α_{1B}	Epinephrine	*Cirazoline*	Spiperone
		Norepinephrine	*Methoxamine*	CEC
			Phenylephrine	*Prazosin**
	α_{1C}	Epinephrine	*Cirazoline*	*WB 4101*
		Norepinephrine	*Methoxamine*	CEC
			Phenylephrine	*5-Methylurapidil*
				*Prazosin**
	α_{1D}	Epinephrine	*Cirazoline*	*WB 4101*
		Norepinephrine	*Methoxamine*	CEC
			Phenylephrine	*Prazosin**
Adrenergic α_2	α_{2A}	Epinephrine	Oxymetazoline	
			UK 14304	
			Guanabenz	
	α_{2B}	Epinephrine	*UK 14304*	*Prazosin**
			Guanabenz	ARC 239
	α_{2C}	Epinephrine	*UK 14304*	*Prazosin**
			Guanabenz	ARC 239
				BAM-1303
				Rauwolscine
	α_{2D}	Epinephrine	*UK 14304*	BRL 44408
			Guanabenz	
Adrenergic β	β_1	Norepinephrine	Norepinephrine	CGP20712A
			Xamoterol	Betaxolol
				Atenol
	β_2	Epinephrine	Procaterol	ICI 118551*
				Butaxamine
				α-Methylpropranolol
	β_3	Epinephrine	BRL37344	

TABLE 1   G-Protein Coupled Receptors Grouped by Type of Ligand[a] (continued)

Receptor	Sub-type	Endogenous Ligand	Exogenous Agonist	Exogenous Antagonist
Muscarinic	M_1	Acetylcholine	*McN-A343* Pilocarpine Bethanechol	Pirenzepine* Telenzepine* *4-DAMP**
	M_2	Acetylcholine	*Bethanechol*	Methoctramine AFDX 116 Gallamine Himbacine
	M_3	Acetylcholine	L-689660 *Bethanechol*	Hexahydrosiladifenidol *p*-Fluorohexahydro- siladifenidol *4-DAMP**
	M_4	Acetylcholine	*McN-A343* *Bethanechol*	Tropicamide Himbacine *4-DAMP**
	M_5	Acetylcholine		*4-DAMP*
Serotonin 5-HT$_1$	5-HT$_{1A}$	Serotonin	8-OH-DPAT* *5-CT** R(+)-UH-301	WAY100135 *Spiperone*
	5-HT$_{1B}$	Serotonin	CP93129 *5-CT**	*Isamoltane*
	5-HT$_{1D}$	Serotonin	LY694247 *5-CT** *Sumatriptan** BRL 56905	GR127935 *Isamoltane*
	5-HT$_{1E}$	Serotonin	*5-CT** RU 24969	*Methiothepin*
	5-HT$_{1F}$	Serotonin	*5-CT** *Sumatriptan**	*Methiothepin*
Serotonin 5-HT$_2$	5-HT$_{2A}$	Serotonin	*α-Methyl-5-HT* *DOI* *DOB*	*Ketanserine** *LY53857* Ritanserin *Spiperone* *Mesulergine**
	5-HT$_{2B}$	Serotonin	*α-Methyl-5-HT* *DOI*	*LY53857* *Ketanserine** SB200646
	5-HT$_{2C}$	Serotonin	*α-Methyl-5-HT* *DOB*	*LY53857* *Ketanserine** SB200646 *Spiperone* *Mesulergine**
Serotonin misc.	5-HT$_4$	Serotonin	BIMU8 Renzapride SC-53116	GR113808* SB204070 SDZ 205-557
	5-HT$_6$	Serotonin	*LSD** *5-CT*	*Methiothepin* *Amoxipine*
	5-HT$_7$	Serotonin	*LSD**	
Histamine	H_1	Histamine	2-(*m*-Fluorophenyl)- histamin	Pyrilamine* Mepyramine Triprolidine

TABLE 1 G-Protein Coupled Receptors Grouped by Type of Ligand[a] (continued)

Receptor	Sub-type	Endogenous Ligand	Exogenous Agonist	Exogenous Antagonist
	H_2	Histamine	Dimaprit Impromidine Amthamine	Ranitidine Cimetidine Tiotidine
	H_3	Histamine	R-α-Methyl-histamine Imetit Immepip	Thioperamide Iodophenpropit* Clobenpropit
Melatonin	ML-1	Melatonin	2-Iodomelatonin* N-propionyl-melatonin 6-Chloromelatonin	Luzindole

Amino Acid Derivatives

Receptor	Sub-type	Endogenous Ligand	Exogenous Agonist	Exogenous Antagonist
Glutamate	$mGlu_1$ and $mGlu_5$	Glutamate	*ACPD* *L-CCG-1*	MCPG
	$mGlu_2$ and $mGlu_3$	Glutamate	*ACPD* *L-CCG-1*	
	$mGlu_4$ and $mGlu_6$	Glutamate	L-AP-4 *ACPD* *L-CCG-1*	
γ-amino-butyric acid (GABA)	$GABA_{B1α,}$ $B1β, B1γ$	GABA	Baclofen* 3-APPA* 3-APMPA	Phaclofen *CGP 35348*
	$GABA_{B2}$	GABA	3-APPA*	CGP 35348

Peptides and Proteins

Receptor	Sub-type	Endogenous Ligand	Exogenous Agonist	Exogenous Antagonist
Somatostatin	SSTR2	Somatostatin	MK 678* SMS 201-955*	
	SSTR3	Somatostatin	BIM 23056	
	SSTR4	Somatostatin	L-362855	
Opioid	m	Enkephalin	DAMGO* Sufentanyl PL 017 Morphine	β-FNA CTAP CTOP
	$δ_1$	Enkephalin	DPDPE* DADLE	BNTX DALCE Naltrindole ICI-174864

Nucleotides

Receptor	Sub-type	Endogenous Ligand	Exogenous Agonist	Exogenous Antagonist
Purinoceptors	A_1	Adenosine	N^6-cyclopentyl-adenosine N^6-cyclohexyl-adenosine	CPT *CPX**
	A_{2a}	Adenosine	CGS-21600	CP 66713 KF 17837 8-(3-Chlorostryryl) caffeine CGS-15943

**TABLE 1   G-Protein Coupled Receptors Grouped by
Type of Ligand[a] (continued)**

Receptor	Sub-type	Endogenous Ligand	Exogenous Agonist	Exogenous Antagonist
	A_{2b}	Adenosine	NECA	CPX
				Alloxazine
	A_3	Adenosine	APNEA	1-ABOPX
			N[6]-benzyl NECA	
	P_{2T}	Adenosine	2-Methylthio-ADP	ATP
			FLP 66096	

Eicosanoids

Cannabinoid		Anandamide	THC	
			WIN 55,22-2*	
			Levonantradol	
			CP 55940*	

[a] Exogenous ligands in *italics* are not selective for the given subtype, those with * are commonly used to radioactively label receptors.

III. TYROSINE KINASE RECEPTORS

Tyrosine kinase receptors are characterized according to their intrinsic tyrosine kinase activity. The physiological response of tyrosine kinase receptors is produced by the binding of a ligand to the receptor and subsequent activation of the kinase domain. Phosphorylation of effector molecules by the receptors results in a poorly understood cascade whose participants are beginning to be elucidated (for reviews see References 13 and 14). This superfamily of receptors functions within organisms to promote cellular growth and differentiation.[13] A given receptor can act to promote growth in one cell type and arrest growth and induce differentiation in another cell type.[14] Therefore, ligands are not necessarily limited in their action on different cell types.

A. STRUCTURE AND FUNCTION

Tyrosine kinase receptors possess structurally related regions apart from the tyrosine kinase domain. Each receptor contains three domains that are compartmentally separated: an extracellular domain that is responsible for ligand recognition, a single-pass transmembrane domain, and an intracellular domain with tyrosine kinase activity.[14] Phosphorylation sites that are used in signal transduction, as will be discussed below, are also present on the cytoplasmic domain.[14] Further comparisons of structure require dividing the family into four distinct subgroups (Figure 3). Subgroup I consists of a single polypeptide that contains two characteristic cysteine-rich extracellular domains that recognize ligands such as epidermal growth factor (EGF) which stimulates the growth of epidermal and epithelial cells. Subgroup II receptors are heterotetramers linked by disulfide bonds. Each receptor contains two copies of an extracellular ligand-binding α polypeptide and of a transmembrane tyrosine kinase β polypeptide chain. Examples of ligands for this subgroup are insulin, which regulates the blood glucose levels by stimulating

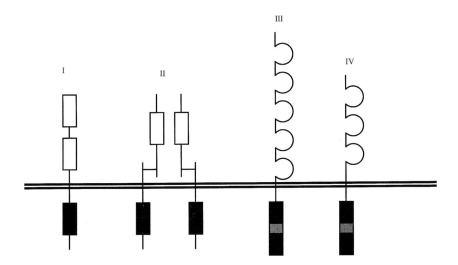

FIGURE 3 Subgroups of tyrosine kinase receptors. Open rectangles represent cysteine-rich domains, semicircles are immunoglobulin-like domains, filled rectangles are tyrosine kinase domains, and shaded squares are kinase inserts. (Adapted from Ulrich, A. and Schlessinger, J., *Cell*, 61, 203, 1990. With permission.)

glucose and amino acid cellular uptake, and insulin-like growth factor (IGF) which stimulates proliferation of fat and connective tissue cells. The remaining subgroups, III and IV, consist of monomeric proteins with five and three immunoglobulin-like domains, respectively. Subgroup III and IV also are unique in that intervening sequences are added within the cytoplasmic kinase domain.[13] An example of a ligand for subgroup III is platelet-derived growth factor (PDGF), which stimulates proliferation of neural glial cells, while that of a subgroup IV ligand is fibroblast growth factor, which induces proliferation of fibroblast and endothelial cells and formation of mesoderm. Despite these structural diversities, the tyrosine kinase receptors all share a common mechanism of activation: dimerization and phosphorylation.[15]

Transmission of an extracellular signal, such as ligand binding, to the intracellular domain in many receptors system occurs by a conformational change transmitted across the membrane (i.e., the ionotropic receptor, see Section VI). However, with only one sequence traversing the membrane in tyrosine kinase receptors it is difficult for an intracellular conformational change to occur in response to an extracellular ligand. Accordingly, Schlessinger and Ullrich have postulated that receptor dimerization is a universal mechanism for initiation of message transduction in receptors possessing a single transmembrane domain.[14] Dimerization offers a mechanism for single-pass receptors to transfer signals to intracellular domains and has been shown to be used by tyrosine kinase receptors. The necessity of dimerization in tyrosine kinase receptor activation was elegantly demonstrated by coexpression of wild-type and mutant EGF receptors.[16] Defective EGF receptors suppress the activity of wild-type EGF receptors. The mutation is dominant due to heterodimers of mutant and wild-type EGF receptors being inactive. The exact mechanism that causes receptor dimerization is not fully understood, but different methods of dimerization are demonstrated below for three of the receptor subtypes.

In subclass I tyrosine kinase receptors, when a ligand binds to the receptor it causes a conformational change in the extracellular domain. The conformational change facilitates dimerization of receptor molecules which activates the intra-cellular kinase domain.[13] Subclass II receptors are normally present as dimers due to the disulfide bridge between proteins. Ligand binding, such as by insulin, to each of the binding sites causes a conformational change that activates the kinase domain.[13] Finally, subclass III receptors recognize ligands that are dimers them-selves, such as platelet-derived growth factor (PDGF). Since the ligand can bind two receptors it will itself facilitate dimerization, causing activation of the intra-cellular kinase.[14] A single growth hormone peptide is able to bind two receptors and cause dimerization. PDGF, on the other hand, makes things a little more complicated as it consists of two associated molecules, AA, BB, and AB. Further-more, there are also two types of receptors, α and β, that are present at the plasma membrane which recognize A and B, respectively. AA or BB will cause homo-dimerization of receptors, while AB will cause heterodimerization. Each set of dimerized receptors displays differential binding characteristics.[14] Therefore, receptor dimerization allows a further level of fine-tuning receptor-ligand interac-tions.[14] The exact nature of the receptor-receptor interaction is not known . How-ever, regardless of the mechanism of dimerization, the net result is activation of the tyrosine kinase and the resultant message transduction.

Once activated by dimerization, the intrinsic tyrosine kinase will phosphory-late itself and other tyrosine kinase receptors. Autophosphorylation of receptors is an essential step for further activation of cytosolic proteins. Effector proteins that are regulated by tyrosine kinase receptor phosphorylation will only become associated with phosphorylated receptors. The tyrosine phosphorylated regions of tyrosine kinase receptors function as high-affinity binding sites for cytosolic pro-teins with src homology 2 (SH2) domains.[14] SH2 domains are conserved regions that recognize specific tyrosine phosphorylated regions.[14] Each effector that is regulated by tyrosine kinases receptors contains a SH2 domain that will recognize specific tyrosine kinase receptors that have been autophosphorylated. To date, three examples of effectors containing SH2 domains that are activated by tyrosine kinase receptors have been discovered: phospholipase C-γ, phosphatidylinositol 3-kinase, and ras GTPase-activating protein.[17-19] The end result of the signal transduction varies from increased Na^+/H^+ exchange (EGF and TGFα), Ca^{2+} influx (e.g., EGF), to increased glucose and amino acid transport (e.g., insulin).[13]

As mentioned when discussing G-protein receptors, ligand stimulation often causes subsequent down-regulation of the receptor it binds. For tyrosine kinase receptors, down-regulation is mediated by internalization.[13] However, only certain receptors subtypes are degraded after internalization. Upon entry within the cell, insulin receptors are quickly recycled. EGF receptors differ in that they are degraded upon internalization. After EGF is bound to its receptor, the ligand-receptor complexes aggregate and are internalized in coated pits. Once internal-ized, the receptors are degraded within lysosomes. The kinase domain has been shown to be instrumental for targeting EGF receptors to lysosomes, since mutant EGF receptors lacking kinase activity are internalized but quickly recycled to the membrane.[20] Another common mechanism of receptor regulation is phosphorylation which, as mentioned, is a step in the normal activation of tyrosine kinase receptors. Where tyrosine kinase receptors differ from others is that they are controlled by autophosphorylation. Tyrosine kinase receptors are also regulated by phosphorylation of serine and threonine residues.[13] Protein kinase C (PKC), which is a downstream product of phospholipase C activation, phosphorylates conserved residues on tyrosine kinase receptors, leading to reduced ligand affinity.[13] Therefore, through internalization and reduction of ligand affinity, tyrosine kinase receptor activity can be down-regulated.

B. ENDOGENOUS LIGANDS

The four subgroups of tyrosine kinase receptors all recognize peptides as endogenous ligands. Accordingly, the synthesis of nonpeptide agonist or antagonist is extremely difficult. Therefore, in the treatment of disorders such as insulin-dependent diabetes, human insulin from recombinant bacteria is used.

IV. DNA BINDING RECEPTORS

The superfamily of DNA binding receptors is unique in that the ligand-receptor complex binds to response elements on chromosomal DNA to modulate transcription.[1] While many other superfamilies of receptors modulate transcription, such as G-protein coupled receptors and tyrosine kinase receptors, they do so using soluble intermediate molecules such as cAMP and IP_3. Intermediate molecules are used by these receptors in part because their ligands can not cross the plasma membrane of cells. DNA binding receptors possess their unique DNA binding property due to their ligands — steroids, thyroid hormone, vitamin D_3 (cholecalciferol), and vitamin A (retinol) — being lipid soluble. Upon diffusing into cells, steroids will associate with their receptors in the cytoplasm and then be translocated into the nucleus, while vitamins bind their receptors directly in the cell nucleus.[1]

Besides vitamin A and D_3, which are important for vision, development, growth, reproduction, and homeostasis and control of phosphorus and calcium metabolism, the glucocorticoids, mineralocorticoids, thyroid hormone, and sex steroids are recognized by DNA binding receptors. Sex steroids (i.e., estrogen, progesterone, and testosterone) are vital for sexual differentiation and functioning, while glucocorticoids (cortisol) and mineralocorticoids (aldosterone) are needed to maintain homeostasis of glucose and mineral metabolism. Thyroid hormone, which is derived from tyrosine and iodine, acts to increase the level of general metabolism. As a whole, these ligands act on DNA binding receptors throughout the body to produce cell- and tissue-specific increases or decreases in transcription.

A. STRUCTURE AND FUNCTION

DNA binding receptors all share discrete domains that are separable by function into the following domains: ligand-binding, DNA-binding, nuclear localization, and transcriptional modulating (Figure 4).[1] The DNA binding domain is the most

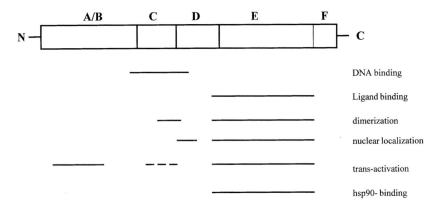

FIGURE 4 Domains of DNA binding receptor. Boldface letters are the conventional designations for the different domains. Lines indicate which regions of the receptor sequence have been demonstrated to act in the given function of the receptor. (Adapted from Luisi, B.F., Schwabe, J.W.R., and Freddman, L.P., *Vitam. Horm.*, 49, 1, 1994. With permission.)

conserved throughout the superfamily and is made up of two zinc finger-like structures that can act as a true domain, binding DNA in the absence of the other domains.[1] These two structures each consist of four cysteine residues that bind a zinc ion in a similar but distinct manner of a zinc finger.[21-23] The zinc ion acts to orient the protein so that an amphipathic α-helical region will position itself in the major groove of the DNA.[1] The remaining functional domains — the ligand binding, dimerization, nuclear localization, transactivation and HSP-90 binding domains — are all contained in a conserved segment of the DNA binding receptor, which is proximal to the C-terminus (E, Figure 4).[1]

The steroid receptors, which are primarily cytoplasmic, are associated with HSP-90 prior to stimulation by their complementary ligands. Upon stimulation, HSP-90 dissociates and the receptor is translocated to the nucleus.[1] Nuclear receptors, such as the vitamin receptors, will bind their ligands only after the ligands have diffused into the nucleus (Figure 5). Dimerization of both classes of receptors may or may not precede DNA binding,[24,25] however, when one receptor has bound to DNA, binding of the second receptor is cooperative in that its affinity is increased by two orders of magnitude.[1] The cooperativity may occur through protein-protein as well as protein-DNA interactions.[26-28] The element of DNA that is bound by DNA binding receptors is referred to as the response element (RE). DNA binding receptors bind consensus REs of specific DNA sequences that consist of palindromic (e.g., steroids) or direct (e.g., thyroid hormone) repeats that are separated by short sequences of random DNA, referred to as half-sites.[1] Recognition of RE by specific receptors occurs through recognition of the half-site spacing, with slight specificity offered by the consensus sequence.[27] The dimerization region or D-box of DNA binding receptors determines the half-site spacing and therefore the RE that is recognized by each receptor.[29,30] As demonstrated, the physiological actions of each DNA binding receptor differ but the mechanisms by which these controls are exerted are similar.

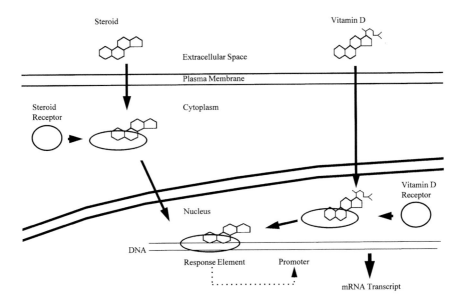

FIGURE 5 Functioning of vitamin D and steroid receptors in the modulation of mRNA transcription.

   The properties and regulation of the glucocorticoid receptor (GR) have been the most extensively examined and offer an excellent example of DNA binding receptors. Activated GR can act to stimulate or repress the transcription of specific genes. As mentioned, stimulation occurs through GR binding to glucocorticoid RE (GRE) and subsequent interaction of GR with the transcriptional machinery to induce transcription via a poorly understood mechanism. Stimulation of transcription in a number of genes depends not only on activated receptors and a RE but also on the presence of cAMP, cAMP RE (CRE), and CRE binding protein (CREB). The phosphoenolpyruvate carboxykinase gene promoter contains both a GRE and CRE and requires a GR and CREB protein-protein interaction for full activation.[31,32] Repression, on the other hand, has been shown to occur through two separate mechanisms that do not necessarily involve a RE. First, GR can repress the expression of pro-opiomelanocortin (POMC) in the anterior pituitary through binding to a negative GRE.[33] The negative GRE is in the 5' region of the POMC gene and overlaps with a CCAAT box sequence. Accordingly, it been postulated that repression occurs through occlusion of the binding of positive elements.[34,35] The second mechanism of repression by GR has been shown to not involve a GRE. Using chimeric receptors, it was demonstrated that GR interferes with the expression of the glycoprotein alpha subunit through interactions with CREB.[36,37] The GR-CREB interaction may be direct or through an intermediary protein.[36] Regardless, GR does not bind to the promoter sequence of the glycoprotein alpha subunit.[37] While these examples demonstrate actions of GR in modulating transcription there are also a number of examples demonstrating phosphorylational control of GR function.

   GR initially binds glucocorticoids in the cytoplasm of the cells before translocation into the nucleus. Ligand binding initiates receptor phosphorylation that is required for the translocation of the receptor-ligand complex into the nucleus.[38] Prior to ligand binding GR are phosphorylated at a basal level. Upon activation the receptor becomes hyperphosphorylated[39] and moves into the nucleus. Phosphorylation also is responsible for increasing the affinity of GR for the GRE on DNA. Protein Kinase A will up-regulate DNA binding of GR and of the DNA binding domain of GR alone.[40] As seen with other receptor superfamilies, phosphorylation offers a diverse range of control.

The DNA binding superfamily is unique in that it can directly modulate transcription of genes; however, it can also initiate second-messenger systems such as increased intracellular Ca^{2+},[41] making it a truly multifunctional family. The DNA binding capacity offers the superfamily access to many genes without the interference of second-messenger molecules. Removing the amplification of the signal provided by second-messengers allows very specific responses. Finally, the direct interaction between receptor and gene can also be regulated by the normal second-messenger pathways, allowing further refinement of the signal.

B. ENDOGENOUS AND EXOGENOUS LIGANDS

Any ligand for a DNA binding receptor must have the capacity to cross the plasma membrane so that the receptor can bind the ligand. In order to accomplish this, all DNA binding receptor ligands are amphipathic molecules. Perhaps due to this limitation, the majority of DNA binding ligands are derived from cholesterol, such as the steroids and vitamin D. The two exceptions are vitamin A and thyroid hormone. Vitamin A is an isoprenoid alcohol derived from β-carotene while thyroid hormones, thyroxine or triiodothyronine, are formed by joining two iodinated tyrosine residues.

V. CYTOKINE RECEPTORS

Cytokines are a class of molecules that act to regulate proliferation and differentiation of many cell types, including hemopoietic and neuropoietic cells.[42] As a superfamily, individual cytokines have the potential to produce many different effects and different cytokines possess overlapping effects[43] (Table 2). While cytokines mostly recognize distinct receptors, there are commonalities both in receptor binding and signal transduction mechanism. Cytokine receptors acts as homo- or heterodimers of β subunits with or without a high-affinity α subunit.[44] Accordingly, the naming of receptors for each cytokine has not proven simple and remains eclectic.

A. STRUCTURE AND FUNCTION

Discussing the function of cytokine receptors can prove to be difficult since little is known about how the receptors initiate the changes produced by cytokines. A better approach is to discuss the effects of different cytokines, which are of course mediated by their receptors. Cytokines are known for their redundancy of function[43] (Table 2). However, transgenic experiments have revealed functions unique to individual cytokine receptors.[45] It was also demonstrated that development of the immune or hemopoietic system does require each cytokine.[43] The sequence homology between cytokine proteins, which is approximately 30%,

TABLE 2   Actions of Cytokines and the Multiples Cytokines Producing Those Effects

Function	Prototype for Function	Others Producing Similar Effects
Activated T cell proliferation	IL-2	IL-4, IL-6, IL-7, IL-10, TNF-α/β
Costimulation of CTL with IL-2	IFN-γ	IL-4, IL-6, IL-7, IL-10
B cell growth and differentiation	IL-4	IL-1, IL-2, IL-3, IL-5, IL-6, IL-10, TNF-α/β
Macrophage MHC class II expression	IFN-γ	IL-3, IL-4, GM-CSF, M-CSF, TNF-α/β
Macrophage tumoricidal function	IFN-γ	IL-4, IL-7, GM-CSF
Neutrophil activation	G-CSF	IL-3, GM-CSF, TNF-α/β
GM progenitor growth and differentiation	GM-CSF	G-CSF, M-CSF, IL-3
Costimulation of IL-3-dependent mast cell production	IL-4	IL-9, IL-10

would suggest that there are few overlaps in receptor binding. However, despite sequence differences, structural analysis of cytokines does predict a common helical framework and disulfide bonds with similar C-terminal domains.[42] Additionally, the neurokines can be distinguished by a D2 neurospecific motif.[46] The structural similarity of cytokines is reflected by a complementary structural similarity in their receptors. They contain binding domains of approximately 200 amino acids with 4 characteristic cysteine residues as well as a WSxWS cytokine receptor motif.[47] Receptor complexes consist of an optional α component and two β components which dimerize on ligand binding. The β component can form hetero- or homodimers, depending on the receptors involved[44] (Figure 6). The associations of cytokine receptor components are just beginning to be understood, however, good examples are offered by interleukin-6 (IL-6) and leukemia inhibitory factor (LIF).

IL-6 and LIF are members of a subfamily of cytokines that also includes ciliary neurotrophic factor (CNTF) and oncostatin M.[44] IL-6 binds to a unique α receptor (IL-6Rα; Figure 6) that possesses a low affinity for the ligand but does not carry out signal transduction. The second component of the receptor complex, gp 130β, does not contribute to ligand recognition, but does initiate tyrosine phosphorylation upon dimerization through a separate protein, JAK, by a poorly understood mechanism.[44] On the other hand, LIF and CNTF, which carry out the same actions in the nervous system (LIF also acts in the periphery) bind to the same β components, gp 130β and LIF-Rβ. CNTF differs in requiring an α component, CNTFα, allowing selection between the CNTF and LIF[44] (Figure 6). The need for an α component for LIF binding is eliminated due to the LIF binding capacity of LIF-Rβ. As with gp 130β homodimerization, LIF-Rβ and gp 130β heterodimerization forms an active receptor complex and initiates autophosphorylation of JAK.[44] The subsequent steps of signal transduction are not known; however, phosphorylation has been deemed necessary due to kinase inhibitors eliminating cellular responses to cytokines.[48] Further research into the signaling mediated by cytokine receptors is needed in order to fully understand their function.

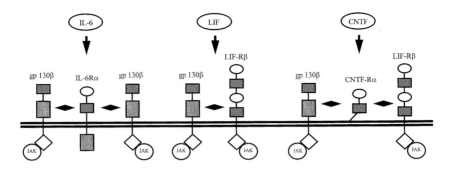

FIGURE 6 Cytokine receptor and ligand interactions for IL-6, CNTF, and LIF, demonstrating dimerization with and without component. Striped boxes indicate cysteine domains, shaded rectangles are WSxWS domains, diamonds are box 1 and box 2 sequences, and circles are potential JAK interactions.[44,46]

B. ENDOGENOUS LIGANDS

The endogenous ligands for cytokine receptors, the cytokines themselves, are dissimilar in primary sequence but similar in structure. Bazan has demonstrated that there is a common theme of a four-amphipathic-helix bundle that forms a similar helical framework.[42] The sequence similarity between cytokines is mostly held within these helical bundles. The four α-helices pack together in an antiparallel bundle with hydrophobic amino acids hidden within the center.[42] These structural similarities between cytokines allow the sharing of receptor components mentioned previously.

VI. IONOTROPIC RECEPTORS

Ionotropic receptors are characterized by their function, which is the opening of an ion channel in the membrane upon activation. They are distinguished from other gated channels such as the voltage-gated channel in that a ligand is required to bind in order for the channel to open. However, there are members of the ionotropic receptor family that also contain voltage gates, notably the N-methyl-D-aspartate (NMDA) subtype of glutamate receptors.

The ionotropic receptor family seemingly is simple since there are only six known members with relatively basic ligands and function. However, the complexity can be seen on closer examination. Each member of the family has a number of genes encoding the five subunits of the complete receptor complex. Most subunits of the pentameric complex have a number of subtypes, each conferring different ligand affinity or ion conductance. Furthermore, alternative splicing mRNA and editing offer the final complexity.

The necessity for such a large degree of complexity can be understood since the family is responsible for the majority of the central nervous system's fast excitatory and inhibitory inputs, as well as mediating communication at neuromuscular junctions. Specifically, glutamate, the predominant excitatory neurotransmitter in the vertebrate nervous system, is involved in neuronal plasticity as well as learning and memory.[49,50] Cellular damage associated with hypoxia, hypoglycemia, ischemia, and epilepsy has also been shown to involve glutamate.[51,52] Accordingly, glutamate is also implicated in the pathology seen in Alzheimer's and Parkinson's diseases and AIDS encephalopathy.[53]

On the other hand, GABA$_A$ receptors, which recognize γ-amino butyric acid (GABA), mediate most of the inhibitory synaptic transmission, along with glycine receptors. Benzodiazepines, barbiturates, steroids, general anesthetics, and alcohol all potentially exert their actions through GABA$_A$ receptors.[54] Glycine receptors are closely related to GABA receptors and, as mentioned, also mediate inhibitory synaptic transmission. Their actions have been localized primarily to the spinal cord and brain stem[55] and are implicated in stroke, epilepsy, and anxiety.[56]

Acetylcholine binds both to nicotinic and muscarinic receptors. The muscarinic subgroups are complexed to G-proteins but the nicotinic subgroup gates a sodium/potassium channel. It is one of the most-studied receptors, partially due to its high expression at the neuromuscular junctions where it mediates the contraction of muscles by motor neurons.[56] One of the members of the serotonin receptor family, the 5-HT$_3$ receptor, is distributed primarily in mesocorticolimbic structures.[58] It has been implicated in schizophrenia due to its localization, its mediation of dopamine release in the nucleus accumbens, and its blockade by clozapine.[58-60] Additionally, 5-HT$_3$ receptors are located within the pain and nausea areas of the brain, where 5-HT$_3$ antagonists are thought to reduce pain associated with migraines and nausea associated with chemotherapy or anesthesia.[61] Finally, certain purinoceptors recognize adenosine triphosphate and are involved in nociception[62,63] and fast synaptic transmission between neurons and between autonomic nerves to smooth muscles mediating events such as vasoconstriction.[64,65]

A. STRUCTURE AND FUNCTION

The function of ionotropic receptors is to open an ion channel in response to a neurotransmitter. The ionotropic receptors allow the flow of cations, as in the case of acetylcholine, 5-HT$_3$, glutamate, and purinoceptors, which leads to depolarization of the postsynaptic membrane. GABA and glycine receptors differ in that they allow the inward flow of an anion, chloride, which will hyperpolarize the postsynaptic membrane and inhibit subsequent action potentials. The functioning of ionotropic receptors gets more complex, as in the case of the NMDA subtype of glutamate receptors. NMDA receptors have a high permeability for calcium ions, which is relatively unusual among the cationic ionotropic receptors.[66] The inflow of calcium ions will activate a cascade of protein kinases and lead to the induction of long-term potentiation (LTP), a process believed to be central to memory mechanisms.[67] However, the simple recognition of glutamate by NMDA receptors is not sufficient for the induction of LTP, as the NMDA receptor also contains a voltage gate which is usually blocked by magnesium ions. Therefore, prior flow of cations through the AMPA subtype of glutamate receptor is required to partially depolarize the membrane and remove the magnesium ion that is acting at the voltage gate from the NMDA receptor.[68] As illustrated, ionotropic receptors mediate actions as mundane as producing action potentials to as elaborate as LTP, depending on the receptor and system involved.

Structurally, ionotropic receptors have been studied extensively. The majority of the work has been on nicotinic acetylcholine receptors. The acetylcholine receptor complex is formed by a pentamer of four different subunits in a $\alpha_2\beta\gamma\delta$ stoichiometry.[68] The subunit structure is that of a type I channel, in that both the N and C terminus face the extracellular matrix.[69] The receptor extends 65 Å extracellularly and 15 Å intracellularly in a 20 Å cylinder that narrows in the membrane.[71] The structure of the membrane spanning region has produced some controversy of late. Originally, from hydrophobicity plots each subunit was determined to contain four antiparallel α-helices,[72] but this has been put in question by recent results. Of the four proposed transmembrane units, termed M1-M4, only the α-helical nature of M2 has been experimentally supported.[73] The remaining transmembrane structures seem to be β-structures that could form a β-barrel surrounding the pore-forming α-helical M2 subunit,[56,72-74] which is supported, along with other research, by a 9 Å resolution image of the receptor. Regardless, of the exact nature of the transmembrane sequences, a possible mechanism of channel opening has been revealed by the 9 Å resolution image of the receptor in both the closed[74] and open[71] state. Acetylcholine, which binds the two α subunits, causes molecular changes that are transmitted to the structures within the membrane by small rotations between subunits.[71] The M2 helices then transfer these rotations to the gate-forming sidechains, creating an open pore.[71] The structural determination of acetylcholine receptors may not be applicable directly to the other ionotropic receptors but rather can offer a guide since there is a large amount of sequence homology.[54] However, ionotropic purinoceptors are the exception, as they are not related to other ligand-gated ion channels[64] but rather each subunit consists of two transmembrane domains forming a pore that resembles the potassium channel.[65]

As alluded to before, the considerable diversity of a receptor family that consists of only six members is derived from multiple subtypes and alternate splicing and editing of the mRNA of those subtypes. Glutamate receptors offer a good example of all these diversification mechanisms. There are three subtypes of glutamate receptors: NMDA, α-amino-3-hydroxy-methyl-4-isoxazolepropionate (AMPA), and kainate, which are characterized by the agonist that selectively activates each receptor. Within the AMPA subtypes, four different genes have been cloned, GluR1-4. Of these GluR2 mRNA is selectively edited to change the calcium ion permeability and rectifying properties.[75] A single base pair is altered so that the codon encodes arginine (CGG) instead of glutamine (CAG).[75] The conversion occurs by a deamination of the adenosine base pair to a inosine by a double-stranded RNA-specific adenosine deaminase.[76,77] The enzyme will recognize the desired CAG sequence due to a double-stranded RNA structure formed between it and an intronic complementary site.[76] The end result is a change in the functional characteristic of the GluR2 subunit that can be regulated.

In addition to RNA editing, GluR2 can also be expressed as either flip or flop isoforms, depending on alternate mRNA splicing.[78] The isoforms differ in that each splices out a different exon in forming the final mRNA transcript. Expression of either the flip or flop exon has been demonstrated to produce different functional characteristics.[78] This final form of diversification adds to those previously mentioned, forming a complex group of receptors built up from six families.

TABLE 3 **Ionotropic Receptors**[a]

Receptor	Subtype	Endogenous Ligand	Exogenous Agonist	Exogenous Antagonist
Acetylcholine	Nicotinic	Acetylcholine	Nicotine	d-Tubocurraine
			Cytisine	Lophotoxin
			Methylisoarecolone	α-Bungarotoxin*
			Anatoxin-A	Conotoxin MI
GABA	GABA$_A$	γ-Amino-butyric acid	Isoguvacine	Bicuculine*
			Musciomol*	SR-95531*
Glutamate	NMDA	Glutamate	N-methyl-D-aspartic-acid	AP-5*
				CPP*
				CGS-19755*
				MK-801*
				PCP*
	AMPA	Glutamate	AMPA*	NBQX
			S(-)-5-fluorowillardine	GYKI 52466
			Quisqualic acid	*CNQX**
	Kainate	Glutamate	Kianate acid	*CNQX**
			Domoic acid	
Glycine	—	Glycine		MLD 100, 458
				MLD 102, 288
Purinoceptor	P$_{2Z}$	ATP	3'-O-(benzoyl)-benzoyl-ATP	2-Methylthio-L-ATP*
	P$_{2T}$	ADP	2-methylthio-ADP	2-Chloro-ATP
			FLP 66096	
Serotonin	5-HT$_3$	Serotonin	2-Methyl-5-HT	Zacopride
			1-(m-Chlorophenyl)-biguanide*	Ondasetron
			5-HTQ	Tropisetron
				Granisetron

[a] Exogenous ligands in italics are not selective for the given subtype, those with * are commonly used to radioactively label receptors.

B. ENDOGENOUS AND EXOGENOUS LIGANDS

A list of the endogenous and exogenous ligands recognized by ionotropic receptors is given in Table 3. A noticeable feature of the endogenous ligands is that they are all amino acid or amino acid derivatives, with the exception of ATP which is recognized by purinoceptors. However, even in the case of ATP, it is a widely available molecule and used in general metabolism.

VII. CELL ADHESION RECEPTORS

Interactions between cells and between cells and extracellular matrix (ECM) are of fundamental importance in cellular growth and differentiation. These interactions are mediated by cell surface proteins of diverse structure and function. The proteins involved in such interactions were initially believed to carry out their function simply by homologous or heterologous binding and therefore they were named cell adhesion molecules (CAMs). However, recent investigations have revealed that homologous binding of CAMs between two cells or binding of a cellular CAM to ECM is not a passive event, but results in the activation of a variety of intracellular signal transduction pathways that is the basis of change in cell dynamics. In appreciation of the signal transducing properties of the CAMs, it is perhaps more apt to term these molecules as cell adhesion receptors (CARs). The range of functions subserved by CARs includes embryogenesis, neuronal connectivity, immune cell response during infection and inflammation, and cancer cell metastasis.

Adhesion receptors generally belong to one of the following four families: immunoglobulin (Ig) superfamily, cadherins, integrins, and selectins (see Figure 7). In most cases CARs are membrane-bound single peptide chain glycoproteins with one transmembrane domain and a short cytoplasmic tail, although different structural variants are also known.

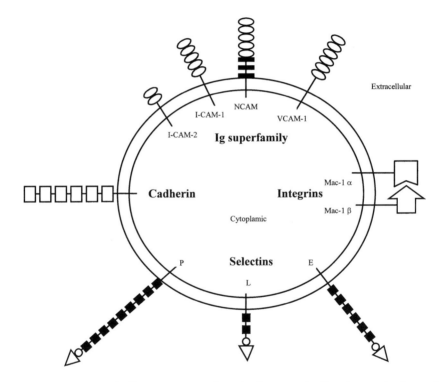

FIGURE 7   Structure of adhesion receptors from the four superfamilies. Ig superfamily: ovals represent Ig domains; filled rectangles, FN II repeats. Selectins: filled squares, complement regulatory protein domains; open circles, EGF-like domains; open triangles, carbohydrate. Cadherin: open squares, extracellular repeats with multiple glycosylation sites.

A. STRUCTURE AND FUNCTION

The Ig superfamily comprises one of the largest groups of CARs, the classic example of which is the neural cell adhesion molecule (NCAM).[79] Ig family members contain one or more Ig-like C or V domains with a characteristic disulfide bond in their extracellular N-terminal region. In addition some Ig superfamily members contain fibronectin type III (FNIII) repeats in the N-terminal. The Ig superfamily of

CARs are most widely distributed in the nervous system and leukocytes. There are many subclasses of the Ig superfamily in the nervous system,[80] such as

1. Small molecules with one of two Ig domains (e.g., P0, Thy1, and Ox2),
2. Molecules with three Ig domains and a glycosyl phosphatidylinositol (GPI) membrane anchor (e.g., the limbic system-associated membrane protein (LAMP), opioid binding CAM (OBCAM), and neurotrimin,
3. Large molecules containing five to nine Ig domains (e.g., myelin associated glycoprotein (MAG), SC1/DM-GRASP, and telencephalin,
4. Molecules with five to six Ig domains and two to five FNIII repeats (NCAM/fasciclin II, L1/NgCAM/NILE/neuroglian, F3/F11/contactin, and TAG 1/axonin 1), and
5. Molecules with Ig domains plus domains other than FNIII repeats (e.g., neurocan, trk, trkB, and trkC); however, the role of this particular subgroup of the Ig superfamily in cell-cell adhesion is not clear.

Most neural CARs of the Ig superfamily interact with homologous binding, e.g., NCAM on one cell binds to NCAM on another cell; however, other types of cis-cis interactions between two cell adhesion molecules on the same cell are also possible, e.g., NCAM on one cell may laterally interact with L1 on the same cell, which may affect L1-L1 homologous interaction between two apposing cells.

There are many molecules of the Ig superfamily involved in adhesion and signal transduction in cells involved in inflammatory and immune functions such as vascular endothelial cells and various types of leukocytes, as well as cancer cells.[81] For example, endothelial cells express intercellular adhesion molecule 1 (ICAM-1; CD54), ICAM-2 (CD102), ICAM-3 (CD50), vascular cell adhesion molecule 1 (VCAM-1; CD106), PECAM-1, and mucosal addressin cell adhesion molecule 1 (MAdCAM-1). Low levels of ICAM-1 and VCAM-1 are expressed in resting endothelium, however, ICAM-2 seems to be constitutively present. T-lymphocytes express several Ig superfamily molecules including CD2, CD4, or CD8, ICAMs 1 and 2, and T-cell receptor. The Ig superfamily molecules of this system engage mainly in heterologous interactions with different classes of CARs (in particular, members of the integrin family) present on other cells. Cancer cells express many characteristic antigens on their surface that belong to the Ig super-family, e.g., carcinoembryonic antigen (CEA), and deleted in colon carcinoma (DCC) associated with colon cancer. Recently, a number of transmembrane tyrosine phosphatases belonging to the Ig superfamily have been identified that are involved in cell adhesion and behavior.

As alluded to earlier, changes in cellular function and behavior that results from the interactions between CARs are caused, in large part, by activation of signal transducing second-messenger pathways within participating cells.[82] Some of the critical information on the signal transduction events by Ig superfamily members has come from studies on NCAM-NCAM, or L1-L1 interactions.[83] For example, it has been shown that binding and activation of these molecules can result in a variety of second-messenger responses including altered intracellular $[Ca^{2+}]$, pH, and cyclic AMP (cAMP) levels. Ca^{2+} influx through L- and N-type calcium channels is the obligatory step in neurite outgrowth promoted by NCAM and L1. It has recently been shown that activation of fibroblast growth factor receptor (FGFR) tyrosine kinase following NCAM-NCAM interaction may be one of the first steps in the signal transduction mechanism. The scheme of events that is emerging suggests that FGFR activates phospholipase-γ, resulting in the production of diacylglycerol (DAG). DAG lipase in turn converts DAG into arachidonic acid (AA) which directly activates N- and L-type calcium channels. Interestingly, a CAM homology domain (His-Ala-Val motif within a 20-amino-acid domain) present in neuronal CAMs (NCAM, L1, and N-cadherin) has been found in the extracellular domain of FGFR, suggesting the possibility of a direct interaction between neuronal CAMs and FGFR.[84]

The second superfamily of adhesion receptors, the cadherins, like NCAM, participate in homophilic cell-cell interactions.[85] However, unlike NCAM or other members of the Ig superfamily, cell adhesion mediated by cadherins requires the presence of calcium. Cadherins also differ from the Ig family in not having an IgG fold. There are over a dozen cadherins known; for example, N-cadherin (most abundant in nervous tissue and present in differentiated neurons and glia), E-cadherin (epithelium, liver), and P-cadherin (placenta). The cadherins are transmembrane proteins with an extracellular N-terminal having 5 tandem repeats, and about a 150-amino-acid-long cytoplasmic C-terminus. Calcium ions are thought to stabilize cadherin-cadherin interaction by binding to a charged region in the extracellular domain of the protein. The cytoplasmic domain of cadherins interacts with the cytoskeleton via proteins called catenins. The signal transduction events following cadherin-cadherin interactions are not fully elucidated; however, as described for NCAM, recent results point to an involvement of the FGFR and calcium influx in the process of neurite outgrowth promoted by N-cadherin. Cadherins play important roles in cell adhesion during development and morphogenesis. In the nervous system, for example, N-cadherin is expressed from the time of neural tube induction and plays a critical role in neural tube formation. A loss of cadherins on tumor cells can lead to an invasive and malignant phenotype.

The third superfamily adhesion receptors are the integrins, which mediate adhesion between the cell and extracellular matrix (ECM), Ig superfamily adhesion molecules (e.g., ICAMs, VCAMs), and even viruses.[86,87] The integrins thus act as receptors for ECM and counter-receptors for Ig superfamily proteins. Integrin receptors are heterodimeric transmembrane glycoproteins composed of an α and a β subunit linked in a noncovalent but stable structure. Currently there are at least 15 α and 8 β subunits known that can combine in different ways to form over 20 integrins of differing affinities towards ECM and other proteins. Each α β combination imparts unique ligand specificity to the integrins, e.g., α1, β1 integrin binds both collagen and laminin equally well, but α5, β1 integrin has a relatively high specificity for fibronectin. The β1 integrins are expressed at low levels on lymphocytes and their number increases after cell stimulation. The predominant integrins in all leukocytes are in the β2 subfamily (leukocyte cell adhesion molecule or LeuCAM).

Integrins transduce information from outside of the cell to the inside and cause reorganization of the cytoskeleton partly by interactions of cytoplasmic tails of integrin subunits with talin, vinculin, and α-actinin. Integrin-mediated signal transduction appears to involve tyrosine phosphorylation of a cytoplasmic tyrosine kinase,

TABLE 4   Ligands of Selected Cell Adhesion Receptors

Cell Adhesion Receptor	Ligand
Ig Superfamily	
NCAM	NCAM, L1
L1	L1, NCAM
MAG	MAG
ICAM-1, 2, and 3	α 1 β 2 integrin (LFA-1)
VCAM-1	α 4 β 1, α 4 β 7 integrins
Cadherins	
N-Cadherin	N-Cadherin
E-cadherin	E-cadherin
P-cadherin	P-cadherin
Selectins	
E-selectin	Sialylated Lewis X (SLe^x)
	Sialylated Lewis a (SLe^x)
P-selectin	SLe^x, SLe^a
L-selectin	Sialylated, sulfated sugar related to SLe^x, SLe^a, CD34, MAdCAM-1
Integrins:	
α 1 β 1	Collagen (col), laminin (LN)
α 2 β 1	Col, LN
α 3 β 1	Fibronectin (FN), col, LN, kalinin (KN), α 3 β 1
α 4 β 1	LN, VCAM-1
α 5 β 1	FN, IGF-BP-1, HIV-Tat
α 6 β 1	LN, KN
α 7 β 1	LN
α 1 β 2	ICAM-1,2,3
α M β 2	Fibrinogen (FG), ICAM-1, iC3b, factor X
α IIb β 3	FN, vitronectin (VN), FG, von Willebrand factor (vWF), thrombospondin (TS)
α V β 3	VN, FG, vWF, TS, osteopontin, adenovirus, HIV-1 Tat
α R β 3	FN, VN, col, FG, vWF
α 6 β 4	LN, KN
α V β 5	FN, VN, adenovirus
α V β 6	FN
α 4 β 7	FN, VCAM-1, MAdCAM-1

pp125[FAK] (125 kDa focal adhesion-associated kinase) and activation of the MAP kinase pathway;[88,89] however, the exact details are not fully understood. Interestingly, integrins can also transduce signals from inside of the cell to the outside; changes inside of cell can lead to "activation" of integrins, possibly by conformational change following modification of the cytoplasmic domain. This usually results in increased affinity of a ligand for the integrin, as has been demonstrated for LFA-1.

The final superfamily of adhesion receptors are the selectins, which are cell adhesion receptors involved in adhesion of leukocytes to endothelium during inflammation.[90] The selectins are transmembrane glycoproteins with N-terminal domain similarity to calcium-dependent animal lectins. The lectin-like domain is followed by an EGF-like domain and a complement of 2 to 9 regulatory protein repeats. Currently, three different members of this family are recognized. E-selectin (CD62E/ELAM-1) is synthesized by endothelial cells in response to stimulation by cytokines such as interleukin-1 and tumor necrosis factor. E-selectin interacts with carbohydrate ligands on leukocytes, which is important during tissue immune response. P-selectin (CD62P) is present in platelets and endothelial cells in a latent form and is expressed on the cell surface upon cell activation by thrombin, histamine, and peroxides. P-selectin then binds to ligands on leukocytes to permit their adhesion to endothelium at sites of inflammation. In contrast to E- and P-selectins, L-selectin (CD62L, lymphocyte homing receptor) is constitutively expressed by leukocytes and plays a role in lymphocyte binding to endothelium in the peripheral lymph node. It also recognizes MAdCAM-1 and mediates lymphocyte rolling. The identification of ligands for selectins is not complete yet (some of the ligands are listed in Table 4); however, the sialyl Lewis X motif, present on glycoproteins and glycolipids, appears to be a general ligand for the three selectins.

B. ENDOGENOUS LIGANDS

The endogenous ligands for the four superfamilies are listed in Table 4. As mentioned, many of the ligands, such as the cadherins, interact through homophilic binding. However, the integrins are quite different in that they are not involved in homophilic binding but rather each different subtype binds to a number of extracellular molecules. Exogenous ligands for adhesion receptors have not been synthesized. However, there are a number of blocking antibodies available that can act as antagonists.

REFERENCES

1. Luisi B.F., Schwabe J.W.R., and Freddman L.P., The steroid/nuclear receptors: From three-dimensional structure to complex function. *Vitam. Horm.,* 49, 1, 1994.
2. Strader C.D., Fong T.M., Graziano M.P., and Tota M.R., The family of G-protein-coupled receptors. *FASEB J.,* 9, 745, 1995.
3. Gudermann T., Nurnberg B., and Schultz G., Receptors and G proteins as primary components of transmembrane signal transduction: Part 1. G-protein-coupled receptors: structure and function. *J. Mol. Med.,* 73, 51, 1995.
4. Hoflack J., Trumpp Kallmeyer S., and Hibert M., Re-evaluation of bacteriorhodopsin as a model for G protein-coupled receptors. *Trends Pharmacol. Sci.,* 15, 7, 1994.
5. Wang C.D., Buck M.A., and Fraser C.M., Site-directed mutagenesis of α_{2A}-adrenergic receptors: identification of amino acids involved in ligand binding and receptor activation by agonists. *Mol. Pharmacol.,* 40, 168, 1991.
6. Lefkowitz R.J., Cotecchia S., Samana P., and Costa T., Constitutive activity of receptors coupled to guanine nucleotide regulatory proteins. *Trends Pharmacol. Sci.,* 14, 303, 1993.
7. Chung F.Z., Wang C.D., Potter P.C., Venter J.C., and Fraser C.M., Site-directed mutagenesis and continuous expression of human β-adrenergic receptors. *J. Biol. Chem.,* 263, 4052, 1988.
8. Zhu S.Z., Wang S.Z., Hu J., and El-Fakahany E.E., An arginine residue conserved in most G-protein-coupled receptors is essential for the function of the M_1 muscarinic receptor. *Mol. Pharmacol.,* 45, 517, 1994.
9. Montmayeur J.P., Guiramand J., and Borelli E., Preferential coupling between dopamine D_2 receptors and G-proteins. *Mol. Endocrinol.,* 7, 161, 1993.

10. Liu Y.F., Jakobs K.H., Rasenick M.M., and Albert P.R., G protein specificity in receptor-effector coupling. *J. Biol. Chem.,* 269, 13880, 1994.
11. Nurnberg B., Gubermann T., and Schultz G., Receptors and G proteins as primary components of transmembrane signal transduction: Part 2. G proteins: structure and function. *J. Mol. Med.,* 73, 123, 1995.
12. Premont R.T., Inglese J., and Leftkowitz R.J., Protein kinases that phosphorylate activated G protein coupled receptors. *FASEB J.,* 9, 175, 1995.
13. Ullrich A., and Schlessinger J., Signal transduction by receptors with tyrosine kinase activity. *Cell,* 61, 203, 1990.
14. Schlessinger J. and Ullrich A., Growth factor signaling by receptor tyrosine kinases. *Cell,* 9, 383, 1992.
15. Barbacid M., The Trk family of neurotrophin receptors. *J. Neurobiol.,* 25, 1386, 1994.
16. Kashles O., Yarden Y., Fischer R., Ullrich A., and Schlessinger J., A dominant negative mutation suppresses the function of normal epidermal growth factor receptors by heterodimerization. *Mol. Cell. Biol.,* 1, 1454, 1991.
17. Margolis B., Rhee S.G., Felder S., Mervic M., Lyall R., Levitzki A., Ullrich A., Zibersterin A., and Schlessinger J., EGF induces phosphorylation of phospholipase C-II: a potential mechanism for EGF receptor signaling. *Cell,* 57, 1101, 1989.
18. Varticovski L., Druker B., Morrison D., Cantley L., and Roberts T., The colony stimulating factor-1 receptor associates with and activates phosphatidylinositol-3-kinase. *Nature,* 342, 699, 1989.
19. Molloy C.J., Bottaro D.P., Flemming T.P., Marshall M.S., Gibbs J.B., and Aaronson S.A., PDGF induction of tyrosine phosphorylation of GTPase activating protein. *Nature,* 342, 711, 1989.
20. Honegger A.M., Schmidt A., Ullrich A., and Schlessinger H., Separate endocytic pathways of kinase-defective and active EGF receptor mutant expressed in the same cells. *J. Cell Biol.,* 110, 1541, 1990.
21. Hard T., Kellenbach E., Boelens R., Maler B.A., Dahlman K., Freedman L.P., Carlstedt-Duke J., Yamamoto K.R., Gustafsson J.-A., and Kaptein R., Solution structure of the glucocorticoid receptor DNA binding domain. *Science,* 24, 157, 1990.
22. Luisi B.F., Xu W.X., Otwinowski Z., Freedman L.P., Yamamoto K.R., and Sigler P.B., Crystallographic analysis of the interaction of the glucocorticoid receptor with DNA. *Nature,* 352, 497, 1991.
23. Lee M.S., Kliewer S.A., Provencal J., Wright P.E., and Evans R.M., Structure of the retinoid X receptor a DNA binding domain: A helix required for homodimeric DNA binding. *Science,* 260, 1117, 1993.
24. Wrange O., Eriksson P., and Perlmann T., The purified activated glucocorticoid receptor is a homodimer. *J. Biol. Chem.,* 264, 5253, 1989.
25. Tsai S.Y., Carlsetedt-Duke J., Weigel N., Dahlman K., Gustafsson J.-A., Tsai M.-J., and O'Malley B.W., Molecular interactions of the steroid hormone receptor with its enhancer element: evidence for receptor dimer formation. *Cell,* 55, 361, 1988.
26. Hard T., Dahlman K., Carstedt-Duke J., Gustafsson J.-A., and Rigler R., Cooperativity and specificity in the interactions between DNA and the glucocorticoid receptor DNA-binding domain. *Biochemistry,* 29, 5358, 1990.
27. Schwabe J.W., Chapman L., Finch J.T., and Rhodes D., The crystal structure of the estrogen receptor DNA-binding domain bound to DNA: how receptors discriminate between their response elements. *Cell,* 75, 567, 1993.
28. Eriksson P. and Wrange O., Protein-protein contacts in the glucocorticoid receptor homodimer influence its DNA binding properties. *J. Biol. Chem.,* 265, 3535, 1990.
29. Dalhman-Wright K., Grandien K., Nilsson S., Gustafsson J.A., and Carlstedt-Duke J., Protein-protein interactions between DNA-binding domains of nuclear receptors: influence on DNA-binding. *J. Steroid Biochem. Mol. Biol.,* 45, 239, 1993.
30. Zilliacus J., Wright A.P., Norinder U., Gustafsson J.A., and Carlstedt-Duke J., Determinations for DNA-binding site recognition by the glucocorticoid receptor. *J. Biol. Chem.,* 267, 24941, 1992.
31. Angrand P.O., Coffinier C., and Weiss M.C., Response of the phosphoenolpyruvate carboxykinase gene to glucocorticoids depends on the integrity of the cAMP pathway. *Cell Growth Differ.,* 5, 957, 1994.

32. Imia E., Miner J.N., Mitchell J.A., Yamamoto K.R., and Granner D.K., Glucocorticoid receptor-cAMP response element-binding protein interaction and the response of the phosphoenolpyruvate carboxykinase gene to glucocorticoids. *J. Biol. Chem.*, 268, 5353, 1993.

33. Drouin J., Charron J., Gagner J.P., Jeannotte L., Nemer M., Plante R.K., and Wrange O., Pro-opiomelanocortin gene: a model for negative regulation of transcription by glucocorticoids. *J. Biol. Chem.*, 35, 293, 1987.

34. Drouin J., Trifiro M.A., Plante R.K., Nemer M., Eriksson P., and Wrange O., Glucocorticoid receptor binding to a specific DNA sequence is required for hormone-dependent repression of pro-opiomelanocortin gene transcription. *Mol. Cell. Biol.*, 9, 5305, 1989.

35. Nakai Y., Usui T., Tsukada T., Takahashi H., Fukata J., Fukushima M., Senoo K., and Imura H., Molecular mechanism of glucocorticoid inhibition of human proopiomelanocortin gene transcription. *J. Steroid Biochem. Mol. Biol.*, 40, 301, 1991.

36. Stauber C., Altschmied J., Akerblom I.E., Maron J.L., and Mellon P., Mutual cross-interference between glucocorticoid receptor and CREB inhibits transactivation in placental cells. *New Biologist*, 4, 527, 1992.

37. Chatterjee V.K., Madison L.D., Mayo S., and Jameson J.L., Regression if the human glycoprotein hormone alpha-subunit gene by glucocorticoids: evidence for receptor interactions with limiting transcriptional activators. *Mol. Endocrinol.*, 5, 100, 1991.

38. Boulikas T., Phosphorylation of transcription factors and control of the cell cycle. *Crit. Rev. Eukaryotic Gene Expression*, 5, 1, 1995.

39. Orti E., Hu L.M., and Munck A., Kinetics of glucocorticoid receptor phosphorylation in intact cells. Evidence for hormone-induced hyperphosphorylation after activation and recycling of hyperphosphorylated receptors. *J. Biol. Chem.*, 268, 7779, 1993.

40. Rangarajan P.N., Umesono K., and Evans R.M., Modulation of glucocorticoid receptor function by protein kinase A. *Mol. Endocrinol.*, 6, 1451, 1992.

41. Merelli F., Stojilkovic S.S., Iisa T., Krsmanovic L.Z., Zheng L., Mellon P.L., and Catt K.J., Gonadotrophin-releasing hormone-induced calcium signaling in clonal pituitary gonadotrophs. *Endocrinology*, 131, 925, 1992.

42. Bazan J.F., Neuropoietic cytokines in the hematopoietic fold. *Neuron*, 7, 197, 1991.

43. Paul W.E., Pleiotropy and redundancy: T cell-derived lymphokines in the immune response. *Cell*, 57, 521, 1989.

44. Stahl N. and Yancopoulos G.D., The alpha, betas and kinases of cytokine receptor complexes. *Cell*, 74, 587, 1993.

45. Huang S., Hendriks W., Althage A., Hemmi S., Bluethmann H., Kamijo R., Vilcek J., Zinkernagel R.M., and Aguet M., Immune response in mice that lack the interferon-γ receptor. *Science*, 259, 1742, 1993.

46. Hall A.K. and Rao M.S., Cytokines and neurokines: related ligands and related receptors. *Trends Neurosci.*, 15, 35, 1992.

47. Bazan J.F., Structural design and molecular evolution of a cytokine receptor superfamily. *Proc. Natl. Acad. Sci. U.S.A.*, 87, 6934, 1990.

48. Campbell G.S., Christian L.J., and Carter-Su C., Evidence for involvement of the growth hormone receptor-associated tyrosine kinase in actions of growth hormone. *J. Biol. Chem.*, 268, 7427, 1993.

49. Collingbridge G.L. and Singer W., Excitatory amino acids receptors and synaptic plasticity. *Trends Pharmacol. Sci.*, 11, 90, 1990.

50. Nicoll R.A., Kauer J.A., and Malenka R.C., The current excitement in long-term potentiation. *Neuron*, 1, 97, 1988.

51. Choi D.W., Glutamate neurotoxicity and diseases of the nervous system. *Neuron*, 1, 23, 1988.

52. Dingledine R., McBain C.J., and McNamara J.O., Excitatory amino acids receptors in epilepsy. *Trends Pharmacol. Sci.*, 11, 34, 1990.

53. Cunningham M.D., Fernkany J.W., and Enna S.J., Excitatory amino acid receptors: a gallery of new targets for pharmacological intervention. *Life Sci.*, 54, 135, 1994.

54. Smith G.B. and Olsen R.W., Functional domains of GABAA receptors. *Trends Pharmacol. Sci.*, 16, 162, 1995.

55. Betz H., Kuhse J., Fischer M., Schmieden V., Laube B., Kuryotov A., Langosch D., Meyer G., Bormann J., and Rundstrom N., Structure, diversity and synaptic localization of inhibitory glycine receptors. *J. Physiol.*, 88, 243, 1994.

56. Methot N., McCarthy M.P., and Baeziger J.E., Secondary structure of the nicotinic acetylcholine receptor: implication for the structural models of a ligand-gated ion channel. *Biochemistry,* 33, 7709, 1994.

57. Kehne J., Braon B.M., Harrison B.L., McCloskey T.C., Palfreyman M.G., Poirot M., Salituro F.G., Siegel B.W., Slone A.L., Vangiersbergen P.L.M., and White S.H., MLD 100, 458 and MDL 102, 288 — two potent and selective glycine receptor antagonist with different functional profiles. *Eur. J. Pharmacol.,* 284, 109, 1995.

58. Wang R.Y., Ashby C.R., and Zhang J.Y., Modulation of the A10 dopamine system — electrophysiological studies of the role of the 5-HT3-like receptors. *Behav. Brain Res.,* 73, 7, 1995.

59. Mylecharance E.J., Ventral tegmental area 5-HT receptors — mesolimbic dopamine release and behavioral studies. *Behav. Brain Res.,* 73, 1, 1995.

60. Kilpatrick G.J., Hagan R.M., and Gale J.D., 5-HT3 and 5-HT4 receptors in terminal regions of the mesolimbic system. *Behav. Brain Res.,* 73, 11, 1995.

61. Ferrari M.D., 5-HT3 receptors antagonist and migraine therapy. *J. Neurol.,* 238, S53, 1991.

62. Lewis C., Neidhart S., Holy C., North R.A., Buell G., and Suprenant A., Coexpression of P2X2 and P2X3 receptor subunits can account for ATP-gates currents in sensory neurons. *Nature,* 377, 432, 1995.

63. Chenn C.C., Akopian A.N., Sivilotti L., Colquhoun D., Burnstock G., and Wood J.N., A P2X purinoceptor expressed by a subset of sensory neurons. *Nature,* 377, 385, 1995.

64. Suprenant A., Buell G., and North R.A., P2X receptors bring new structure to ligand-gated ion channels. *Trend Neurosci.,* 18, 224, 1995.

65. Valera S., Hussy N., Evans R.J., Adami N., North R.A., Suprenant A., and Buell G., A new class of ligand-gated ion channel defined by P2X receptor for extracellular ATP. *Nature,* 371, 516, 1995.

66. Schoepfer R., Monyer H., Sommer B., Wisden W., Sprengel R., Kuner T., Lomeli H., Herb A., Kohler M., Burnashev N., Gunther W., Ruppersberg P., and Seeburg P., Molecular biology of glutamate receptors. *Progr. Neurobiol.,* 42, 353, 1994.

67. Grant S.G.N., O'Dell T.J., Karl K.A., Stein P.L., Soriano P., and Kandel E.R., Impaired long-term potentiation, spatial learning, and hippocampal development in fyn mutant mice. *Science,* 258, 1903, 1992.

68. O'Connor J.J., Rowan M.J., and Anwyl R., Tetanically induced LTP involves a similar increase in the AMPA and NMDA receptor components of the excitatory postsynaptic current. Investigation of the involvement of mGlu receptors. *J. Neurosci.,* 15, 2013, 1995.

69. Bertazzon A., Conti-Tronconi B.M., and Raftery M.A., Scanning tunneling microscopy imaging of Torpedo acetylcholine receptor. *Proc. Natl. Acad. Sci. U.S.A.,* 89, 9632, 1992.

70. Green W.N. and Millar N.S., Ion-channel assembly. *Trends Neurosci.,* 18, 280, 1995.

71. Unwin N., Acetylcholine receptor channel imaged in the open state. *Nature,* 373, 37, 1995.

72. Fischbarg J., Cheung M., Li J., Iserovich P., Czegledy F., Kuang K., and Garner M., Are most transporters and channels beta barrels? *Mol. Cell. Biochem.,* 140, 147, 1994.

73. Gorne-Tschelnokow U., Strecker A., Kaduk C., Naumann D., and Hucho F., The transmembrane domains of the nicotinic acetylcholine receptor contain α-helical and β structures. *EMBO,* 13, 338, 1994.

74. Unwin N., Nicotinic acetylcholine receptor at 9Å resolution. J. Mol. Biol., 229, 1101, 1993.

75. Egebjerg J., Kukekov V., and Heinemann S.F., Intro sequence directs RNA editing of the glutamate receptor subunit GluR2 coding sequence. *Proc. Natl. Acad. Sci. U.S.A.,* 91, 10270, 1994.

76. Melcher T., Maas S., Higuchi M., Keller W., and Seeburg P.H., Alpha-amino-3-hydroxy-5-methylisoxazole-4-propionic acid receptor GluR-B pre-mRNA in vitro reveals site-selective adenosine to inosine conversion. *J. Biol. Chem.,* 270, 8566, 1995.

77. Dabiri G.A., Lai F., Drakas R.A., and Nishikura K., Editing of the gluR-B ion channel RNA in vitro by recombinant double-stranded RNA adenosine deaminase. *EMBO,* 15, 34, 1996.

78. Rampersad V., Elliot C.E., Nutt S.L., Foldes R.L., and Kamboj R.K., Human glutamate receptor hGluR3 flip and flop isoforms: cloning and sequencing the cDNAs and primary structure of the proteins. *Biochim. Biophys. Acta,* 1219, 563, 1994.

79. Rutishauser U., Adhesion molecules of the nervous system. *Curr. Opin. Neurobiol.,* 3, 709, 1993.

80. Brummendorf T. and Rathjen F.G., Axonal glycoproteins with immunoglobulin- and fibronectin type III-related domains in vertebrates: structural features, binding activities, and signal transduction. *J. Neurochem.,* 61, 1207, 1993.

81. Rosales C. and Juliano R.L., Signal transduction by cell adhesion receptors in leukocytes. *J. Leukocyte. Biol.,* 57, 189, 1995.
82. Rosales C., O'Brien V., Kornberg L., and Juliano R.L., Signal transduction by cell adhesion receptors. *Biochim. Biophys. Acta,* 1242, 77, 1995.
83. Doherty P. and Walsh F.S., Signal transduction events underlying neurite outgrowth stimulated by cell adhesion molecules. *Curr. Opin. Neurobiol.,* 4, 49, 1994.
84. Williams E.J., Furness J., Walsh F.S., and Doherty P., Activation of the FGF receptor underlies neurite outgrowth stimulated by L1, N-CAM, and N-Cadherin. *Neuron.,* 13, 583, 1994.
85. Takeichi M., The cadherin cell receptor family: roles in multicellular organization and neurogenesis. *Progr. Clin. Biol. Res.,* 390, 145, 1994.
86. Reichardt L.F. and Tomaselli K.J., Extracellular matrix molecules and their receptors: functions in neural development. *Annu. Rev. Neurosci.,* 14, 531, 1991.
87. Hynes R.O., Integrins: versatility, modulation, and signalling in cell adhesion. *Cell,* 69, 11, 1992.
88. Kornberg L., Earp H.S. Parsons J.T., Schaller M., and Juliano R.L., Cell adhesion or integrin clustering increases phosphorylation of a focal adhesion-associated tyrosine kinase. *J. Biol. Chem.,* 267, 23439, 1992.
89. Cheng Q., Kinch M.S., Lin T.S., Burridge K., and Juliano R.L., Integrin-mediated cell adhesion activates mitogen-activated protein kinases. *J. Biol. Chem.,* 269, 26602, 1994.
90. Tedder T.F., Steeber D.A., Chen A., and Engel P., The selectins: vascular adhesion molecules. *FASEB J.,* 9, 866, 1995.

Chapter **2**

VISUALIZATION OF STEROID RECEPTORS: EXPERIMENTAL APPROACHES IN THE CENTRAL NERVOUS SYSTEM

Maryvonne Warembourg
Pierre Poulain

CONTENTS

33

I. INTRODUCTION

Steroid hormones exert profound effects upon the structure and functions of the central nervous system such as the neuroendocrine function and reproductive behavior. Our knowledge concerning the anatomical localization of steroid receptors in target organs was initially provided by radioautography, using a radioactive steroid as a marker for the receptor. In early steroid radioautography, presented in this chapter, animals are injected with radioactive steroid and neural tissue is processed to prevent or minimize translocation of material because the steroid might diffuse away from the site of accumulation. It is presumed that the accumulated radioactivity visualized in the light microscope results from the localization of radiolabeled ligand binding to a protein receptor in the target cell.

More recently, development of specific antibodies to steroid receptors has made it possible to detect the receptor molecules themselves, utilizing immunocytochemical techniques, both by light and electron microscopy. These methods allow detection of receptors independently of their binding activity. It is important to identify in the brain the specific neurons involved in each physiological response to steroid hormones. This identification is based on the knowledge of the constituents synthesized in the steroid receptor-containing cells such as neuropeptides and neurotransmitters, and on the description of their axonal projections. Double-label immunocytochemistry allows us to visualize within a single cell both the steroid receptor and another immunocytochemically recognizable element. This is of importance when working with a heterogeneous cell population. Moreover, the approach that combines single or double immunocytochemical staining with the axonal tracing methods allows analysis of the interconnections between steroid-sensitive neurons and neuropeptidergic or neurotransmitter systems. Tissue preparation in combined procedures must achieve a compromise between processing protocols permitting optimal visualization of immunocytochemically demonstrable compounds and demonstration of neuronally transported fluorescent labels. This chapter provides a current view of available procedures and their applications in the brain.

II. LOCALIZATION OF STEROID RECEPTORS BY LIGHT MICROSCOPIC RADIOAUTOGRAPHY

A. GENERAL CONSIDERATIONS

For more than two decades, radioautography has remained one of the principal techniques for studying the localization of steroid receptors. In the well-established genomic model of steroid action, the steroid-receptor complex binds to a specific sequence of DNA to affect target cells. Such cells are identified by the accumulation of steroid in their nuclei (Figures 1 to 3). In the brain, radioautographic studies of steroid binding have almost exclusively been accomplished by injecting a radioactive steroid into the systemic circulation of living animals, allowing the requisite time for nuclear accumulation (1 to 2 h). The principal steroid-producing organs, the gonads and adrenals, are removed with respect to the class of steroid receptor studied, 8 to 12 days prior to the experiment. This reduces levels of endogenous hormone that would compete with the radioactively labeled ligand for cellular binding sites and make impossible the quantitative interpretations of radioautographic results.

B. RADIOAUTOGRAPHIC PROCEDURE

The main radioisotope used was a low-energy-emitting isotope, tritium (^{3}H), which only affects portions of the film quite close to the label and thereby gives a better spatial resolution to detect silver grains over the nuclei rather than over the cytoplasm (Figures 1 to 3). Tritiated steroids of high specific activity are obtainable from commercial sources in ethanol-toluene solution.

The steroids are diffusible in most histological procedures. Therefore, radioautographic methods used for the study of cellular and subcellular localization of steroid receptors avoid the steps in which diffusion may take place.[1,2] This involves using unfixed, unembedded, frozen sections (see Protocol 1) applied to dry emulsion-coated slides (see Protocols 2 to 4).

PROTOCOL 1: ISOTOPE ADMINISTRATION, BRAIN REMOVAL, AND FREEZING

1. Prior to the experiment, evaporate to dryness the labeled compound in a stream of dry N_2 gas and dissolve it in 10% ethanol isotonic saline.
2. Inject the labeled steroid into the animals subcutaneously with doses in the range of 0.1 to 1.0 μg per 100 g body weight.
3. Sacrifice the animals at different time intervals after the injection, mostly at 1 or 2 h — the time of maximal differential retention of steroids in target tissues.
4. Remove the brain as quickly as possible.
5. Excise tissue samples and mount them on a tissue holder.
6. Immerse the tissue block and holder in liquified propane cooled by liquid nitrogen and transfer them into the cryostat previously cooled to –20°C.

PROTOCOL 2: PREPARATION OF EMULSION-COATED SLIDES BY DIPPING[3]

1. Check the darkroom conditions and equipment:
   - The darkroom must be completely light-tight with a safelight compatible with nuclear emulsion such as Kodak filter Wratten no. 2, temperature (20°C), humidity (50 to 70%).
   - A water bath at 43°C.
   - Dipping containers to dilute emulsion.
   - Cleaned histological slides.
   - Drying racks for slides.
   - Light-tight slide boxes and black tape.
2. Place a container with emulsion and a container with distilled water in the water bath for at least 30 min.
3. Dilute the emulsion (Ilford K_5, Kodak NTB_2 or NTB_3) with distilled water (to 1 volume of emulsion add 1 volume of water).

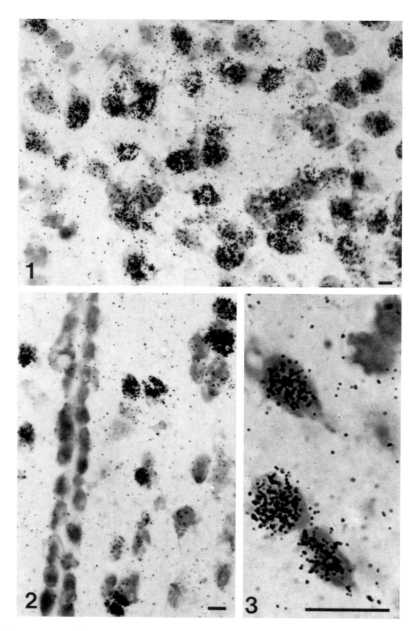

FIGURES 1, 2, and 3 Photomicrographs of radioautograms from various brain regions of an ovariectomized guinea pig 1 h after injection of ^{3}H-estradiol-17β. The target neurons show a concentration of silver grains over the cell nuclei in the nucleus of the stria terminalis (Figure 1), the medial preoptic nucleus (Figure 2), and the arcuate nucleus (Figure 3). Bar = 15 μm.

4. Mix gently to avoid air bubbles.
5. Dip a clean slide into the emulsion by gentle down and up movement and check that the emulsion is free from bubbles. If many bubbles are present, wait a few minutes and dip another clean slide.

6. Hold the emulsion-coated slide in a vertical position for several seconds to allow excess emulsion to drain into paper tissue and wipe the back of the slide.
7. Place the slide in a vertical position on the drying rack until it is dry.
8. Store the emulsion-coated slides into black plastic slide boxes with desiccant and place under refrigeration.

PROTOCOL 3: THAW-MOUNTING OF FROZEN SECTIONS ON EMULSION-PRECOATED SLIDES

1. Move the cryostat into a darkroom.
2. Cut sections, usually 10 μm, under the safelight.
3. Take out of a desiccator box an emulsion-precoated slide and lower it carefully toward the knife until the section adheres to it.
4. Place the slides in light proof boxes containing desiccant and tape the boxes closed with black tape.
5. Expose the slides at −18°C in a freezer between 3 and 12 months. The exposure time depends on the specific activity of the radioactively labeled steroid, the dose, the time interval between application and tissue preparation, the thickness of the section, and the sensitivity of the emulsion.

PROTOCOL 4: PHOTOGRAPHIC PROCESSING AND STAINING

1. Remove the slide box from the refrigerator and warm to room temperature in order to avoid precipitation of moisture on the sections when opening.
2. Develop the preparations in Kodak D19 for 3 min at 20°C in the darkroom.
3. Wash them briefly in 1% aqueous solution of acetic acid and then in distilled water.
4. Fix them in 30% sodium thiosulfate for 6 min at 20°C.
5. Rinse them carefully in water.
6. Stain them with a methyl-green-pyronin staining solution at pH 4.2 for 20 min.
7. Dehydrate them quickly with absolute ethanol, clear in toluene, and mount with a synthetic mounting medium.

C. LABELING CONTROLS
1. Controls for Radioautographic Artifacts

When the unfixed sections are applied to the nuclear emulsion, mechanical and chemical interactions between the sections and the emulsion may occur. The chemical interactions that cause artifacts are referred to as positive or negative chemography.[1,4] Positive chemography is the production of latent images by a chemical activation. Negative chemography is the disappearance of latent images formed during exposure to radiation, often associated with high humidity. The possibility of artifacts must be ruled out in each study by preparation of appropriate control sections from a nonradioactive brain:

- Take sections from a nonradioactive brain and treat them in the same way as the radioactive sections. No grain accumulation under the tissue should be visible after developing.
- Deliberately expose to light the sections mounted on emulsion-precoated slides. In the absence of negative chemography, the emulsion appears entirely black after developing.

2. Controls for Specificity of Binding

The radioactivity detected in the nucleus of a cell may not be part of the originally injected ligand, but a metabolite. Identification of the chemical nature of the radioactivity in the tissue at the time of the preparation of the radioautogram is necessary for the interpretation of the findings. Tissue extraction with radiochemical assays and radioautographic competition with unlabeled antagonists can be used to establish specificity. Nonspecific binding is determined by incubating some adjacent control sections in the presence of a high concentration of unlabeled ligand (100 to 1000 times that of the labeled ligand). Such competition should displace labeled ligand from almost all of the high-affinity receptor sites, but prevent little of the label from binding to the low-affinity nonspecific sites.

III. LOCALIZATION OF STEROID RECEPTORS BY IMMUNOCYTOCHEMISTRY

A. GENERAL CONSIDERATIONS

The purification of steroid hormone receptors has allowed the development of highly specific anti-receptor antibodies. Application of these antibodies to immunocytochemistry requires a clear demonstration that they recognize the receptor in the tissue section. Polyclonal and monoclonal antibodies that recognize the estrogen receptor (ER),[5-8] progesterone receptor (PR),[9-12] androgen receptor,[13-15] and glucocorticoid receptor (GR)[16-19] protein have recently been prepared. Immunocytochemical visualization of steroid receptors is consistenly observed in tissue sections fixed (see Protocol 5) and exposed to subsequent reactions, either as free-floating sections (see Protocol 6) or when mounted on slides sections (see Protocols 7 and 8). The steroid receptor is visualized with standard immunocytochemistry methods such as immunofluorescence (see Protocol 11), biotin-streptavidin complexing (see Protocol 12), or peroxidase anti-peroxidase (PAP) (see Protocol 13). Signals can be enhanced by an intensification of the reaction by nickel (see Protocols 14 and 15) or by glucose oxidase-nickel (see Protocol 16). Although the techniques described here are applied to neural tissue, they also detect steroid receptors in other tissues. For each class of steroid receptor, the immunostaining appears in specific regions of the central nervous system and varies among the species. In the rodent brain, the limbic system is known for its high GR content, whereas receptors for the sex steroids prevail in the hypothalamus. Moreover, the intensity of immunostaining depends on the hormonal status of the experimental animals, which may lead to negative staining results. The ER immunostaining is maximal after ovariectomy, but prolonged treatments with estradiol benzoate decrease the number of ER-positive neurons, as does the immunostaining intensity in their nuclei. On the other hand, an estrogenic treatment of ovariectomized animals induces the PR and is absolutely necessary to visualize it. The GR immunostaining is detected in the intact animal, but is abolished by adrenalectomy and restored rapidly by corticosterone treatment. Taken together, these data support the immunocytochemical method of choosing an animal model possessing the highest levels of steroid receptors. Immunocytochemical studies report a localization of GR both in the cytoplasm and in the nucleus of neurons and glia,[20-22] whereas the sex steroid immunoreactivity is found almost exclusively in the nucleus of target neurons after gonadectomy (Figures 4 to 6).[23-26]

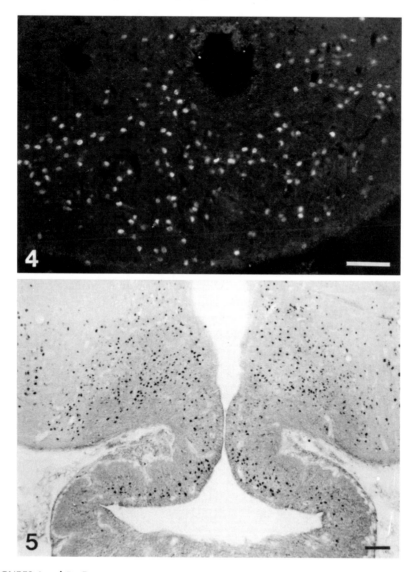

FIGURES 4 and 5 Progesterone receptor immunoreactivity in frontal sections of the guinea pig arcuate nucleus after immunofluorescence (Figure 4) and biotin-streptavidin peroxidase staining (Figure 5). Bar = 100 μm.

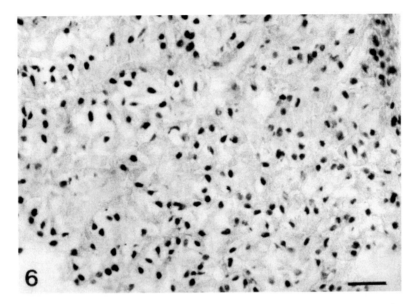

FIGURE 6   Progesterone receptor immunoreactivity in the monkey anterior pituitary after PAP staining with nickel intensification of DAB. Bar = 41 μm.

B. LIGHT MICROSCOPIC VISUALIZATION
1. Tissue Fixation

Tissue fixation is immensely important in immunocytochemical procedures, as the morphological characteristics and the immunoreactivity of the desired antigen must be preserved while allowing penetration of antibodies. Fixatives composed mainly of precipitating reagents such as Carnoy's, Helly's, and ethanol prevent receptor steroid staining, whereas the cross-linking fixatives such as formaldehyde, glutaraldehyde, and acrolein preserve the immunoreactivity of the receptor.[27] We failed to locate gonadal steroid receptors in completely unfixed sections of frozen neural tissue.[23] The method involving fixation by perfusion gives better results than fixation by immersion before or after sectioning of frozen brain tissue. Although a paraformaldehyde-glutaraldehyde-picric acid fixative or a paraformaldehyde-acrolein fixative[28,29] can be used for a light microscopy, we recommend the paraformaldehyde picric acid fixative (see Protocol 5). Before perfusion of fixative, rinsing the vascular system with saline solution removes all the blood from the animal and thus might avoid a nonspecific reaction.

PROTOCOL 5: FIXATION BY PERFUSION

1. Anesthetize the animal. All of the most commonly used drugs (ether, ketamine, xylazine, pentobarbital) seem to give satisfactory results.
2. Perfuse the animal transcardially using an intraventricular catheter attached to a peristaltic pump with 50 to 100 ml of heparinized saline (0.9% NaCl in water) followed by 500 ml (for a guinea pig of 400 g) of cold 4% paraformadehyde

and 15% satured picric acid dissolved in 0.1 M sodium phosphate buffer (PB), pH 7.4. In order to prevent the loss of the tissue antigens only a small volume of rinsing solution should be perfused in as short a time as possible.
3. Remove the brain from the cranium and dissect the regions of interest.
4. Immerse the blocks in the same fixative for 4 to 6 h at 4°C.

2. Sectioning

Neuroanatomical studies of the central nervous system that use immunocytochemical procedures usually require that tissue sections be prepared without embedding in paraffin, plastic, or celloidin since these procedures reduce or abolish many forms of antigenic activity. Although small blocks of fixed, unembedded tissue can be cut on a vibratome (see Protocol 6), cutting frozen sections in a freezing stage (see Protocol 7) or on a cryostat (see Protocol 8) is the most practical alternative for many studies.

PROTOCOL 6: VIBRATOME SECTIONS[30]

1. Wash the tissue block and store it in 0.1 M PB at 4°C.
2. Cut 40- to 60-μm-thick sections with a vibratome cutting system filled with ice-cold 0.1 M PB.
3. Collect brain sections and wash in 0.1 M PB.
4. Transfer sections, before immunostaining to 0.1 M PB containing 10% sucrose and freeze-thaw treat to enhance tissue penetration of the antisera.[30]

PROTOCOL 7: FREEZING MICROTOME SECTIONS[24,26]

1. Transfer the tissue blocks to a 20% sucrose 0.1 M PB solution until they sink.
2. Cut 40- to 60-μm thick sections with a freezing microtome.
3. Mount sections on slides coated with chrome-alum or store sections in an ethylene glycol-based cryoprotectant solution[31] for 1 to 2 weeks at –20°C prior to processing for immunocytochemistry.

PROTOCOL 8: CRYOSTAT SECTIONS[23,25]

1. Transfer the tissue pieces to a 0.1 M PB solution containing 15 to 20% sucrose until they sink.
2. Freeze the tissue embedded in Tissue Tek OCT compound (Miles Inc. Elkhart, NJ, U.S.) by immersion in liquid nitrogen-cooled isopentane. The frozen blocks can be stored at –80°C until sectioning.
3. Mount frozen tissue blocks on cryostat chucks with OCT compound.
4. Cut 5- to 15-μm coronal sections at –20°C.
5. Thaw-mount sections on slides coated with chrome-alum and store at –80°C until processed for immunocytochemistry.

3. Immunocytochemical Staining Procedures

There is a wide range of immunocytochemical methods and enhancement procedures now in common use, and the choice is usually dependent on the experience of the researcher. Among immunocytochemical techniques available to visualize antibodies that are bound to a tissue section, indirect staining procedures (fluorescent conjugate, PAP, enzyme conjugate) are supposed to be more sensitive than direct procedures. Although fluorescence techniques provide a higher resolution in light microscopy, enzyme markers are mainly used because of the stability of the chromogen. The immunocytochemical reactions can be carried out either on free-floating sections[24,26,30] or on sections mounted on slides.[23,25] Most procedures using thick sections include treatment with sodium borohydride[32] to eliminate excess aldehydes from tissue (see Protocol 9).

PROTOCOL 9: SODIUM BOROHYDRIDE PRETREATMENT

1. Incubate sections for 20 min in 1% sodium borohydride-PB.
2. Wash sections in 0.1 M PB (3 × 15 min).

In our laboratory, immunofluorescence and immunoperoxidase procedures used for single PR detection in the central nervous system (Figures 4 to 6) by means of monoclonal mouse immunoglobulin (Ig) antibodies (Mi 60, LET 64, and LET 126; gift of Prof. E. Milgrom) raised against rabbit uterine PR, are described as follows.

PROTOCOL 10: WASHING OF SECTIONS

1. Wash cryostat sections in 0.01 M sodium phosphate-buffered saline (PBS) pH 7.4 for 30 min (3 × 10) before processing for immunocytochemistry. These washes are carried out after each subsequent incubation, unless otherwise stated, for 20 min (2 × 10).

PROTOCOL 11: IMMUNOFLUORESCENCE TECHNIQUE (FIGURE 4)

After washing, incubate sections with:

1. 5% nonimmune serum of the animal species in which the secondary antibody is raised (sheep in this procedure) in PBS for 20 min. This reduces the nonspecific binding of primary antibody.

2. The anti-PR antiserum (2.5 μg/ml) in PBS for 24 to 68 h at 4°C in a humid atmosphere.
3. Biotinylated sheep anti-mouse Ig (species-specific antibodies) 1:200 in PBS for 90 min at 4°C.
4. Fluorescein isothiocyanate (FITC)-streptavidin 1:50 in PBS or Texas-red streptavidin 1:200 for 90 min at 4°C.
5. Coverslip sections with glycerine-PBS (3:1, v/v).
6. Examine sections under a fluorescent microscope equipped with appropriate filters of 450 to 490 nm wavelengths for identification of FITC and 530 to 560 nm for identification of Texas-red.

PROTOCOL 12: BIOTIN-PEROXIDASE-STREPTAVIDIN TECHNIQUE (FIGURE 5)

1. Incubate sections as described in Protocol 11, steps 1 to 3.
2. Apply streptavidin-horseradish peroxidase conjugate 1:100 in PBS to sections for 1 h at room temperature.
3. Flood the slides with 0.05% 3,3'-diaminobenzidine tetrahydrochloride (DAB) in this buffer, pH 7.6, and 0.05% hydrogen peroxide to reveal peroxidase staining. The appropriate time of incubation has to be judged by visual observation of the sections as they are turning brown; this usually takes 8 to 10 min. To enhance the signal, nickel intensification methods of the DAB reaction (see Protocols 14 to 16) should be used.
4. Wash sections, dehydrate in ethanol followed by toluene, and cover with a synthetic mounting medium (Eukitt) under a coverslip. The material is routinely examined without counterstaining with a light microscope.

PROTOCOL 13: PEROXIDASE-ANTIPEROXIDASE METHOD (FIGURE 6)

After washing, incubate with:

1. Normal goat serum 5% in PBS for 20 min.
2. The anti-PR antiserum (2.5 μg/ml) in PBS for 24 to 68 h at 4°C in a humid atmosphere.
3. Goat anti-mouse Ig 1:60 for 1 h at room temperature.
4. Mouse PAP 1:100 in PBS for 1 h at room temperature.
5. Incubate sections with chromogen solution, without (as described in Protocol 12; step 3) or with (see Protocols 14 to 16) intensification.
6. Mount as described in Protocol 12, step 4.

PROTOCOL 14: NICKEL AMMONIUM SULFATE INTENSIFICATION OF REVELATION SYSTEM[33] (FIGURE 6)

1. Incubate sections as described in Protocol 12 (steps 1 and 2) or Protocol 13 (steps 1 to 4).
2. Wash sections in 0.1 M acetate buffer, pH 6.0.
3. Incubate sections for 6 to 10 min in 0.1 M acetate buffer, pH 6.0, containing:
    2.5% nickel ammonium sulfate
    0.05% DAB
    0.01% hydrogen peroxide
4. Mount as described in Protocol 12, step 4.

PROTOCOL 15: NICKEL SULFATE-IMIDAZOLE INTENSIFICATION OF REVELATION SYSTEM[26]

1. Incubate sections as described in Protocol 12 (steps 1 and 2) or Protocol 13 (steps 1 to 4).
2. Prepare acetate-imidazole buffer (175 mM sodium acetate, 10 mM imidazole, pH 7.2).
3. Wash sections twice in acetate-imidazole buffer.
4. Incubate sections for about 20 min with chromogen solution at pH 6.5 containing:
    NiSO$_4$ 100 mM
    sodium acetate   125 mM
    imidazole   10 mM
    DAB   0.05 %
    and add hydrogen peroxide to a final concentration of 0.01%.
5. Transfer to fresh acetate-imidazole buffer to stop the reaction. Imidazole is a nitrogenous base that enhances peroxidase activity. PR immunoreactivity displays a dark blue color.
6. Mount as described in Protocol 12, step 4.

PROTOCOL 16: GLUCOSE OXIDASE-NICKEL INTENSIFICATION OF REVELATION SYSTEM[34]

1. Incubate sections as described in Protocol 12 (steps 1 and 2) or Protocol 13 (steps 1 to 3).
2. Rinse in 0.1 M acetate buffer pH 6.0.
3. Incubate in glucose oxidase-DAB-nickel solution for about 20 min
    A: nickel ammonium sulfate 2.50 g
       0.2 M acetate buffer (pH 6.0) 50 ml
    B: DAB   50-70 mg
       distilled water   50 ml

Mix A and B solutions before use and add:

B-D-glucose	200 mg
ammonium chloride	40 mg
glucose oxidase	0.5 to 1 mg

Check the staining with a microscope: PR immunoreactivity should be dark blue.

4. Rinse in acetate buffer.
5. Mount as described in Protocol 12, step 4.

4. Immunocytochemical Controls

In order to study the specificity of the immunocytochemical staining procedure, sets of sections are incubated with increasing dilutions of anti-PR antibody up to a dilution at which immunoreactivity is not detectable.

A number of essential checks are required in order to eliminate false positives and to demonstrate the specificity of the reaction. For each control experiment, control sections are run in parallel with sections processed as usual for immunocytochemistry. The procedures used to visualize the specificity of the antibodies can be as follows:

- Replace the primary mouse anti-PR antibody by either mouse receptor-unrelated monoclonal antibody used at the same concentration or by the monoclonal anti-PR antibody pretreated (16 to 500 pmol of receptor per microgram of IgG, 16 h at 4°C) before the immunostaining procedure with highly purified PR (7 nmol bound hormone per milligram of protein).[23,35]
- Substitute primary antibody with diluted normal mouse serum or omit primary antibody and perform all other processing steps as usual.
- Incubate with primary antibody followed by secondary and/or tertiary immunoglobulins raised in an inappropriate species.
- Incubate with primary antibody and omit the secondary antibody or DAB.

In all the above-mentioned cases, specific immunostaining is absent.

C. ELECTRON MICROSCOPIC VISUALIZATION

Ultrastructural immunocytochemistry reveals the intracellular distribution of steroid receptors, their specific association with different organelles and nuclear structures (Figures 7 and 8), as well as their trafficking mechanisms. For electron microscopy analysis of neural tissue samples, the same protocols are performed as those applied for light microscopic visualization of receptors regarding vibratome sections (see Protocol 6) and immunocytochemical staining (see Protocols 12 and 13). However, fixation is somewhat different. Glutaraldehyde is known to provide an excellent structure preservation. Thus, the animals are perfused as described in Protocol 5 with paraformaldehyde (1.5 to 4%) and a high concentration of glutaraldehyde (1 to 1.5%) in 0.1 M PB pH 7.4, and postfixed in the same fixative.[30,36,37] Following visualization of immunoreactivity with DAB chromogen or post-DAB intensification (see Protocol 17), the sections of interest are processed for electron microscopy (see Protocol 18).

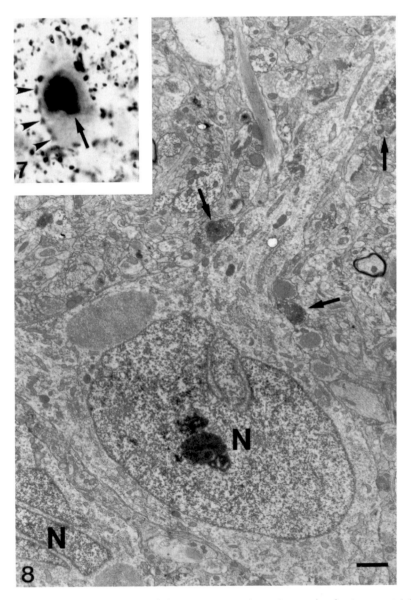

FIGURES 7 and 8   Light (Figure 7) and electron (Figure 8) photomicrographs of an immunostaining for the progesterone receptor, associated exclusively with the cell nucleus (arrow on Figure 7; N on Figure 8), in the monkey ventromedial nucleus. These photographs also illustrate an example of a double immunoperoxidase staining (not described in this chapter) by showing somatostatin-immunopositive boutons surrounding the progesterone receptor-containing neuron (arrowheads on Figure 7; arrows on Figure 8). Original magnification of Figure 7: × 250; bar = 1 μm. The photomicrographs are a generous gift of Dr. C. Leranth.

PROTOCOL 17: SILVER-GOLD POSTINTENSIFICATION OF THE DAB REACTION

For the step-by-step protocol of silver enhancement see Reference 38.

1. Incubate sections with 10% thioglycolic acid for 2 h to reduce tissue argyrophilia.
2. Intensify sections in the special physical developer described by Gallyas[39] and allow to react until the brown DAB turns black.
3. Tone the silver with gold to partly replace the silver grains by more stable gold particles.

PROTOCOL 18: PROCESSING FOR ELECTRON MICROSCOPY[30,36,37]

1. Postfix sections in 1% osmium tetroxide in 0.1 M PB pH 7.4 for 20 min.
2. Dehydrate through graded ethanol solutions (include 1% uranyl acetate as a contrasting agent in the 70% ethanol solution).
3. Flat-embed into Epon resin.
4. Fix some of sections of regions of interest on cylindrical araldite blocks.
5. Cut ultrathin sections and collect them on Formvar-coated single-slot grids.
6. Counterstain the sections with lead citrate and uranyl acetate or leave unstained.
7. View with an electron microscope.

D. APPLICATIONS OF COMBINED TECHNIQUES
1. Demonstration of Two Antigens

By means of double labeling, the neuropeptide or neurotransmitter content of steroid receptor-containing neurons can be determined. The purpose of the immunocytochemical double staining is to visualize two different antigens in one tissue on the basis of two different fluorescent compounds or two different chromogens. Numerous combined techniques are available for this, both in light and electron microscopy (Figures 7 and 8).[36] This chapter discusses the protocols used in our laboratory to demonstrate via light microscopy the presence of two steroid (PR and ER) receptors in the same neurons,[25] as well as the coexistence of PR with different neuropeptides[40-42] or neurotransmitters.[43] The perfusion, fixation, sectioning parameters, dilution of anti-PR antibody, and washes are the previously described techniques which permit optimal immunocytochemical visualization of the steroid receptors. All the protocols described herein use primary antibodies raised in different species. Immunoreactivity of two antigens can be detected by various revelation methods: using a secondary antibody conjugated either with two fluorescent markers (see Protocol 19) or with a fluorescent reagent and an enzymatic marker (peroxidase) (see Protocol 20), or with an enzymatic marker (peroxidase) (see Protocol 21) revealed by two different chromogens (DAB and aminoethylcarbazole). The double immunoperoxidase staining procedures involve sequential incubations of tissue sections (see Protocols 20 and 21).

PROTOCOL 19: DOUBLE IMMUNOFLUORESCENCE METHOD

In our laboratory, the following protocol was applied to visualize neurotensin in PR-containing neurons[42] (Figures 9 and 10):

1. Incubate sections with a mixture containing 10% normal sheep serum and 10% normal monkey serum for 30 min.
2. Incubate sections with a mixture containing primary antibodies (mouse monoclonal anti-PR and rabbit polyclonal anti-neurotensin) for 68 to 70 h at 4°C in a humid atmosphere.
3. Visualize the neurotensin immunoreactivity by incubating the sections with fluorescein-conjugated donkey anti-rabbit Ig 1:50 for 90 min at 4°C.
4. Visualize the PR immunoreactivity by incubating the sections with biotinylated sheep anti-mouse Ig, then with Texas-red streptavidin (see Protocol 11, steps 3 to 5).
5. Coverslip and examine sections as described in Protocol 11, steps 5 and 6.

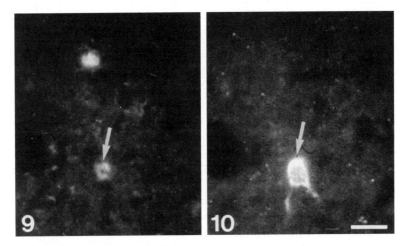

FIGURES 9 and 10 Photomicrographs of the same frontal section through the guinea pig ventrolateral nucleus after a double immunofluorescence technique with mouse antiserum to progesterone receptor (Figure 9) and rabbit antiserum to neurotensin (Figure 10). Observation of neurons was made by switching from the filter block for identification of Texas-red to the filter block for identification of FITC. Observe a neurotensin-immunoreactive cell which is also PR-immunoreactive (arrows). Bar = 20 μm.

PR-immunoreactive cells display a red fluorescence which is present only in the nucleus, whereas neurotensin immunoreactive cells show green fluorescence in the cytoplasm. By switching filters, coexistence can be directly established. There is no overlap in excitation wavelength between Texas-red and FITC, and Texas-red does not shine through when the FITC-induced immunofluorescence is analyzed. This procedure has also been applied to visualize PR and ER in the same neurons, using a monoclonal mouse antibody against PR and a monoclonal rat antibody against ER.[25] However, a convincing demonstration of coexistence of two receptors in the same neuron requires color photographs since both fluorophores occupy a nuclear position. The sequential immunofluorescence and immunoperoxidase staining (see Protocol 20) is suitable for demonstrating two distinct antigens in the same compartment of a cell (Figures 11 and 12).[25]

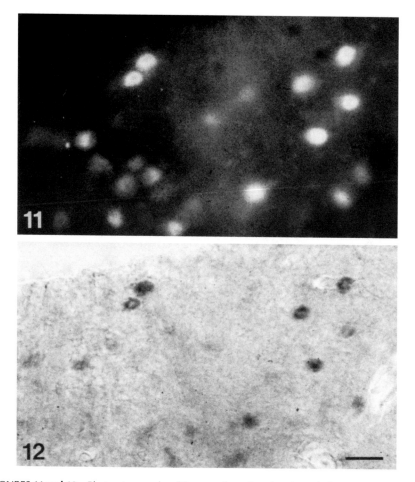

FIGURES 11 and 12 Photomicrographs of the same frontal section through the guinea pig arcuate nucleus after immunofluorescence with antiserum to the estrogen receptor (Figure 11) and, after photography, immunoperoxidase staining with antiserum to the progesterone receptor (Figure 12). The majority of cell nuclei are doubly labeled. Bar = 22 μm.

PROTOCOL 20: IMMUNOFLUORESCENCE AND PEROXIDASE-ANTIPEROXIDASE METHOD

For immunofluorescence (see Protocol 11), incubate sections with:

1. Primary antiserum to the first antigen (either PR or ER).
2. Biotinylated sheep anti-mouse or anti-rat Ig.
3. FITC-streptavidin or Texas-red streptavidin.
4. Wet-mount sections and observe by fluorescence microscopy.
5. Remove coverslips after photography.

For the immunoperoxidase reaction (see Protocol 13), incubate sections with:

6. Primary antiserum to the second antigen (either PR or ER).
7. Mouse or rat PAP.
8. Chromogen solution, with or without intensification (see Protocols 14 to 16).
9. Compare results to previous photographs.

PROTOCOL 21: DOUBLE IMMUNOPEROXIDASE METHOD

In our laboratory, the following protocol was applied to demonstrate neuropeptide Y (NPY) in PR-containing neurons (Figure 13):[41]

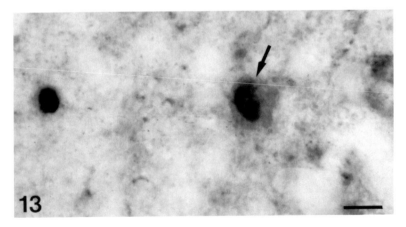

FIGURE 13 Photomicrograph of a frontal section of the guinea pig arcuate nucleus after double immunoperoxidase technique with mouse antiserum to the progesterone receptor and rabbit antiserum to NPY. Two progesterone receptor-immunoreactive cells with dark staining of the nucleus can be observed, one cell containing a cytoplasmic NPY immunoreactivity (arrow). Bar = 9 μm.

1. Incubate sections as described in Protocol 12 with:
   a. Primary antiserum directed against PR.
   b. Biotinylated sheep anti-mouse IgG.

    c. Streptavidin peroxidase.

    d. Chromogen solution with ammonium sulfate intensification (see Protocols 14 to 16).

2. Rinse sections in a methanol/hydrogen peroxidase series[44] to inhibit any remaining peroxidase activity as follows:

    a. Methanol 20% in PBS for 2 min.

    b. Increase methanol concentrations in 20% increments until sections are bathed in 100% methanol for 10 min.

    c. Decrease methanol concentrations in a similar step-wise fashion.

    d. Place sections, after the final 20% methanol rinse, in PBS containing 0.3% hydrogen peroxidase for 10 min.

    e. Rinse sections in PBS for 1 h.

3. Incubate sections as described Protocol 13 with:

    a. Primary antiserum directed against NPY in PBS for 24 to 48 h at 4°C.

    b. Goat anti-rabbit Ig 1:80 for 1 h at room temperature.

    c. Rabbit PAP.

    d. Prepare a chromagen solution just before use containing:

        4 mg 3-amino-9-ethylcarbazole.

        1 ml N-N-dimethylformamide.

        14 ml 0.1 M acetate buffer pH 5.0.

    and add extemporaneously 15 µl 33% hydrogen peroxide.

    e. Apply one drop of the solution to the section for 20 min in darkness.

    f. Wash in tap water.

    g. Mount in a hydrophilic medium (Glycergel).

Each tissue antigen can be labeled by a distinct color staining and be separated on the basis of cellular localization. The dual immunoperoxidase method results in a dark blue-to-black-stained nuclei and a red-stained cytoplasm.

Controls for the sequential dual-staining immunocytochemical procedures include:

- Omission of primary antibodies.
- Substitution of primary antibodies by preabsorbed antiserum or diluted normal serum.
- Incubation in the primary antibodies followed by inappropriate secondary and/or tertiary Ig, one at a time, to test the absence of cross-reactivities between the individual immunoreagents.

No specific immunoreactivity is visible in sections following any of the controls.

2. Combination of Steroid Receptor Immunocytochemistry With Axonal Tracing

a. Retrograde Axonal Tracing

The combination of steroid receptor detection with retrograde axonal tracing can determine whether steroid receptor-containing neurons project to a particular field in which the tracer has been injected. Because fluorescent substances are visible without histochemical procedures, they have become widely used for retrograde tracing. Radioautographic detection of steroid receptors can be combined with retrograde tracing using fluorescent dyes.[45] However, association of fluorescent immunocytochemical detection of steroid receptors with fluorescent retrograde tracing is simpler because visualization of the immunostaining and retrograde label is achieved by switching filter systems during light microscopic observation.

The selection of the fluorescent retrograde tracer is based on two main considerations. First, the tracer must provide strong labeling that is stable after exposure to the water phases when the incubation and rinsing procedures of the immunocytochemical processes are performed. Second, there must be a clear separation between the fluorescence of the tracer and the fluorochrome associated to the second antibody. Fluoro-Gold (FG)[46] is very useful in combination with receptor immunocytochemistry when Texas-red is conjugated to streptavidin (see Potocol 11), since UV excitation is used to visualize FG and green excitation is used for Texas-red. Fluoro-Gold gives a strong yellow-gold color to the perikaryon and the proximal dendritic processes of the neuron (Figure 14) that is readily distinguished from the red nuclear labeling (Figure 15). One drawback of the association is that FG also labels the nucleus, but the fluorescence of the nucleus does not shine through the Texas-red filters.

The effectiveness of retrograde axonal tracing mainly depends on the reproducibility of the tracer injection. On the other hand, discrete injections are often needed when small projection fields are studied. FG can be injected either by pressure or by iontophoresis, and both types of injection give good results. We prefer to apply FG by iontophoresis because application by pressure may damage the tissue and imposes minimum limits to the size of the injection. Details for iontophoretic injection of FG have been given.[47,48] In our observations, using FG dissolved in cacodylate[48] permitted more consistent injections because the tips of the pipettes were less frequently blocked during application of the iontophoretic current, due to an excessive increase of the micropipette resistance.

PROTOCOL 22: RETROGRADE AXONAL TRACING FOLLOWED BY IMMUNOCYTOCHEMISTRY

1. Dissolve 2% Fluoro-Gold (Fluorochrome, Englewood, CO, U.S.) in a solution of 0.1 M sodium cacodylate trihydrate filtered through 0.22 μm pores.
2. Fill glass micropipettes (1.5 mm external diameter with inner glass filament for facilitating filling, with tip broken at approximately 40 μm) with the solution. Prepare the micropipettes just before filling (cleaning or treating the tubes does not improve the results).
3. Stereotaxically lower the filled micropipette into the selected area of the brain.
4. Inject by passing constant positive current, 4 s on and 4 s off, through a precision current source. Monitoring the micropipette resistance or using an audible warning that sounds when the maximum load voltage occurs (given, for example, by the A360 high-voltage stimulator manufactured by World Precision Instruments) is useful to ascertain that the micropipette is not blocked.
5. Use 1 to 5 μA for 1 to 5 min, depending on the injection site diameter required. A standard injection with 5 μA for 3 min results in a site diameter

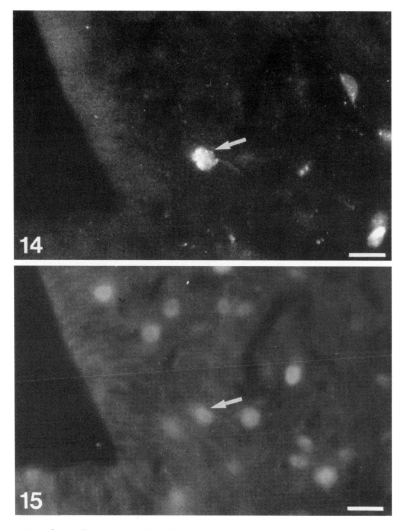

Figures 14 and 15 Photomicrographs of the same frontal section after retrograde axonal transport of FG (Fluoro-Gold) (Figure 14) and immunocytochemistry with antiserum to the progesterone receptor (Figure 15). A doubly labeled neuron is indicated by arrows. Fluoro-Gold has been injected in the preoptic area and the fluorescent neurons are observed in the arcuate nucleus of the guinea pig. Observation of cells was made by switching from the filter block for identification of FG to the filter block for identification of Texas-red. Bar = 20 μm.

of approximately 600 μm. The maximal value that can be used without micropipette blockage is around 25 μA/min.

6. Survival interval of 3 to 5 days
7. Use Protocol 5 for fixation of the brain.
8. Use Protocol 8 for cryostat sections.
9. Examine sections under a fluorescent microscope equipped with filters of 340 to 380 nm to observe the neurons retrogradely labeled with FG.

10. Perform immunofluorescence techniques to visualize the steroid receptor with the complex biotin-Texas-red-streptavidin, taking Protocols 10 and 11 as examples.

11. Examine sections under a fluorescent microscope equipped with filters of 340 to 380 nm wavelenghts for identification of FG and 530 to 560 nm for identification of Texas-red.

It is often practical to observe the injection site separately, in order to verify its localization on counterstained slides. For this, the piece of fixed brain containing the injection site is removed after cryoprotection in the sucrose solution (see Protocol 8, step 1) and the following protocol is used.

PROTOCOL 23: OBSERVATION OF THE INJECTION SITE

1. Cut sections at 80 μm on a freezing microtome.
2. Collect on ice-cold 0.1 M PB.
3. Mount on slices coated with chrome-alum.
4. Dry in air at room temperature in the dark for 6 to 24 h.
5. Dehydrate in ethanol 50% (1 min), ethanol 100% (30 s), clear in xylene (2 x 5 min) and coverslip with Fluoromount (Gurr). Sections can be stored in a cold room in the dark for months.
6 *Counterstaining*: after step 4, rehydrate in cold distilled water (1 to 5 min), stain in cold 0.0001% acridine orange for 30 s to 2 min[49] or in 0.05% methylene blue for 5 to 30 s,[50] then wash in cold distilled water (3 × 10 s). Go back to steps 4 and 5.

b. Anterograde Axonal Tracing

The combination of steroid receptor detection with anterograde axonal tracing allows us to know whether a particular structure in which the tracer has been injected establishes direct contacts with the neuron possessing steroid receptors. Dextran-amines are axonal tracers that display a very high degree of sensitivity.[51] Anterograde labeling with dextran-amines results in a detailed morphology of the fiber's trajectories, varicosities, and presumptive terminal boutons. We have observed that the fluorescent tetramethylrhodamine-conjugated dextran-amine Fluoro-Ruby (FR)[52,53] is compatible with the immunocytochemical detection of steroid receptors at the light microscopic level. The intensity of the labeled fibers is unaffected by exposure to the water phases during receptor immunostaining. Choosing FR offers the advantage that it can be visualized with the same filter combination used to observe the red color of the stained nucleus when Texas-red is conjugated to streptavidin (see Protocol 11). It is obviously unnecessary to use markers with different colors for the labeled fibers and for the stained nucleus, because the two markers are localized in different neuronal compartments (Figure 16). The drawback inherent to this approach is that the evaluation of putative contacts at the light microscopic level is hampered by the fact that the postsynaptic target is identified by the presence of nuclear staining while the cytoplasm is unlabeled.

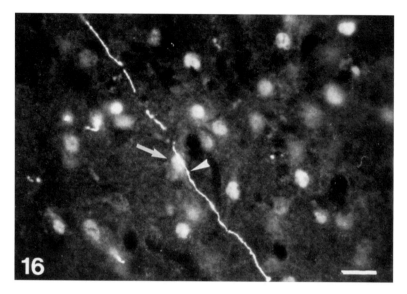

Figure 16   Low-power view of a fiber labeled after anterograde axonal transport of Fluoro-Ruby, associated with a progesterone receptor-immunoreactive nucleus (arrow). Observe varicosities (arrowhead) signalling boutons *en passant*. The fiber originates from the lateral septum and the immunoreactive neurons are located in the preoptic area of the guinea pig (courtesy of F. Varoqueaux). Observation is made through the filter block for identification of Texas-red. Bar = 20 μm.

PROTOCOL 24: ANTEROGRADE AXONAL TRACING FOLLOWED BY IMMUNOCYTOCHEMISTRY

1. Dissolve 10% Fluoro-Ruby (MW 10,000, Molecular Probes, Eugene, OR, U.S.) in a solution of 0.01 M PB pH 7.4, filtered through 0.22 μm pores.
2. Fill glass micropipettes (1.5 mm external diameter with inner glass filament for facilitating filling, with tips broken at approximately 40 μm) with the solution.
3. Stereotaxically lower the filled micropipette into the selected area of the brain.
4. Inject by passing constant positive current, 4 s on and 4 s off, through a precision current source. A standard injection with 6 μA for 15 min results in an injection site diameter of approximately 150 μm. Make multiple injections to cover a larger site.
5. Survival interval of 6 to 11 days.
6. Use Protocol 5 for fixation of the brain.
7. Use Protocol 8 for cryostat sections.
8. Perform an immunofluorescence technique to visualize the steroid receptor with the complex biotin-streptavidin-Texas-red, taking Protocols 10 and 11 as examples.
9. Observe sections under a fluorescent microscope equipped with wavelength filters of between 530 to 560 nm for identification of FR and Texas-red.
10. Refer to Protocol 23 to separately examine the injection site.

IV. RESULTS AND DISCUSSION

For over 25 years, tritium-labeled steroids have been used in binding studies of steroid hormones and radioautography has remained the principal method for the study of the distribution of steroid receptors in the brain.[54-59] *In vivo* radioautographic methods have provided evidence for the nuclear localization of hormone in target cells. These methods give sufficient resolution but require relatively large quantities of isotope to occupy the receptors, and metabolism of labeled steroids may occur. The availability of highly specific antibodies directed against steroid receptors has provided an effective tool for the direct visualization by immunocytochemical techniques of the steroid receptors within individual cells. Immunocytochemistry is faster than radioautography. Another advantage of immunocytochemistry is that receptors can be studied at the subcellular level.

This chapter provides detailed descriptions of the most frequently used fixation, sectioning, intensification techniques, and immunostaining procedures at the light microscopic level. Furthermore, it presents some examples of double immunostaining techniques, which are essential to specify the neurotransmitter or neuropeptide content of the steroid receptor-containing neurons, and some examples of the combination of immunocytochemistry with axonal tracing methods.

The analysis of the efferent projections of the neurons that possess steroid receptors requires the use of retrograde axonal tracing methods. Many studies combining retrograde tracing and radioautographic detection of receptors (reviewed in Reference 60) or retrograde tracing and immunofluorescence techniques[61-65] have extended our knowledge of the relationships between neurons possessing gonadal steroid receptors. The triple combination of steroid receptor immunoreactivity, transmitter immunocytochemistry, and retrograde axonal transport methods may even extend the analysis. As it is difficult to localize three labels with specific fluorescent illumination, double fluorescent analysis should be associated to PAP detection of the transmitter.[62]

The combination of anterograde axonal tracing and detection of steroid receptors promises fruitful future developments. Indeed, the detection of afferent connections to neurons containing steroid receptors is a very important issue in determining the neuroanatomical substrates by which information from different parts of the brain access these neurons. By using the popular anterograde tracer *Phaseolus vulgaris* leukoagglutinin, it was recently possible to trace axonal projections to ER immunoreactive neurons.[66-67] This chapter presents some basic instructions about fluorescent axonal tracing methods which are suitable for light-microscopic neuronal mapping. Concerning anterograde axonal tracing, the simple light-microscopic approach described in this chapter is very useful in providing the first evidence of putative relationships between a given area of the brain and steroid receptor-containing neurons. Nevertheless, demonstrating that labeled fibers indeed establish synaptic relationships with the identified neuron requires the use of the electron microscope. For this purpose, biotinylated dextran-amine[68] should be used instead of FR to obtain a reaction product visible at the electron microscopic level.

In conclusion, this review details the procedures used to determine the cells possessing steroid receptors, to elucidate their chemical identity, and to prove their connections by using neuroanatomical tracing methods. The association of these techniques is of great use in better understanding the regulations exerted by the steroid hormones in the brain.

REFERENCES

1. Rogers, A. W., *Techniques of Autoradiography*, Elsevier, Amsterdam, 1979, 429.
2. Warembourg, M., Detection of diffusible substances. *J. Microsc. Biol. Cell.*, 27, 277, 1976.
3. Kopriwa, B. M. and Leblond, C. P., Improvements in the coating technique of radioautography. *J. Histochem. Cytochem.*, 10, 269, 1961.
4. Rogers, A. W. and John, P. N., Latent image stability in autoradiographs of diffusible substances, in *Autoradiography of Diffusible Substances*, Roth, L. J. and Stumpf, W. E., Eds., Academic Press, New York, 1969, 51.
5. Greene, G. L., Nolan, C., Engler, J. P., and Jensen, E. V., Monoclonal antibodies to human estrogen receptor. *Proc. Natl. Acad. Sci. U.S.A.*, 77, 5115, 1980.
6. Greene, G. L. and Jensen, E. V., Monoclonal antibodies as probes for estrogen receptor detection and characterization. *J. Steroid Biochem.*, 16, 353, 1982.
7. Greene, G. L., Sobel, N. B., King, W. J., and Jensen, E. V., Immunochemical studies of estrogen receptors. *J. Steroid Biochem.*, 20, 51, 1984.
8. King, W. J. and Greene G. L., Monoclonal antibodies localize estrogen receptor. *Nature*, 307, 745, 1984.
9. Logeat, F., Vu Hai, M. T., Fournier, A., Legrain P., Buttin, G., and Milgrom, E., Monoclonal antibodies to rabbit progesterone receptor: Cross-reaction with other mammalian progesterone receptor. *Proc. Natl. Acad. Sci. U.S.A.*, 80, 6456, 1983.
10. Loosfelt, H., Logeat, F., Vu Hai, M. T., and Milgrom, E., The rabbit progesterone receptor. Evidence for a single steroid-binding subunit and characterization of receptor mRNA. *J. Biol. Chem.*, 259, 14196, 1984.
11. Logeat, F., Pamphile, R., Loosfelt, H., Jolivet, A., Fournier, A., and Milgrom, E., One-step immunoaffinity purification of active progesterone receptor. Further evidence in favor of the existence of a single steroid-binding subunit. *Biochemistry*, 24, 1029, 1985.
12. Lorenzo, F., Jolivet, A., Loosfelt, H., Vu Hai, M. T., Brailly, S., Perrot-Applanat, M., and Milgrom, E., A rapid method of epitope mapping. Application to the study of immunogenic domains and to the characterization of various forms of rabbit progesterone receptor. *Eur. J. Biochem.*, 176, 53, 1988.
13. Chang, C., Whelan, C. T., Popovich, T. C., Kokontis, J., and Liao, S., Fusion proteins containing androgen receptor sequences and their use in the production of poly and monoclonal antiandrogen receptor antibodies. *Endocrinology*, 123, 1097, 1989.
14. Prins, G. S., Birch, L., and Greene, G. L., Androgen receptor localization in different cell types of the adult rat prostate. *Endocrinology*, 129, 3187, 1991.
15. Takeda, H., Chodak, G., Mutchnik, S., Nakamoto, T., and Chang, C., Immunohistochemical localization of androgen receptors with mono and polyclonal antibodies to androgen receptor. *J. Endocrinol.*, 126, 17, 1990.
16. Krozowski, Z. S., Rundle, S. E., Wallace, C., Castell, M. J., Shen, J. H., Dowling, J., Funder, J. W. and Smith, A. I., Immunolocalization of renal mineralocorticoid receptors with an antiserum against a peptide deduced from the complimentary deoxyribonucleic acid sequence. *Endocrinology*, 125, 192, 1989.
17. Gametchu, B. and Harrison, R. W., Characterization of a monoclonal antibody to the rat liver glucocorticoid receptor. *Endocrinology*, 114, 274, 1984.
18. Okret, S., Wikström, A. C., Wrange, Ö., Andersson, B., and Gustafsson, J. A., Monoclonal antibodies against the rat liver glucocorticoid receptor. *Proc. Natl. Acad. Sci. U.S.A.*, 81, 1609, 1984.
19. Eisen, L. P., Reichman, M. E., Thompson, E. R., Gametchu, B., Harrison R. W., and Eisen, H. J., Monoclonal antibody to the rat glucocorticoid receptor. *J. Biol. Chem.*, 260, 11805, 1985.
20. Ahima, R. S. and Harlan, R. E., Charting of type II glucocorticoid receptor-like immunoreactivity in the rat central nervous system. *Neuroscience*, 39, 579, 1990.

4. Boil 2 h in distilled water.
5. Immerse overnight in denatured 95° ethanol.
6. Rinse through three to four consecutive baths of distilled water; leave in distilled water until subbing.
7. Prepare gelatin solution (2% gelatin, 0.05% chromium-potassium sulfate) as follows: heat distilled water to 70°C, drop gelatin (from porcine skin, type A, 375 Bloom, Sigma) into hot water and dissolve by stirring, allow to cool down to around 25°C under continuous stirring, add the chromium-potassium sulfate (Merck) and allow it to dissolve. Chromium-potassium sulfate is used here as a texture agent; this compound is denatured by heat.
8. Filter and avoid bubbles.
9. Submerse slide racks into the gelatin solution (immersion during a few seconds). N.B.: always use extemporaneously prepared solution.
10. Drain vigorously any excess of subbing solution over absorbent paper.
11. Dry out overnight in dust-free oven at 37°C.
12. Store in closed boxes at 4°C.

Once positioned on the gelatinized glass slide, each section must be defrosted on the slide in order to adhere properly. This can be achieved either by collecting the frozen tissue sections on slides taken at room temperature, or by first placing the section on a cold slide and then heating the slide locally by finger pressure under the section (touch-mounting). The latter method yields better control over section placement. It also allows identical collection of several sections on a single slide: touch-mounting should then be achieved once for all sections of the slide, but should not be delayed by more than 10 min (to avoid tissue alteration). Once defrosted, slide-mounted sections must be refrozen, either at once or after desiccation under vacuum at 4°C during several hours (2 h at least, overnight at most, but similarly for the whole set of slides included in one given experiment). Vacuum desiccation of slide-mounted sections at 4°C before refreezing significantly improves adhesion of the tissue to the glass slide, while not hampering histological preservation (see Figure 3A and C). This can be achieved by collecting mounted sections in slide boxes on ice and then placing filled boxes in a vacuum chamber; once vacuum is attained close the air admission tap and put the bell into a fridge or cold room; after several hours open the bell, close the slide box, and transfer it into the freezer.

The slides must be kept frozen until the radioligand binding assay; avoid defrost/freeze cycles at any rate. Slide-mounted sections must be kept in the freezer for at least 24 h before assay in order to ensure proper adhesion.

IV. RADIOLIGAND BINDING ASSAY ON FROZEN TISSUE SECTIONS

Radioligand binding must be assayed on tissue sections in experimental conditions as similar as possible to those previously found to ensure optimal ligand-receptor interactions in tissue homogenates.

Key parameters to be kept identical between homogenates and sections are pH and ionic composition of assay buffer, temperature, and duration of tissue incubation with radioligand. However, the use of tissue sections for radioligand binding assays involves several specific requirements.

A. EXPERIMENTAL SETUP

Incubation of tissue sections with radioligand can be performed through one of the two following procedures.

1. Drop Incubation

This consists of depositing a drop of incubation mixture onto each section, the supporting slides being placed horizontally. The drop must be prevented from evaporating, especially at room temperature, for instance by placing the slides in humid boxes during the incubation period. In particular, the drop must be prevented from diffusing away from the tissue section: horizontality of slides should be watched carefully and, in case preincubation is required (see below), each slide must be wiped with absorbing paper all around each tissue section before depositing the drop.

2. Bath Incubation

This consists of immersing the section-holding slides within a jar containing the incubation mixture. This method warrants constancy of mixture composition throughout the incubation period much more reliably than the previous one, and should therefore be preferred in the case of pharmacological/biochemical assessments. However, this method is more radioligand-consuming than the other. Disposable cyto-mailers (i.e., commercially available plastic boxes for four or five microscope slides, OSI, France) can advantageously be used for bath incubations rather than conventional histological dishes, thus minimizing volumes of radioligand-containing mixtures and avoiding contamination of glass dishes.

Preincubation and rinsing of slides can be performed in conventional histology glassware.

B. SECTION DEFROSTING

Sections must be taken out of the freezer at the time of the binding assay; indeed, air-drying involves protein tanning and denaturation which can induce alterations of receptor binding characteristics. However, before being dipped into the first aqueous bath of the binding protocol, slides must be allowed to stay at room temperature enough time for any condensation moisture to vanish from tissue sections (usually 2 to 3 min for slides taken out of a –20°C freezer, or 5 to 6 min out of a –80°C freezer); otherwise tissue sections tend to become unstuck from supporting slides. Sections must be submitted to binding assay as soon as they look dry, and then never dry out until the end of the entire procedure.

C. PREINCUBATION

Incubation of tissue sections with radioligand must often be preceded by a preliminary washing of sections in assay buffer; without such "preincubation," specific complexation of radioligand with tissular receptors is decreased or even suppressed. The preincubation step is believed to allow dissociation of endogenous ligand from tissular receptors, and should therefore be long enough for such disso-ciation to occur. Protocols for radioligand labeling of most hydrophilic hormone and neurotransmitter receptors nowadays involve preincubations of 30 min. Even though some exceptions do not deserve preincubation (labeling of D2-like dopaminergic receptors with ^{125}I-iodosulpride) the crucial importance of this step should not be underestimated. For instance, the absence of somatostatin binding sites initially reported in rat hypothalamus as a remarkable mismatch with the distribution of somatostatin-secreting neurons, was later revealed as artifactual by merely increasing the preincubation step from 15 min to 1 h.[5,6] It should be kept in mind, though, that lengthening the stay of unfixed tissue in aqueous solution is damaging for histolog-ical preservation and adherence of slide-mounted sections.

D. INCUBATION WITH RADIOLIGAND

Ionic composition and pH of assay buffer (see Protocol 3) can be directly transposed from optimization trials on tissue homogenates. In the case of widely hypotonic optimals, however, the osmolarity of the assay buffer can be increased without much incidence on receptor binding by adding sucrose.

PROTOCOL 3: SOME ISOSMOTIC BUFFERS

- Tris-base 0.17 M.
- Tris-base 0.05 M, sucrose 0.25 M.
- Media for cell cultures.

Nonspecific binding of radioligands to tissue sections can be reduced by sup-plementing the assay buffer with either 2% bovine serum albumin, or 0.1% poly-ethyleneimine.

Radioligands should be used as freshly made as possible. Peptidic radioligands must be protected against tissular proteases by adding bacitracin (up to 0.1%) to the assay buffer. Radioligands sensitive to oxidation (like aminergic molecules) can be protected by the antioxidant 0.1% ascorbic acid (ex-temporaneously dissolved in assay buffer, the pH of which must be readjusted afterward).

E. RINSING

Incubation of tissue sections with radioligand can be ended by merely removing either the drop from the slide or the slide out of the jar. Subsequent elimination of the remaining unbound radioligand by rinsing in conditions that limit dissociation of receptor-bound molecules, though, is more difficult than with suspensions of tissue particles. In most cases, slide-mounted 20-μm-thick sections are efficiently rinsed through 3 consecutive baths of fresh buffer at 4°C, totaling a 10-min period. For ligand-receptor systems reaching equilibrium in less than 15 min, the total rinsing duration should be decreased; radioligand labeling of glutamate receptors thus involves a 30-s-rinse.[7] The key points, whatever the total duration, are (1) to allow different consecutive baths, and (2) to ensure the same global time of rinsing for all slides, since dissociation of bound radioligand is likely to occur in the meanwhile.

In any case, cold buffer rinses should be always followed by a quick dip of slides into ice-cold distilled water, in order to avoid salty and proteic deposits on the sections after drying; such deposits can indeed generate artifacts in subsequent radioautography.

F. DRYING

Rinsed slides must be placed subvertically on racks, under a stream of cold air. Drying must be as rapid and dust-free as possible. Slides must be thoroughly dry before starting radioautography.

G. POST-FIXATION FOR LIQUID EMULSION WET RADIOAUTOGRAPHY

Since receptor labeling by radioligand binding relies upon a reversible equilibrium in aqueous solution, the mere dipping of thus radiolabeled sections into liquid photographic emulsion (to be used around 40°C) would result in dissociation of specifically bound radioligand and artifactual redistribution of radioactive molecules within tissue. For wet radioautography purposes, bound radioligand must be irreversibly cross-linked to receptors immediately after the rinsing of radiolabeled sections. The cross-linking agent to be used depends upon the chemical structure of the radioligand — ultraviolet irradiation for probes endowed with an azido group,[8] or immersion in 4% glutaraldehyde containing 50 mM phosphate buffer at pH 7.4 for 30 min at 4°C for probes containing at least one –NH_2 group.[9]

In either case, post-fixation must be followed by dehydration-delipidation of tissue sections in order to avoid lipid-induced quenching of radioactive disintegrations, in accordance with Protocol 4.

PROTOCOL 4: DEHYDRATION-DEFATTING OF TISSUE SECTIONS

Ethanol 70°, 2 × 5 min.
Ethanol 95°, 2 × 5 min.
Ethanol 100°, 2 × 5 min.
Xylene, 2 × 15 min.
Ethanol 100°, 2 × 5 min.
Ethanol 95°, 2 × 5 min.
Ethanol 70°, 2 × 5 min.
Distilled water, 3 × 5 min.
Air-dry in oven at 37°C overnight.

Although quantitative,[10] this radioautographic method is mainly used for qualitative purposes, eventually in combination with immunocytochemistry and for neuronal tracing (see Chapter 6).

V. FILM RADIOAUTOGRAPHY

The film to appose onto radiolabeled sections must be chosen among the commercially available types in accordance with the radioactive isotope of the probe used. Long-term storage of films before radioautographic use generates background, which can be lessened and delayed by storage at 4°C; even so, storage duration should not exceed several months.

All films for radioautography can be safely handled in a darkroom with inactinic lights used for photographic papers. Check, though, the nature of inactinic filters recommended by companies commercializing the films. *Film emulsion must be apposed directly in contact with the radiolabeled tissue sections,* since radioelement-emitted radiations are absorbed by the plastic substrate of films and even by air. Make sure therefore that: (1) the emulsion side of the radioautographic film is positively identified; and (2) all slides to be apposed to one film sheet make up a regular plane surface; all slides must be the same thickness and must be carefully arranged with respect to the others so as to avoid any piling up, even away from tissue sections per se.

Exposure of radioautographic films onto labeled tissue sections must be achieved in cassettes ensuring both perfect darkness and tight contact of the film onto the sections without the least relative displacement. The commercially available cassettes meeting these demands are made of lead which, in addition, warrants radiation proofness; these expensive items can be replaced by self-made wooden cassettes. Exposure time must be empirically determined, so that optical densities of the images to be analyzed fall within a range of values related to radioactivities by a linear function (Figure 1). Radioautographic films must always be developed routinely according to a standard protocol (see Protocol 5).

PROTOCOL 5: DEVELOPMENT OF TRITIUM-SENSITIVE FILMS (HYPERFILM, AMERSHAM)

At 21°C:

1. 4 min in Kodak D-19 developer (used as prepared from commercial powder according to Kodak's instructions).
2. Brief rinse through a running-water bath.
3. 10 min in Kodak Rapid Fixer (used as prepared from stock according to Kodak's instructions).
4. 20 min rinsing under tap water.
5. Air drying at room temperature.

N.B.: the emulsion side of films is very sensitive to the least mechanical touches until dry (unlike photographic papers). Even when dried, film radioautographs can still be damaged by mechanical stresses; films must therefore be sheltered for storage and handled with care.

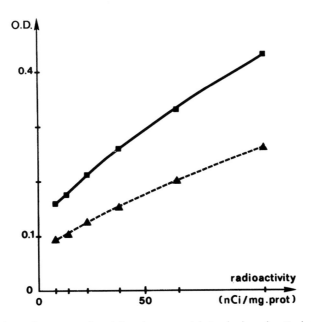

FIGURE 1   Relation between radioactivity of commercial standards and optical density (O.D., arbitrary units) of corresponding radioautographic film images after 38 h (▲) and 13 d (■) exposure in light-proof cassettes.

VI. MEASUREMENT OF RADIOACTIVITY CONTENT OF TISSUE SECTIONS BY DENSITOMETRY ON RADIOAUTOGRAPHS

A. METHODOLOGICAL PRINCIPLE

The parameter of radioautographic images that relates quantitatively to the radioactivity content of the sample is the number of silver grains per unit area, which in turn determines the optical density (O.D.) of the radioautographic image. Optical density can be measured indirectly through the proportion of light absorbed by the image from a source of known intensity, according to the physical law: O.D. = –log I/I_o, where "I_o" represents the intensity of light incident to the image and "I" the light intensity recovered beyond the image.

B. COMPUTER-ASSISTED DENSITOMETERS

Several firms commercialize computerized systems that allow radioactivity determination from radioautographic images by: (1) measuring light intensity transmitted by the radioautogram placed over a light source; (2) converting this measure into optical density by reference to the light source intensity; and (3) converting O.D. into radioactivity by reference to a standard curve established by the images of a series of known radioactivity samples that had been radioautographed along with the sections to analyze. One of these systems (Biocom, Les Ulis, France) is illustrated in Figure 2 as an example.

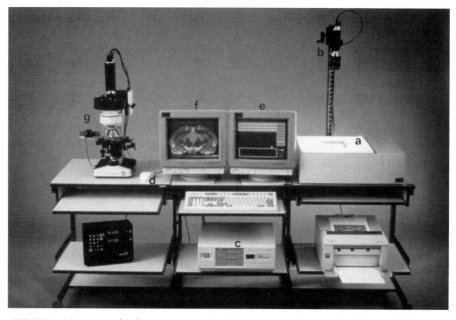

FIGURE 2   One example of a computer-assisted densitometer for analysis of film or light micro-scopic radioautograms (Biocom, France). The different components of the system are designated by small letters as follows: **a**, light source to expose the radioautographic film to be analyzed; **b**, CCD video camera, equipped with a macrophotographic lens and connected to **c**, personal computer; **d**, computer mouse; **e**, monitor for software instructions; **f**, "control" monitor displaying the image received by the video camera either untreated or after computer-assisted digitalization; **g**, microscope equipped with another CCD video camera connected to the computer, for light microscopic densitometry and morphometry (see Reference 10 for details).

For the present purpose, the light source intensity must be kept constant as a function of time without heating the films too much. Such requirement has been fulfilled in the Biocom system by placing a translucent diffuser screen over fluorescent tubes plugged into an electronic device that converts the domestic 50-Hz alternating current into a 40-kHz current. Light intensity is measured in each pixel of a CCD video camera with a 512×512 field as a voltage (V) delivered by a condenser proportionately to the number of afferent photons, according to the equation:

$$V = k \times \text{number of photons} \times i$$

where k is a proportionality constant and i the current provided to the video camera. Elementary voltage values are then affected arbitrary values by the computer soft-ware, ranging within a linear scale between 0 (minimal light, i.e., dark, no light admission to the camera) and 255 (maximal light). Validity of this intensity scaling relies drastically upon correct calibration of the system at the beginning of each measurement session, in order to avoid saturation of the CCD video camera at low and high light intensities; also, automatic gain control of the CCD camera must be disconnected. Following this pixel-by-pixel measurement of afferent light intensity, subfields of the radioautographic image used to deduce the radioactivity content of the corresponding section area, can be delineated over the control screen (Figure 2) by using the computer-coupled mouse. The software then *averages* light intensities of all the pixels included within the delineated area into one mean value that represents the light intensity transmitted by this area of the radioautograph.

C. AREA DELINEATION

This step is highly critical in the presently addressed experimental approach because it can generate dramatic artifacts even out of perfect radioautographic material. Basically, since the software determines the optical density of the delineated area via averaging the values of all pixels within it, the area of the radioautographic image used to determine optical density must correspond to a portion of the tissue section whose radioactivity concentration can be assumed similar to that of the anatomical structure studied. The first criterion to meet, therefore, is that the densitometric area must be equal to, or contained within, the perimeter of the targeted structure in the plane of section examined. This requirement can be fulfilled by several techniques:

1. Delineate the histological boundary of the targeted structure on the Nissl-counterstained radiolabeled section, and apply this area onto the corresponding radioautographic image. Such an operation is allowed by the morphometric functions usually included in densitometric softwares.
2. Delineate inside the radioautographic image of the targeted structure the largest area wherein radioautographic labeling appears homogeneous.
3. Use small geometrically predefined areas (like squares or circles), repeated as many times as can be fit within anatomical limits of the targeted structure.

These methods are commonly used in neuropharmacology. For a comparison, we applied all three techniques on the same radioautographic film image generated from a 20-μm-thick section of rat olfactory bulb wherein D2-like receptors of dopamine had been labeled by 0.1 nM ^{125}I-iodosulpride (Figure 3E).

The amount of specifically bound ^{125}I-iodosulpride within glomerular layer of olfactory bulb differs significantly among the three methods used:

- With method 1: 4749 ± 336 nCi/mg.prot.
- With method 2: 5585 ± 329 nCi/mg.prot.
- With method 3: 6640 ± 244 nCi/mg.prot.

The lowest radioactivity value was obtained with method 1 which is obviously due to "edge effects." On the one hand, the transfer of anatomical delineation from the counterstained preparation onto radioautographic image has macroscopically limited accuracy; on the other hand, the anatomical boundary of the glomerular layer is not necessarily orthogonal to the plane of the section, which can result in tissue sampling bias being specially misleading at the limit between the glomerular layer and unlabeled neighboring structure.

Method 3 yields a mean radioactivity value with low S.E.M., which is probably due to the number of measurements (15) for a single structure. However, all these

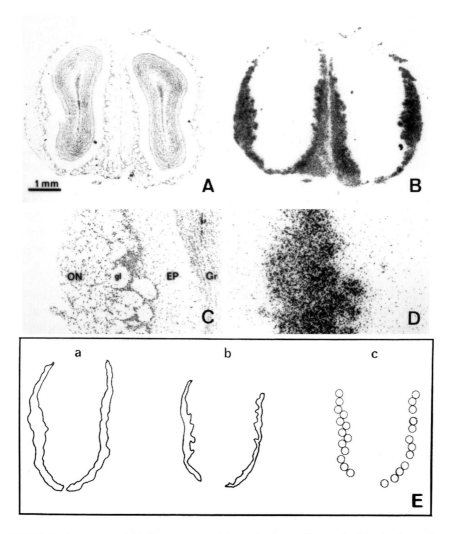

FIGURE 3 Assessment of D2-like receptors of dopamine in rat olfactory bulb by *in vitro* radio-autography. **A, C**: coronal sections (20 μm thick) of frozen rat olfactory bulbs prepared as described in Section III and counterstained with cresyl-violet after *in vitro* labeling with ^{125}I-iodosulpride (0.1 nM). **B, D**: film radioautograms generated from the section (B) or sector (D) photographed, respectively, in **A** and **C**. **C and D** represent one enlarged sector of the peripheral olfactory bulb. **E**: protocols for densitometric quantification of specific radioautographic labeling in the glomerular layer of olfactory bulb (see text Section VI.C). Abbreviations: EP, external plexiform layer; gl, one glomerulus within the glomerular layer; Gr, granular layer; ON, olfactory nerve layer.

measurements range between 117.96 and 154.54 (arbitrary unit values of initial densitometric measurements); such dispersion is probably due to the small number of grains included in each area.

Figure 4 illustrates an example of one competition kinetic curve yielded by application of method 2.

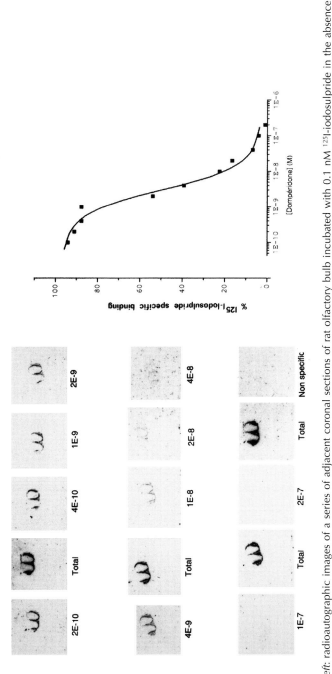

FIGURE 4 *Left*: radioautographic images of a series of adjacent coronal sections of rat olfactory bulb incubated with 0.1 nM ^{125}I-iodosulpride in the absence ("Total") or in the presence of either haloperidol at 10 μM ("Non specific") or domperidone at the concentration indicated underneath (in M). *Right*: Competition kinetics plotted from densitometric analysis of the radioautograms reproduced on the left, according to protocol b of Figure 3.

In conclusion, on the basis of our past experience, we advise the use of method 2 for densitometric analysis of radioautographic series intended for pharmacological determinations.

D. STANDARD CURVE FOR ABSOLUTE QUANTIFICATION OF RADIOACTIVITY

The absolute measurement of tissue radioactivity content through densitometric analysis of radioautographs requires calibration of O.D. readings by reference to the densities of radioautographic labeling generated by several samples of known radioactivities. Commercially available standards (Microscales, Amersham) made out of homogeneous radioactive plastic fabrics, precalibrated vs. equivalent tissue-bound radioactivities ("tissue equivalents") and sectioned at the same thickness as tissue slices, should be used preferentially to any homemade products. As illustrated in Figure 1, the resulting standard curves are always saturable, i.e., nonlinear, and depend heavily upon radioautographic exposure time. For purposes of relative radioactivity determinations, therefore, one must ensure that all radioautographic labeling densities be mutually compared and fall within a linear relationship range between radioactivities and radioautographic O.D. The use of radioactivity standards is thus highly recommended, even for so-called semiquantitative assessments of radioautographic labeling.

VII. PRACTICAL TRANSPOSITION OF BIOCHEMICAL PROTOCOLS ONTO SERIES OF TISSUE SECTIONS

Biochemical protocols for receptor characterization by radioligand binding all involve variation of one physical parameter among a series of identical tissue samples. The basic problem is that distinct histological sections cannot be assumed as much identical as repetitive volumes of a tissular homogenate, especially in the case of anatomically heterogeneous organs like nervous systems. Resulting errors in biochemical determinations can be minimized by observing several precautions.

First, any biochemical protocol should be applied onto a series of serially cut sections, thus the total number of conditions to compare for within one assay are spread over the least number of planes of sections. For ensuring multiple determinations in each condition assayed, take one item out of several sets of serial sections, as represented on the table below.

	I	II	III	N
First series	1	2	3	n
Second series	n + 1	n + 2	n + 3	2n
Third series	2n + 1	2n + 2	2n + 3	3n

Each row represents one series of adjacent tissue sections, the increasing numbers corresponding to the chronological order of section-making at the cryostat; the columns correspond to the different conditions assayed in the biochemical protocol. For economy of both radioligand and radioautographic film, all sections of column I can be mounted onto one single slide (see Section III, this chapter).

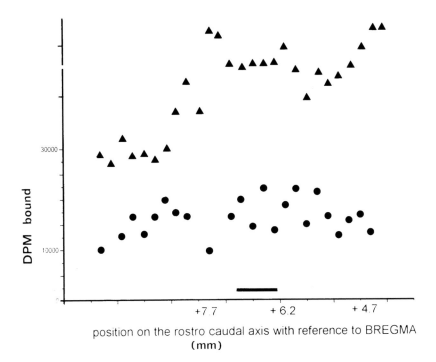

FIGURE 5 Radioautographic distribution of D1-like (●) and D2-like (▲) receptors of dopamine in the olfactory bulb, respectively labeled by ^{125}I-SCH23982 and by ^{125}I-iodosulpride. Optical densities of radioautographic labeling (expressed in dpm) are plotted against the anatomical position of the section along the rostro-caudal axis of the olfactory bulb (expressed in mm from Bregma). The horizontal bar indicates anatomical position of the series of coronal sections used for receptor assessments illustrated in Figures 4 and 6.

Second, the anatomical distribution of binding sites within the organ of study should be mapped as a preliminary step before the biochemical assay per se. Distribution of D1-like and D2-like dopaminergic binding sites along the rostro-caudal axis of the olfactory bulb are thus illustrated in Figure 5 as an example; the series of 20 sections used for the pharmacological result of Figure 4 was deliberately sampled in the region of the olfactory bulb outlined on Figure 5.

Assessment of any given biochemical protocol onto tissue sections therefore requires a specially deviced protocol of cryostat sectioning. Even so, the signal-to-noise threshold of radioligand binding is higher on series of tissue sections than on homogenates and can challenge assay validity. For instance, we failed to characterize the pharmacological profile of D1-like ^{125}I-SCH 23982-specific binding in rat olfactory bulb, while the same protocol yielded positive results with a series of striatum sections (Figure 6).

VIII. CONCLUSION

The biochemical protocols derived from kinetic laws of receptor-ligand interaction can indeed be readily applied *in vitro* to a series of tissue sections. This

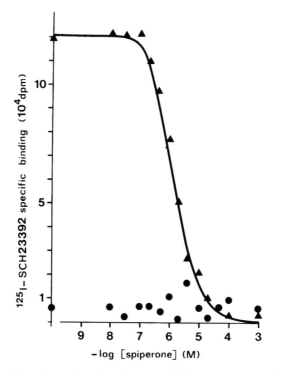

FIGURE 6   Competition of nonradioactive spiperone with ^{125}I-SCH23982 (D1-like) specific binding, as assessed by film radioautography on a series of frozen brain sections in striatum (▲) and in olfactory bulb (●). The two series of sections have been processed together in the same binding experiment.

approach can yield the same type of information as batteries of homogenate samples, at a level of observation much beyond the limits of macroscopical dissection. However, the reliability of such determinations requires rigorous fulfilment of numerous precautions throughout tissue handling, radioautography, and image analysis.

ACKNOWLEDGMENTS

The authors wish to express their gratitude to René Michel and Jean-Paul Robert (Biocom, Lyon, France) for several years of competent and diligent advice in the field of computerized image analysis. We also thank Miss Catherine Berthet for her excellent clerical assistance.

REFERENCES

1. Scatchard, G., The attractions of proteins for small molecules and ions, *Ann. N.Y. Acad. Sci.*, 51, 660, 1949.
2. Taylor, P. and Insel, P.A., Molecular basis of pharmacological selectivity, in *Principles of Drug Action*, 3rd ed., Pratt, W.B. and Taylor, P., Eds., J & A Churchill, London, 1990.

3. Quirion, R., Hammer, R.P., Jr., Herkenham, M., and Pert, C.B., Phencyclidine (angel dust)/σ "opiate" receptor: visualization by tritium-sensitive film, *Proc. Natl. Acad. Sci. U.S.A.*, 78, 5881, 1981.

4. Quirion, R., Comparative localization of putative pre- and post-synaptic markers of muscarinic cholinergic nerve terminals in rat brain, *Eur. J. Pharmacol.*, 111, 287, 1985.

5. Krantic, S., Martel, J.C., Weissmann, D., and Quirion, R., Radioautographic analysis of soma-tostatin receptor subtype in rat hypothalamus, *Brain Res.*, 498, 267, 1989.

6. Krantic, S., Quirion, R., and Uhl, G., Somatostatin receptors, in *Handbook of Chemical Neuroanatomy*, Vol. 11, Björklund, A., Hökfelt, T., and Kuhar, M.J., Eds., Elsevier, Amsterdam, 1992, 321.

7. Greenamyre, J.T., Olson, J.M.M., Penney, J.B., and Young, A.B., Autoradiographic characterization of N-methyl-d-aspartate-, quisqualate- and kainate-sensitive glutamate binding sites, *J. Pharmacol. Exp. Ther.*, 233, 254, 1985.

8. Pasquini, F., Bochet, P., Garbay-Jaureguiberry, C., Roques, B.P., Rossier, J., and Beaudet, A., Electron microscopic localization of photoaffinity-labeled delta opioid receptors in the neostriatum of the rat, *J. Comp. Neurol.*, 326, 229, 1992.

9. Moyse, E., Rostène, W., Vial, M., Leonard, K., Mazella, J., Kitabgi, P., Vincent, J.P., and Beaudet, A., Distribution of neurotensin binding sites in rat brain: a light microscopic radioautographic study using monoiodo(^{125}I)Tyr3-neurotensin, *Neuroscience*, 22, 525, 1987.

10. Moyse, E., Miller, M.M., Rostène, W., and Beaudet, A., Effects of ovariectomy and estradiol replacement on the binding of ^{125}I-neurotensin in rat suprachiasmatic nucleus, *Neuroendocrinology*, 48, 53, 1988.

Chapter **4**

VISUALIZATION OF *IN VITRO* BOUND RADIOLIGANDS BY ELECTRON MICROSCOPY

Marie-Elisabeth Stoeckel
Marie-José Freund-Mercier

CONTENTS

0-8493-2644-3/97/$0.00+$.50
© 1997 by CRC Press LLC

79

I. INTRODUCTION

The radioligand binding technique to tissue sections, first developed by Young and Kuhar,[1] makes possible morphological localization of receptors simultaneously with their pharmacological characterization. This technique is still largely in use for detection of neuropeptide and neurotransmitter receptors, especially in the central nervous system in which the blood-brain barrier impedes access to the receptors of systemically administered radioligands. The method consists in incubating slide-mounted frozen tissue sections in the presence of radioligand under the same conditions as for binding studies on membrane preparations. Localization of the binding sites was originally detected on nuclear emulsion-coated coverslips apposed against these sections. This rather fragile support was soon replaced by radiosensitive film[2] which furthermore permitted quantitative analysis of radioligand binding by densitometry and, consequently, pharmacological studies and detection of functional or experimental alterations of the receptor expression in discrete tissue areas. The resolution of the film technique, however, remains limited to the macroscopic scale. Detection of radioligand binding at the cellular level by light microscopy is possible by processing the sections for conventional historadioautography, provided that the radioligand can be firmly attached to its receptor to avoid its leakage during the procedure. However, the poor preservation of the cellular structures in frozen sections from unfixed tissue limits the exact localization of the radioligand binding sites. Ultrastructural localization of radioligand binding sites is performed by applying a pre-embedding binding technique to samples subsequently processed for electron microscopy. The technique is also suitable for high resolution light microscope radioautography on semithin sections.

II. RADIOLIGANDS

A. GENERAL CONSIDERATIONS

- The ligand used needs previous pharmacological characterization as regards its selectivity and affinity for the receptor studied.
- A prerequisite for the ligand is that it includes at least one free amine (NH_2) radical which permits cross-linking to the receptor site with aldehydes (e.g., glutaraldehyde) involving covalent bonds.[3] This operation is indeed necessary in view of the reversibility of the receptor-ligand interaction, which results in leakage of the ligand during histological processing unless the radioligand is irreversibly attached to the receptor.
- High resolution radioautography requires the ligand to be coupled to a low-energy radioisotope. Tritiated ligands meet this requirement but their specific activity is too low (30 to 50 Ci/mmol) and thus would need exaggerated exposure times (several months to years). ^{125}Iodine is actually the most suitable radioisotope. It has a high specific activity (2000 to 2500 Ci/mmol) and its radiations, involving Auger electrons, have even lower energy than the β radiations of tritium. However, in some cases, e.g., oxytocin or vasopressin, direct ^{125}I coupling by substitution impedes the binding of the peptide to its receptor. In these cases, specific antagonists, which keep their binding abilities when they are radioiodinated, must be used. On the other hand, the short half-life (60 days) of ^{125}I is a disadvantage of this marker.

B. RADIOIODINATION OF THE LIGAND

Most ^{125}I-labeled radioligands are commercially available. Several techniques have been described for radioiodination of peptides. Protocol 1, adapted from Elands et al.,[4] describes a technique which has been shown suitable for labeling of oxytocin and vasopressin selective antagonists.[5] Its application needs imperatively controlled working conditions adapted to manipulation of volatile radioelements.

PROTOCOL 1: RADIOIODINATION OF THE LIGAND

The whole procedure is performed in a radioprotective compartment conveniently equipped with an iodine retention filter and under negative pressure (^{125}INa is volatile and thus very harmful).

1. In an silicon-coated Eppendorf tube, cooled in ice, place successively:
   10 μl of 1 mg/ml oxytocin or vasopressin antagonist solution in methanol and
   10 μl of 1 M KH_2PO_4 buffer, pH 7.4.
   10 μl of 1 mg/ml chloramin T solution in water
   1 mCi ^{125}INa (IMS-300, Amersham).

2. Mix gently.
3. After 2 min, stop the reaction by adding 900 μl trifluoroacetic acid 0.11% in distilled water.

C. PURIFICATION AND STORAGE OF THE RADIOLIGAND

HPLC purification (see protocol 2) of the radioligand is absolutely necessary (1) to eliminate free ^{125}INa, and (2) to separate the monoiodinated peptide from the other forms of the molecule, i.e., the noniodinated peptide which could compete with the radioligand, and the diiodinated compound, often inactive, whose radioactivity can lead to wrong estimations of the specific activity.

PROTOCOL 2: HPLC PURIFICATION OF THE MONOIODINATED RADIOLIGAND

A multipump gradient HPLC system is required. We normally use a C_{18} μBonda-pak reverse-phase column (Waters Associated).

1. Directly apply the radioiodinated ligand to the HPLC system. Buffer A (trifluoroacetic acid 0.11%)/buffer B (60% acetonitrile in trifluoroacetic acid 0.11%).
2. Make a linear gradient from 24 to 60% acetonitrile over 30 min at a flow rate of 1 ml/min.
3. Collect 30-s fractions.
4. Monitor the elution of the radioligand by counting sample aliquots of each tube in a gamma scintillation counter (diluted 1:50,000 in order not to saturate the counter). The tube with the highest radioactivity contains the monoiodinated peptide (which for the oxytocin antagonist occurs at 76 to 77% of B and for the vasopressin antagonist at 72 to 74% of B).
5. Eliminate by vacuum evaporation (Speed Vac) the organic solvents of the tube containing the radioligand.
6. Dissolve in 15 ml Tris-HCl buffer (see Appendix).
7. Distribute in aliquots in silicon-coated tightly stoppered tubes.
9. Store at −180°C (liquid nitrogen container) until needed.

The specific activity of the monoiodinated ligand is calculated taking into account the specific activity of ^{125}INa given by the manufacturer (mCi ^{125}I/μg iodine or Ci ^{125}I/mmol at the activity reference date = day 0).

III. PRELIMINARY TESTS

In contrast with the conventional film radioautographic or historadioautographic techniques (see Chapters 3 and 6) radioligand binding for electron microscopic receptor detection is performed on previously fixed tissues, assuring adequate ultrastructural preservation. The denaturing effect on proteins of the fixatives leads to decreased binding capacity of the receptors, thus determination of fixation conditions which result in maximal labeling intensity of tissue with acceptable ultrastructural preservation is needed. The effective fixative is determined by comparing the binding capacities (i.e., labeling intensities) of frozen sections from tissues fixed with variable concentrations of paraformaldehyde and/or glutaraldehyde, taking as reference unfixed tissue sections. The radioligand binding is determined by conventional film radioautography which permits quantification of the labeling intensity. For oxytocin receptor labeling, 2% paraformaldehyde fixation was an acceptable compromise for both ultrastructural preservation and receptor labeling.

Preliminary tests should also include verification of the effects on labeling intensity of the treatments following application of the binding procedure: cross-linking with glutaraldehyde, postfixation with osmium tetroxide, and dehydration.

- Glutaraldehyde should not notably affect the labeling intensity; moreover, it has a beneficial effect on preservation of the ultrastructure of the tissues.
- Osmium tetroxide treatment, applied after glutaraldehyde for ultrastructural studies, fixes membrane structures which are not preserved by aldehydes. Osmium tetroxide has been reported to provoke loss of proteins, even in aldehyde-fixed tissues,[6] and may thus decrease labeling intensity. The labeling intensity of oxytocin receptors has been found to be only slightly reduced by osmium tetroxide.

A. TISSUE FIXATION

The fixatives tested contain freshly depolymerized paraformaldehyde (used no later than 1 day after preparation) at final concentrations varying between 1 to 4%, glutaraldehyde between 0.1 to 1%, or mixtures of both in variable proportions. All fixatives are prepared in 0.1 M phosphate buffer, pH 7.35 (see Appendix).

Fixation is performed by intracardiac perfusion for 30 min with a 20 ml/min flow rate for a rat weighing 200 to 300 g. After perfusion, the tissues are dissected out and immersed overnight in 0.1 M phosphate buffer containing 20% saccharose.

B. FREEZING

The fixed tissues are frozen by immersion for 1.5 min in isopentane cooled to –40°C with dry ice. Unfixed tissues, dissected out after decapitation of the rats, are similarly frozen. The frozen tissues can be wrapped in an aluminum foil and stored in tightly stoppered vials at –20°C until sectioning.

C. SECTIONING

Cut 20-μm thick frozen sections on a cryostat microtome and thaw-mount on gelatin-coated slides (see Appendix). They can be stored at –20°C until application of the binding procedure.

D. RADIOLIGAND BINDING

The incubation conditions (see Protocol 3) are those used for conventional film radioautography on unfixed tissue sections.

- The concentration of the radioligand, close to the dissociation constant (K_D), is calculated on the basis of the specific activity. The persistence or not of the biological activity of the tellurium-labeled ligand, resulting from the decay of ^{125}I, must be taken into account in this determination.
  - If the tellurium-labeled ligand is biologically inactive, the specific activity of the ligand is constant and the counts per minute per milliliter determined at day 0 are used whatever the age of the ligand.
  - If the tellurium-labeled ligand keeps its biological activity (which is the case for the oxytocin and vasopressin antagonists), the concentration of the radioligand is calculated according to the decay rate of ^{125}I as given on tables by the manufacturer.
- The endogenous ligand is normally eliminated by the preincubation. Its dissociation can be favored by guanyl nucleotides.[7] Adjunction of GTP, however, did not notably increase labeling intensity with oxytocin or vasopressin radioiodinated antagonists.
- Protection against peptidases by adding bacitracin (40 mg/l), chymostatin (2 mg/l), and leupeptin (4 mg/l) to the incubation medium has been recommended.[8] In the case of tritiated oxytocin and vasopressin, such agents did not appreciably modify labeling intensity.
- Low temperature of the incubation medium improves the signal/background ratio. Duration of incubation must last until the binding equilibrium is reached. For oxytocin and vasopressin antagonists, optimal and reproducible labeling is obtained by incubating sections at 4°C for 24 h.

PROTOCOL 3: INCUBATION OF TEST SECTIONS IN THE RADIOLIGAND

Place the slides back-to-back in a Coplin jar, the volume of which is reduced with dental wax in order to reduce the amount of radioligand used.

1. Preincubate for 20 min at room temperature in Tris-HCl buffer (see Appendix).
2. Incubate for 24 h at 4°C in Tris-HCl buffer added with the radioligand.
3. Wash 2 × 5 min in cold (4°C) Tris-HCl buffer.
4. Rinse rapidly with distilled water.
5. Dry in a cold airstream for 12 h.

E. CROSS-LINKING, POSTFIXATION, AND DEHYDRATION

After application of the binding procedure, the slides are successively immersed in:

1. Glutaraldehyde 5% in 0.1 M phosphate buffer (see Appendix) for 60 min.
2. Phosphate buffer 0.1 M, 3 × 5 min.
3. Osmium tetroxide 1% in 0.1 M phosphate buffer for 60 min.
4. Distilled water (rapid rinse).
5. Increasing concentrations of ethanol: 70°, 95°, 2 × 100°, (5 min each).
6. Propylene oxide, 2 × 15 min.

7. Decreasing concentrations of ethanol: 100°, 95°, 70°.
8. Distilled water (rapid rinse).
9. Air dried.

F. FILM RADIOAUTOGRAPHY

Radioautographs from the fixed and the unfixed incubated sections, taken after the incubation and after steps 1, 3, and 6 of Section III.E, are generated by apposing the slides against tritium-sensitive film (Hyperfilm, Amersham) in X-ray cassettes, stored at 4°C. After 4 to 6 days exposure, the films are developed in D19 for 5 min at 18°C. Labeling intensity and signal/background ratio are measured on the film radioautographs by using a computer-assisted optical densitometry system (see Chapter 3).

These controls, much less time-consuming than the effective ultrastructural labeling procedure, permit the choice of optimal conditions for the various steps of the tissue preparation and binding procedure for electron microscopy.

IV. TISSUE PREPARATION

For electron microscopic detection of radioligand binding sites, tissues are fixed with the fixative selected acccording to the procedure described in Section III.A:

- Fixation is performed by intracardiac perfusion.
- The tissues are dissected out and stored in 0.1 M phosphate buffer (see Appendix) pH 7.35 for 1 to 12 h at 4°C.
- Slices, 100 μm thick, are prepared with a vibratome in ice-cooled 0.1 M phosphate buffer.

V. BINDING PROCEDURE

Binding procedure is basically the same whatever the resolution level of detection (macroscopic on film radioautographs, light or electron microscopic).

PROTOCOL 4: INCUBATION OF TISSUE SLICES IN THE RADIOLIGAND

Transfer the slices in 0.1 M phosphate buffer (see Appendix) into flat silicon-coated vials containing 1 to 2 ml reagent and place on a slowly moving agitator. The following reagents are replaced by decantation:

1. Tris-HCl buffer (see Appendix) for 20 min at room temperature.
2. Radioligand-containing Tris-HCl buffer for 24 h at 4°C, under mild agitation.
3. Tris-HCl buffer at 4°C, for 2 × 5 min.

For determination of nonspecific binding, treat some of the slides by adding at step 2, in addition to the radioligand, a large excess of unlabeled ligand (about 10^3 to 10^4 times the concentration of the radioligand).

VI. CROSS-LINKING, POSTFIXATION, EMBEDDING

A. CROSS-LINKING WITH GLUTARALDEHYDE

The radiolabeled slices are treated for 1 h in 5% glutaraldehyde in 0.1 M phosphate buffer (see Appendix) at 4°C, and rinsed in 0.1 M phosphate buffer. At this step, selected areas of the slices are trimmed into tissue blocks of 2 to 4 mm^2 with a razor blade.

B. POSTFIXATION WITH OSMIUM TETROXIDE

The blocks are treated for 1 h in 1% osmium tetroxide in 0.1 M phosphate buffer (see Appendix).

Although it has a crucial effect on general ultrastructural preservation, especially fixation of the membranes, osmium tetroxide postfixation can be omitted in the case of application of post-embedding immunocytological labeling, e.g., using colloidal gold, before the radioautographic procedure. Most immunocytological reactions are indeed impeded by osmium tetroxide treatment.

C. EMBEDDING

The samples are flat-embedded in Araldite-Epon mixture (see Protocol 5). The Araldite-Epon mixture permits large surface sections, but any other epoxide resin is suitable. Flat embedding avoids crinkling of the samples and thus facilitates sectioning.

PROTOCOL 5: EMBEDDING

1. Dehydrate in increasing concentrations of ethanol: 70°, 95° (2 ×), 100° (3 ×), for 10 min each.
2. Rinse in propylene oxide, 2 × 10 min.
3. Immerse in a 50:50 mixture of propylene-oxide and Araldite-Epon embedding medium (see Appendix) for 1 h.
4. Immerse in Araldite-Epon embedding mixture 2 × 1 h at 50°C.
5. Embed in a drop of the Araldite-Epon mixture between a slide and a coverslip, both coated with aluminum foil.
6. Allow to polymerize for at least 48 h at 60°C.
7. Trim the samples and stick them with superglue onto the flattened top of an empty Araldite-Epon block.

These preparations can be used for both light and electron microscopic studies.

VII. RADIOAUTOGRAPHY ON SEMITHIN SECTIONS (LIGHT MICROSCOPY)

Semithin sections of plastic-embedded samples permit radioautographic localization of binding sites by light microscopy with a much higher precision than frozen sections. On the other hand, they constitute the preliminary step for ultrastructural localizations.

A. SECTIONING

Semithin sections 1 to 2 μm thick are cut on an ultramicrotome using glass knives. They are transferred with a wire loop in a drop of distilled water onto a gelatin-coated glass slide (or simply to a clean slide). Several sections (possibly serial) can be placed on the same slide at least 0.5 cm from the edge and 1 cm from the top of the slide.

The slides are dried on a hot plate (90°C), which ensures spreading of the sections and their adherence to the slide. At least two series of slides and some additional test slides should be prepared for development using different exposure times.

B. RADIOAUTOGRAPHIC PROCEDURE

Radioautographic detection is conventionally performed by dipping the slides in melted nuclear emulsion (see Protocol 6). The nuclear emulsion is a suspension of silver halide crystals in gelatin which ideally form an uniform layer on the section. The labeling precision depends directly on the size of the crystals whereas the sensitivity is inversely related. The Auger electrons emitted by ^{125}I induce formation in the overlying crystals of a latent image which catalyzes reduction of the silver halide of the crystal to metallic silver grains by the developer, whereas the fixer eliminates the unreacted crystals.

PROTOCOL 6: EMULSION COATING FOR LIGHT MICROSCOPY

Equipment and reagents

In the darkroom equipped with safelight adapted to the nuclear emulsion used, the following materials and reagents are required:

- Water bath at 43 to 44°C
- Jar to melt (or dilute) the emulsion
- Jolly tube (a specially manufactured tube with minimal volume that reduces the amount of emulsion used for a small series of slides)
- Glass stirring rod
- Nuclear emulsion (Amersham LM-1, ready for use when melted or Ilford K5 to be diluted 1:1)*

---

* Nuclear emulsions for electron microscopy (Amersham EM-1 or Ilford K4) can also been used although the silver grains are of very small size.

- Five clean histological slides
- Light-tight slide boxes containing drying agent in its lid or wrapped in paper tissue.

Procedure

1. Place the emulsion jar in the water bath until emulsion melts (which takes approximately 1 h).
2. Dip clean glass slides to remove the bubbles.
3. Allow to dry in a position close to vertical for 6 h (or overnight).
4. Place the sections in the light-tight boxes.
5. Store at 4°C until development.

The exposure time for semithin sections is approximately twice that of histo-radioautographs of frozen sections. For detection of the oxytocin binding sites in the kidney it lasted between 1 and 2 months. Development (see Protocol 7) of test slides at regular time intervals permits determination of optimal exposure time.

The use of fresh reagents and strict respect for the development time and temperature are prerequisites for correct labeling.

PROTOCOL 7: DEVELOPMENT OF SEMITHIN SECTION HISTORADIOAUTOGRAPHS

Operate in a darkroom equipped with a safety light.
Reagents

Developer D19 Kodak (filtered if necessary).
Distilled water.
Fixer: 30% sodium thiosulfate (commercial fixers, e.g., Hypam, Ilford, can also be used).

Procedure

1. Remove the boxes containing the sections from the refrigerator at least 1 h before development to reach room temperature.
2. Immerse the sections for exactly 5 min in D19 at 18°C.
3. Rinse in distilled water for a few seconds.
4. Immerse in fixer.
5. Wash in running water for at least 30 min.
6. Rinse rapidly 3 times in distilled water.
7. Allow to dry on racks in a dust-free atmosphere.

Staining (see Protocol 8) of the tissue section for localization of the radio-autographic label risks the elimination of emulsion and consequently the silver grains, since staining of tissues in semithin sections needs heating and alkaline solutions to permeate the plastic embedding medium. Protocol 6 describes the method of Descarries and Beaudet[9] which results in adequate staining without altering the radioautographic labeling.

PROTOCOL 8: POSTSTAINING OF SEMITHIN SECTIONS WITH TOLUIDINE BLUE

1. Filter stock solution (1% toluidine blue in 1% sodium tetraborate) and dilute 1:10 with 70° ethanol.
2. Place slides on a hot plate (50 to 60°C) and stain for 45 s with a sufficient amount of dilute solution to cover the whole slide.
3. Rinse and differentiate with 70° ethanol; wipe off excess alcohol and evaporate any residue by heat.
4. Clear the section with a drop of xylene and mount the coverslip with Permount or Eukitt.

VIII. ELECTRON MICROSCOPE RADIOAUTOGRAPHY

For radioautographic labeling at ultrastructural scale, coating of the ultrathin sections with emulsion can be performed either by means of the loop technique (see description in Chapter 7) or by dipping the sections, mounted on a celloidin-coated slide, in the nuclear emulsion. As for light microscope radioautography, the "dipping" technique reported in this section largely derives from that previously described by Descarries and Beaudet.[9]

A. PREPARATION OF THE SECTIONS

PROTOCOL 9: MOUNTING OF SECTIONS ON CELLOIDIN-COATED SLIDES

1. Prepare celloidin coated slides.*
   - Engrave two crosses with a diamond scribe on the backs of the slides, 2.5 cm from the end and 0.5 cm from the lateral edge.
   - Clean slides thoroughly.
   - Prepare a solution of 2% collodion (parlodion) in isoamyl acetate, filter, and transfer into a Borel tube.
   - Dip the slides into the collodion solution and allow to dry for 2 to 3 h at a small angle to vertical, the upside downward.
   - Dip again the lower tip of the slide to a depth of about 5 mm. Dry again.
2. With an ultramicrotome and preferably using a diamond knife, cut white-yellowish sections of the area selected by previous semithin sections.

---

* Celloidin-coated slides can be stored for several months in a tight slide-box.

3. Transfer the sections, floated in a drop of water in a wire loop (or in a heuse cut out of plastic foil), onto the collodion-coated slide over the engraved crosses. Draw off the water drop containing the sections with the tip of a filter paper interposed between the edge of the loop and the slide, allowing sections to adhere (without folds!) to the slide. Care must be taken not to scratch the collodion film during this operation.
4. Contrast the sections by covering them with a large drop of uranyl acetate (5% in distilled water)* for 10 min, rinse with a gentle stream of distilled water; apply a drop of lead citrate (see Appendix), wash again, and allow to dry.
5. Coat the slides with a carbon film in an evaporator. The thickness of the film is the same as that used for stabilization of thin section for conventional ultrastructural observation.**

B. RADIOAUTOGRAPHIC PROCEDURE

The protocol for emulsion coating is similar to Protocol 6, except that special emulsions for electron microscopy, with smaller silver halide crystals, are used: Ilford L-4 (diluted 1:4 in distilled water) or Amersham EM-1 (ready for use).

As for light microscopy, the slides are placed in light-tight slide boxes containing a drying agent which needs to be replaced monthly. Development of test slides at regular intervals permits optimization of exposure times, which are approximately 3 to 5 times those of semithin sections. However, in view of the 60-day half-life of ^{125}I, exposure times of over 3 to 4 months do not notably increase labeling intensity.

Development of electron microscopic radioautographs is performed as described in Protocol 7. The developer, Microdol, filtered before use, for 4 min at 18°C is most frequently used. Use of other developers and their effects on size and shape of the silver filament are discussed in Chapter 7.

During the development procedure, the celloidin films tend frequently to detach from the slides. To avoid this accident, rinsing, fixation, and washing must be performed at the same or slightly lower temperature than development.

After washes in distilled water, the slides are placed on racks to drain but not dry completely in a humid atmosphere, before transferring (see Protocol 10) them onto grids.

C. TRANSFER OF SECTIONS ON GRIDS

PROTOCOL 10: STRIPPING OF THE CELLOIDIN FILM AND PLACEMENT OF THE SECTIONS ON GRIDS

Material required

Crystallizing dish filled to the brim with distilled water, placed on a black plastic tray

---

* Uranyl acetate associated with lead citrate (see Appendix) confers an appreciable contrast. It should, however, not been systematically used since it provokes conspicuous α-tracks in the emulsion.
** Although not an absolute necessity, the carbon coating facilitates layering of the emulsion and avoids destaining of the sections during the radioautographic procedure.

Glass rod
Clean razor blade
Fine forceps
Clean grids 100 to 200 mesh
Parafilm
Small Petri dishes

Procedure

1. Holding the slide by its frosted end, detach the thick edge of the celloidin film with the razor blade and incise the emulsion 2 to 3 mm from the lateral edges of the slide.
2. Slowly lower the slide into the water at a sharp angle, sections facing upward. If the celloidin film does not spontaneously strip off, pull it gently with the fine forceps or with a mounted needle.
3. Place grids, shiny face down, over the sections.
4. Dispose carefully a parafilm, slightly exceeding the size of the slide, over the celloidin film, remove it gently and place it on a filter paper in a Petri dish.
5. Let dry at room temperature.
6. Detach the grids from the film by puncturing around them with the fine forceps.
7. Dissolve the celloidin film by immersing the grids for 1 to 5 min in isoamyl acetate in a watch glass. Allow to dry.

The sections are now ready for observation under the electron microscope.

IX. RESULTS

A. PRELIMINARY TESTS

The effects of tissue fixation with 2% paraformaldehyde perfusion on the distribution and density of oxytocin binding sites in the cortex of rat kidney are shown in the radioautographic films in Figures 1A and B. The specific binding expressed as optical density (OD) × 100 was 28.48 ± 1.54 in unfixed tissue, vs. 27.60 ± 1.42 in paraformaldehyde-fixed tissue. Subsequent treatment of the sections with 5% glutaraldehyde, osmium tetroxide, and final defatting (Figure 1C) only slightly decreased the labeling intensity (OD × 100 = 20.85 ± 1.23).

B. SEMITHIN SECTIONS

Radioautographs from semithin sections provide cellular localization of the radioligand binding sites by light microscopy with a precision which cannot been reached on frozen sections (Figure 2). In our present example, localization of the oxytocin binding sites previously detected on the juxtaglomerular apparatus did not allow clear-cut identification of the component of this very complex formation bearing the binding sites. Radioautographs from semithin sections unambiguously indicated that the macula densa (a differentiation of the distal tubule apposed

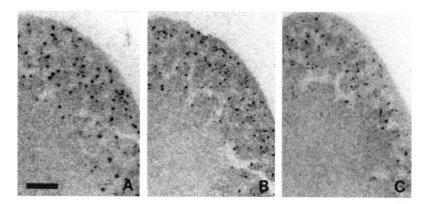

FIGURE 1   Film radioautographs of frozen kidney sections incubated in the presence of 0.03 nM
^{125}I-labeled oxytocin antagonist. A: unfixed; B: fixed with 2% paraformaldehyde before application
of the binding procedure; C: fixed in 2% paraformaldehyde, treated after the binding procedure
with glutaraldehyde, osmium tetroxide, and defatted with alcohol and xylene. Calibration bar =
1 mm.

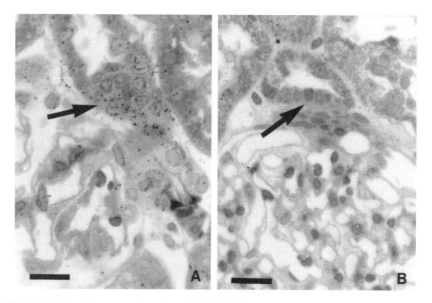

FIGURE 2   Light microscope detection of ^{125}I-labeled oxytocin antagonist on semithin sections.
A: specific labeling concentrated on the macula densa (arrow) of the juxtaglomerular apparatus.
B: Absence of labeling on the macula densa (arrow) after incubation in the presence of 0.03 nM
^{125}I-labeled oxytocin antagonist plus 10 μM oxytocin for detection of nonspecific binding. Cali-
bration bars = 20 μm.

against the vascular pole of the glomerulus) was the only labeled structure (Figure
2A). No labeling occurred when the incubation was performed in the presence of
10 μM oxytocin (Figure 2B), which indicates that labeling was related to specific
oxytocin binding sites.

The low sensitivity of the technique, compared to the conventional historadio-autographic technique (see Chapter 6), however, must be pointed out. High concentrations of binding sites are thus required for unequivocal labeling, even after prolonged exposure times.

C. ELECTRON MICROSCOPY

The sections processed for electron microscope radioautography revealed concentration of silver grains at the basal pole of the cells of the macula densa, closely related to the plasma membrane (Figure 3). In this case labeling was unequivocal and did not require analysis of the silver grain distribution. In more complex tissues, especially in the central nervous system, the limited resolution of the radioautographic method calls for statistical analysis of the silver grain distribution to assess the precise localization of the radioisotope.[10] The precision of radioautographic labeling is dependent on both the energy of the radiations emitted by the radioisotope and the size of the silver crystals of the nuclear emulsion (see Chapter 7).

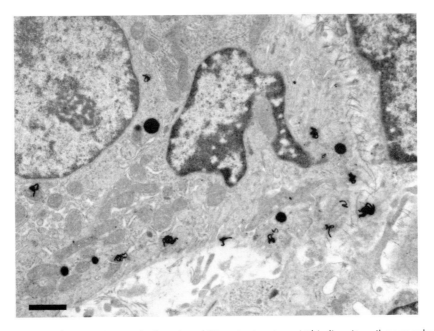

FIGURE 3 Electron microscopic detection of ^{125}I oxytocin antagonist binding sites: silver granules are aligned along the basal plasma membrane of a macula densa cell. Calibration bar = 1 μm.

X. CONCLUSION

The radioautographic detection of radioligand binding sites at the ultrastructural scale constitutes a useful tool in receptor research, despite the complexity of the procedure and the relative imprecision of the labeling obtained. The more recently developed immunocytochemical localization of receptors is certainly easier to perform and results in more precise locations provided the receptor is cloned and antibodies are available and sensitive enough. In fact, the two approaches are not equivalent and immunolabeling demonstrates presence of the receptor protein whereas radioautography localizes binding sites. The main advantage of the radioligand binding procedure consists in the detection of the binding sites under pharmacologically controlled conditions.

ACKNOWLEDGMENTS

We are indebted to Dr. M. Manning for generously providing the oxytocin and vasopressin antagonists. We thank M.J. Klein and E. Waltisperger for technical assistance and J.M. Gachon for photographic work.

APPENDIX

Araldite-Epon Embedding Mixture
Add successively under moderate stirring:

4 ml Araldite M (CY 212)
5 ml Epon (Glycidether 100)
12 ml Hardener DDSA (DBA)
1 ml Accelerator DMP 30

Mix thoroughly, avoiding bubble formation, for 10 min.

Gelatin-Coated Slides

- Dissolve by heating moderately (below 45°C) 1.25 g powdered gelatin and 0.125 g chrome alum in 250 ml distilled water. Filter.
- Dip thoroughly cleaned sections in the solution.
- Let dry.
- Heating of the gelatin-coated slides in an oven at 120°C for 3 h improves adherence of frozen sections.

Lead Citrate

- 16 ml Distilled water
- 4 ml Sodium citrate M solution
- 3 ml Lead nitrate M solution
- 4 ml Sodium hydroxide N solution, freshly prepared.

Stir until the precipitate is completely dissolved and store at room temperature in a stoppered vial for 1 to 2 weeks.

Phosphate Buffer 0.1 M, pH 7.35

- Disodium phosphate $(Na_2HPO_4 \cdot 12\ H_2O)$ 2.9 g
- Monosodium phosphate $(NaH_2PO_4 \cdot H_2O)$ 0.264 g
- Distilled water to 100 ml.

Tris-HCl Buffer, pH 7.4
85 mM Tris-HCl:

- Solution A: Tris base (MW 121.1) 170 mM: 20.587 g/l
- Solution B: hydrochloric acid 170 mM, HCl N: 170 ml/l
- Mix equal parts of solutions A and B, adjust pH at 7.4 with solution A if necessary.
- Add: 5 mM magnesium chloride $(MgCl_2 \cdot 6\ H_2O)$: 1.016 g/l and 0.1% bovine serum albumin: 1 g/l

REFERENCES

1. Young, W. S. and Kuhar, M. J., A new method for receptor radioautography: (3H) opioid receptors in rat brain. *Brain Res.*, 179, 255, 1979.
2. Palacios, J. M., Niehoff, D. L., and Kuhar, M. J., Receptor radioautography with tritium-sensitive film: potential for computer densitometry. *Neurosci. Lett.*, 25, 101, 1981.
3. Hamel, E. and Beaudet, A., Localization of opioid binding sites in rat brain by electron microscopic radioautography. *J. Electron Microsc. Technol.*, 1, 317, 1984.
4. Elands, J., Barberis, C., Jard, S., Tribollet, E., Dreifuss, J. J., Bankowski, K., Manning, M. and Sawyer, W. H., ^{125}I-labeled $d(CH_2)^5$ $(Tyr\ (Me)^2, Thyr^4, Tyr\text{-}NH_2^9)OVT$: a selective oxytocin receptor ligand. *Eur. J. Pharmacol.*, 147, 197, 1988.
5. Stoeckel, M. E. and Freund-Mercier, M. J., Autoradiographic demonstration of oxytocin binding sites in the macula densa. *Am. J. Physiol.*, F310, 1989.
6. Hayat, M. A., *Fixation for Electron Microscopy*, Academic Press, New York, 1978, chap.4.
7. Leroux, P., Gonzalez, B. J., Laquerriere, A., Bodenanat, C., and Vaudry, H., Autoradiographic study of somatostatin receptors in the rat hypothalamus: validation of a GTP-induced desaturation procedure. *Neuroendocrinology*, 47, 533, 1988.
8. Palacios, J. M. and Dietl, M. M., Regulatory peptide receptors: visualization by autoradiography. *Experientia*, 43, 750, 1987.
9. Descarries, L. and Beaudet, A., The use of radioautography for investigating transmittter-specific neurons. In *Handbook of Chemical Neuroanatomy*, Vol. 1., Methods in Chemical Neuroanatomy, Björklund, A. and Hökfelt, I., Eds., Elsevier, Amsterdam, 1983, 286.
10. Beaudet, A., Hamel, E., Leonard, K., Vial, M., Moyse, E., Kitabgi, P., Vincent, J. P., and Rostène, W., Autoradiographic localization of brain peptide receptors at the electron microscopic level. In *Neurotransmitters and Cortical Function*, Avoli, E., Reader T. A., Dykes, R. W., and Gloor, P., Eds., Plenum Press, New York, 1988, 547.

Chapter 5

WHOLE BODY AND MICRORADIOAUTOGRAPHY OF RECEPTORS

Masahisa Shimada
Masahito Watanabe

CONTENTS

I. INTRODUCTION

For relating biochemical, physiologic, and pharmacologic information on receptors to an anatomic background, a morphologic technique might be necessary, although microdissection techniques allow determination of receptor levels in tissues as small as 200 to 500 μm in diameter.[1,2] However, use of an ultramicroscopic technique as a starting point for a morphologic experiment would result in a loss of perspective for the big picture, like the proverbial case of the blind men examining the elephant. The body is made up of many organs, and each individual organ is functionally and morphologically heterogeneous. It is important that all experiments proceed step by step, from the macroscopic to the microscopic level. Radioautography is a powerful tool to visualize receptor localization. Whole-body radioautography is suitable for studying the tissue distribution of radiolabeled ligands in the whole animal. Microradioautography at the light microscopic level is suitable for visualization of radiolabeled ligand bindings at the cellular level.

II. GENERAL REMARKS ON RECEPTOR RADIOAUTOGRAPHY

From the viewpoint of size, radioautography is classified into macro- and microradioautography. Macroradioautography includes whole-body radioautography of animals such as small monkeys, rats, and mice, and organ radioautography of large organs like the brain, liver, and kidney. Radioautographs obtained through this procedure can be observed with the naked eye. Microradioautography, on the other hand, includes radioautography of pieces of organs and tissues or cultured cells. Microradioautography is further divided into light microscopic and electron microscopic radioautography. For macroradioautographic studies, high-energy radioisotopes such as ^{32}P, ^{35}S, and ^{14}C can be used, whereas in microradioautography low-energy radioisotopes such as ^3H and ^{125}I should be used for reasons of resolution.

From the viewpoint of the solubility of radiolabeled ligands and receptor complexes, radioautography is divided into radioautography of diffusible and nondiffusible substances. For example, in radioautographic studies on tritiated steroids by Bogoroch,[3] use of aqueous fixatives such as 4% paraformaldehyde resulted in the loss of most of the radioactivity into the fixative. In such a case, radioautography of diffusible substances is necessary to stabilize the hormone *in situ* throughout the radioautographic procedure. The best method for this purpose is a freeze-drying cryotome and a dry-mounting technique. This technique is described elsewhere.[4-7] However, the procedures described in this chapter to visualize receptors with radioautography are based on the assumption that the radiolabeled ligands and receptor complexes are stable during radioautographic procedures, which include fixation with aqueous fixatives, dehydration by a graded series of ethanols, and the use of wet emulsions except for *in vivo* whole-body radioautography. Therefore, the techniques described here are applicable for peptide and protein ligands because histologic fixatives do not cross-link low molecular weight compounds.

For radioautography, two types of labeling techniques for receptors can be used, *in vivo* and *in vitro*. In *in vivo* receptor radioautography, tissues and organs are removed from experimental animals that have received radiolabeled ligand injections, and then sections are prepared. The sites labeled by the ligand are visualized by radioautography with films or nuclear emulsions.

During *in vitro* procedures, sections of tissues and organs from experimental animals, which have not received a radiolabeled ligand, are incubated with a radiolabeled ligand. The sections are then exposed to films or nuclear emulsions to detect radioactivity from the ligand. Because many factors such as metabolism of ligands and tissue barriers might influence the *in vivo* receptor labeling, the interpretation of their results might be more complicated than that of *in vitro* labeling. However, *in vivo* procedures might better reflect physiologic binding of ligands to their receptors. As has been described by Kuhar,[8] *in vitro* procedures have certain advantages over the *in vivo* procedures: (1) the quantity of radioisotope needed is much smaller than the case with *in vivo* radioautography, and (2) physiologic conditions of the receptor sites can be easily regulated by removing endogenous sources of ligands.

III. MACRORADIOAUTOGRAPHY FOR TISSUE RECEPTOR DISTRIBUTION

A. ROUTINE PROCEDURE FOR WHOLE BODY RADIOAUTOGRAPHY[9-11]

The technique of whole-body radioautography was first developed by Sven Ullberg of Uppsala University in Sweden in 1954.[12] Whole-body radioautography is divided into two categories: *in vivo* whole-body radioautography and *in vitro* whole-body radioautography.

1. *In Vivo* Whole-Body Radioautography

1. After intravenous injection of a radiolabeled ligand into experimental animals, the animals are perfused with Ringer's solution to wash out unbound ligands from the whole-body. Usually, to avoid the nonspecific binding of ligands to the organ and tissue, washing the syringe with trace amounts of serum albumin

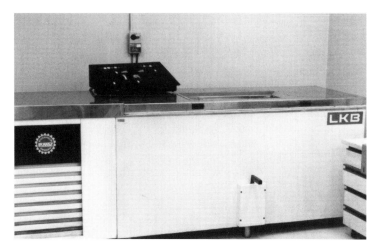

FIGURE 1   A heavy-duty cryomicrotome.

is recommended. Then, 3 min before the perfusion, the mice are anesthetized by intraperitoneal injection of sodium pentobarbital.

2. The region incised for perfusion is covered with 3% carboxymethylcellulose that has been frozen with powdered dry ice. Then, the entire body of the animal is frozen at −70°C in a mixture of dry ice and acetone or dry ice and hexane.

3. The frozen animal is embedded into 6% carboxymethylcellulose on the microtome stage and after equilibrating the block with the temperature of the cryotome (−20°C), whole-body cryosections of 20 µm thickness are prepared by using a heavy-duty microtome. We use a Swedish cryotome microtome (LKB 2250), which allows sectioning of 450 mm × 150 mm tissue blocks (Figure 1). The range of sectioning can be varied from 1 to 999 µm.

4. Large frozen sections containing various organs and tissues are very fragile; therefore, to prevent the section from falling apart, adhesive tape is applied to the cut surface of the frozen animal. Scotch® tape (Type 800, 810, and 8110, 3M Co., St. Paul, MN, U.S.) is most commonly used as the adhesive tape. Good sections can be obtained with the appropriate pressure on the knife edge of the microtome (Figure 2).

5. The sections obtained are freeze-dried in a cryotome or deep freezer. The freeze-dried sections are brought to room temperature in a desiccator containing silica gel.

6. The dry sections are then placed in contact with films, using aluminum plates that are screwed down. Medical X-ray film is useful for ^{14}C- and ^{35}S-labeled ligands. Thin polyester film (Diafoil®, 4 µm thick, Mitsubishi Plastics Co., Tokyo, Japan) can be inserted between the sections and the film to prevent the chemography and the adhesive tape from sticking to the film. However, for ^{3}H- or ^{125}I-labeled ligands, Ultrofilm® (LKB, Bethesda, MD, U.S.), Hyperfilm TM-3H® (Amersham International, Amersham, U.K.), or Konica Macroradioautograph® (Konica Co., Tokyo, Japan) film should be used (they have no protective layer of gelatin), and any inserts

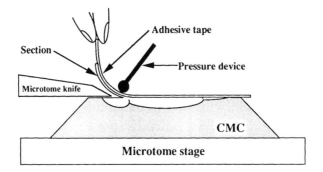

FIGURE 2 Diagram illustrates a method of cryosectioning an animal.

should be avoided. Therefore, before examination using this direct contact method to film, negative and positive chemography should be checked (see discussion of chemography in Section IV.3).

7. After exposure in a cool, dark box, the films are developed and fixed. D19® (Kodak, Rochester, NY, U.S.) is a good emulsion developer because it shows a large range of gray levels and results in good proportionality between the optical density and the radioactivity.[13] Figure 3 shows the distribution of tissue insulin receptors in a male ICR mouse.

There is an alternative method of whole-body radioautography in which the whole-body cryosections on glass slides are used rather than cryosections adhering to adhesive tape. The technique is described later (see Section III.4). The dried cryosections on the glass slides are exposed to X-ray film while under slight pressure. To prevent uneven pressure, the slide-mounted sections are placed in a plastic bag, and the bag is evacuated of air and sealed by a vacuum sealer (SQ-202, Sharp Co. Ltd., Tokyo, Japan) (Figure 4).[14]

2. *In Vitro* Whole Body Radioautography

Without injection of radiolabeled ligands into experimental animals, the animals are perfused with Ringer's solution and prepared into whole-body cryosections as described above. The whole-body sections adhering to the adhesive tape are fixed, washed, and incubated with a radiolabeled ligand in the absence or presence of excess unlabeled ligand. After washing and drying, the sections are ready for exposure against films. On the basis of this technique, it might be possible to study the tissue distribution of radiolabeled ligand binding sites, but no such attempt has been made so far because this technique presents big problems. Adhesive tapes used for whole-body radioautography could be deformed and denatured during fixation, incubation, and washing. Whole-body sections on the glass slide may come off during the processing.

3. Observation of Whole Body Radioautographs

Whole-body radioautographs are usually observed with the naked eye. However, when ^{14}C, ^{35}S, ^{125}I, or ^3H radioisotopes are used, the whole-body radioautographs are also suitable for observation under a low-power microscope. Before the whole-body

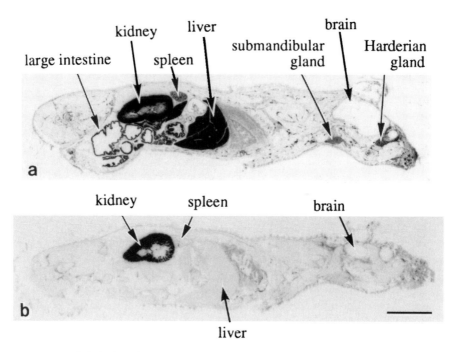

FIGURE 3   Whole-body radioautographs show the distribution of total (a) and nonspecific (b) insulin-binding sites in mice after 3 min of radiolabeled insulin. The mice were injected intravenously with 185 kBq of ^{125}I-insulin (porcine, ^{125}I-Tyr^{A14} insulin dissolved in Ringer's solution at a concentration of 185 kBq/0.4 ml) in the absence (a) or presence (b) of excess (50 µg) unlabeled insulin. Bar = 1 cm. (From Watanabe, M., Hirose, Y., Sugimoto, M., Nakanishi, M., Watanabe, H., and Shimada, M., *J. Recept. Res.*, 12, 13, 1992. With permission.)

radioautographs are observed, organs and tissues appearing in the radioautographs must be identified. Whole-body histologic sections that correspond to the radioautographs are prepared for this purpose.

Usually, sections taken adjacent to the radioautographic specimens are stained. If sections adhering to the tape are used, several problems could occur during staining, although certain stains like Goldner's collagen and very weak Meyer's hematoxylin and eosin give good results.[15] The cellulose acetate backing of Scotch® tape expands and contracts in different solvents and may be stained by several dyes. Organic solvents cannot be used. Observation of the stained section on the tape under a microscope is difficult because under the coverslip the tape crinkles in the embedding medium. The preparation of good quality whole-body sections on glass slides will resolve these problems. Several techniques have been developed in our laboratory.[16-18]

4.   Preparation of Whole-Body Histologic Sections

Dry, thin paper, which is cut nearly the same shape as the animal, is put on the surface of the animal and adhesive tape applied to the entire cut surface as shown in Figure 5.[18] Japanese paper (Yamanashigasen or Inshugasen, obtained from Kyoto Kyukyodo, Teramachi, Nakagyo-ku, Kyoto, 604, Japan) is recommended.[17] The

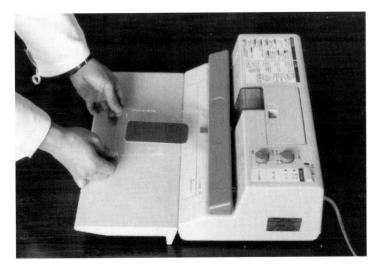

FIGURE 4 Contact of film with a whole-body section mounted on a glass slide. A plastic bag containing a piece of photographic film placed on the slide is evacuated of air by a vacuum sealer. (From Watanabe, M., Hirose, Y., and Shimada, M., *Acta Histochem. Cytochem.*, 24, 167, 1991. With permission.)

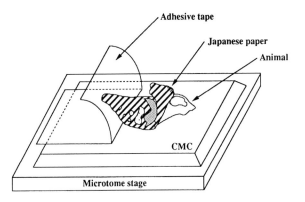

FIGURE 5 Diagram illustrating the method of cryosectioning an animal using dry paper and adhesive tape.

sections are dried in a cryotome and pressed onto glass slides coated with an egg albumin and glycerin mixture. Both the adhesive tape and the paper are then removed from the glass slides. The whole-body sections on the glass slides are fixed and stained. A mouse whole-body section stained with hematoxylin and eosin thus obtained is shown in Figure 6.

B. ROUTINE PROCEDURE FOR ORGAN RADIOAUTOGRAPHY
1. *In Vivo* Organ Radioautography

With whole-body radioautography, radioactivies in various parts of organs and tissues can be estimated and compared. For tubular organs, these techniques are not

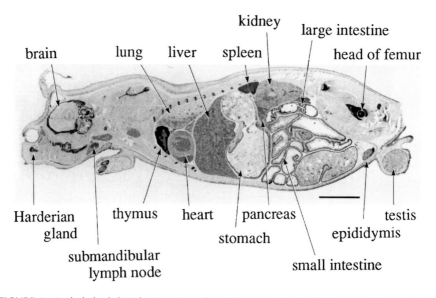

FIGURE 6   A whole-body histologic section of a male mouse stained with hematoxylin and eosin. Bar = 1 cm.

sufficient for such demands in research because a single whole-body section does not display any tubular organ as a whole. For this purpose, radioautography of tubular organs has been established.[19] Insulin binding in the gastrointestinal tract of mice was studied by this method. After injection of ^{125}I-insulin in the presence and absence of excess unlabeled insulin into fed male ICR mice, the mice were perfused with Ringer's solution and fixed with 4% paraformaldehyde solution through the left ventricle. The alimentary tract from the esophagus to the anus was then removed. The alimentary tract was placed on a paraffin plate in the form of a coil and adjusted so that all rings were set on the same plane. After being set on the paraffin plate, the organs were immediately frozen in liquid nitrogen. The paraffin plate was prepared in advance by pouring hot paraffin into a plastic Petri dish. The frozen organ and paraffin plate were together embedded in 3% (w/v) CMC paste in a block (Figure 7a), and refrozen in a dry-ice powder. Then the paraffin plate was removed from the block. The block was turned so that the organ surface was facing upward and then fixed on the microtome stage with 6% (w/v) CMC paste (Figure 7b). With the aid of adhesive tape cryosections of 25 μm thickness were cut, dried, and exposed to the film for macroradioautography (^{3}H-type, Konica) as described in Section III.A.1. The macroradioautographs showing the distribution of insulin-binding sites throughout the alimentary tract of mice are shown in Figure 8.

For the parenchymal organs such as the brain, liver, and kidney, the following organ radioautography is recommended. We studied the distribution of insulin receptors in the mouse liver. After injection of ^{125}I-labeled insulin, with or without excess unlabeled insulin, the animals were perfused with Ringer's solution and fixed with 4% paraformaldehyde solution through the left ventricle. The organs were then removed and immersed in the solution for 2 h. After the immersion, the specimens were dehydrated with a graded series of ethanol and then embedded into paraffin

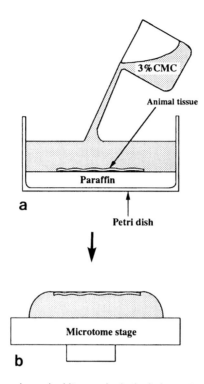

FIGURE 7 Diagram shows the embedding method of tubular organs. (as described in *in vivo* organ radioautography) (From Kuno, N., Watanabe, M., Akagi, N., Sugimoto, M., and Kihara, T., *Acta Histochem. Cytochem.*, 20, 547, 1987. With permission.)

after immersion in xylene. Paraffin sections, 5 μm thick, on glass slides were deparaffinized and brought into contact with the film for macroradioautography. After appropriate exposure, the films were developed and fixed. The macroradioautographs are shown in Figure 9a. In this case, nuclear emulsion will give better results (Figure 9b). The emulsion was applied by the dipping technique (see Section IV.A.2), but exposure time was much longer than that in microradioautography.

2. *In Vitro* Organ Radioautography

In vitro organ radioautography could visualize specific binding sites or receptors through exposure of the thaw-mounted tissue sections on glass slides which are incubated with radiolabeled ligand to photographic film. The brains were removed quickly from the animals without injection of a radiolabeled ligand. The brains were washed with ice-cold physiologic saline to remove blood and then frozen as soon as possible in isopentane, cooled by liquid N_2. During the freezing, time is an important factor to prevent cracking of the organs (e.g., 15 s for mouse brains). Sections 25 μm thick were cut on a microtome at $-15°C$ and thaw-mounted on glass slides. Thinner sections are preferred, but it is more important that they are cut with a uniform thickness throughout the sections. To prevent the section from peeling off the glass slide, the glass slides must be coated with either dichlorodimethylsilane, poly-l-lysine, or gelatin. Although the adhesive power of gelatin is inferior to the

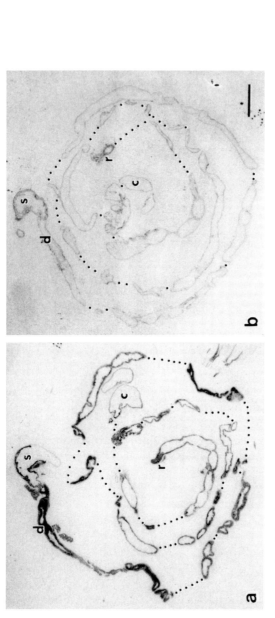

FIGURE 8 Macroradioautographs of the alimentary tracts of mice taken 3 min after intravenous injection of 185 kBq ^{125}I-insulin (porcine, ^{125}I-Tyr^{A14} insulin, receptor grade, specific activity 81.4 TBq/mmol, New England Nuclear, Boston, MA, U.S., dissolved in Ringer's solution at a concentration of 185 kBq/0.4 ml) in the absence (a) or presence (b) of excess (50 μg) unlabeled insulin (porcine, Sigma Chemical Company, St. Louis, MO, U.S.); s, stomach; d, duodenum; c, cecum; r, rectum. Bar = 1 cm.

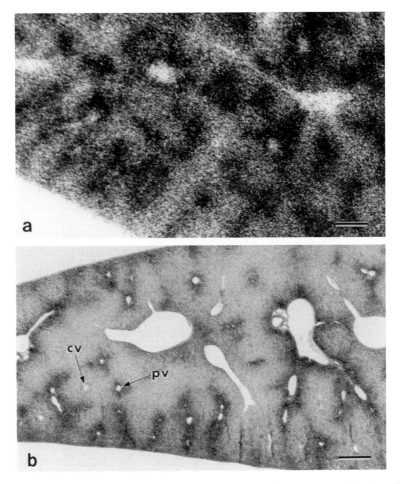

FIGURE 9 Macroradioautographs of the mouse liver 3 min after intravenous injection of 185 kBq ^{125}I-insulin. The paraffin sections were exposed to film (Konica macroradioautograph film) for 5 days (a), and coated with NR-M2 emulsion and exposed for 3 weeks (b); cv, central vein; pv, portal vein. Bar = 500 μm.

other two choices, we use the gelatin coating because the sections can be spread out more easily. The glass slides were subbed in a solution of 0.5% (w/v) gelatin and 0.05% chromium potassium sulfate and then air dried.

The sections obtained must be dried immediately. In our laboratory, the sections were dried for 2 h with a vacuum pump at –20°C. The sections were preincubated with a buffer solution for 15 to 30 min to remove intrinsic ligands and to act as an inhibitor for peptidase. After the preincubation, the sections were incubated with a radiolabeled ligand. Temperature, type of buffer solution, and incubation time are determined by the biochemical characteristics of the receptor-binding experiments with homogenate tissue or with crude membrane fraction. To avoid morphologic deformation, an isotonic buffer solution is usually used. The incubation time is set at maximum binding time for radiolabeled ligands. Because there are several sub-types of receptors in the same tissue section, reaction rates for each binding site must be determined. The stability of the radiolabeled ligand in the binding conditions should be checked.

To decrease nonspecific binding, washing with ice-cold isotonic buffer solution was done after incubation with the radiolabeled ligand. Final washings were done with ice-cold distilled H_2O to remove salts present in the buffer. After being air dried, the sections were exposed to photographic film. Figure 10 shows *in vitro* macroradioautographs of the mouse brain incubated with ^{125}I-insulin with or without cold excess insulin.[20]

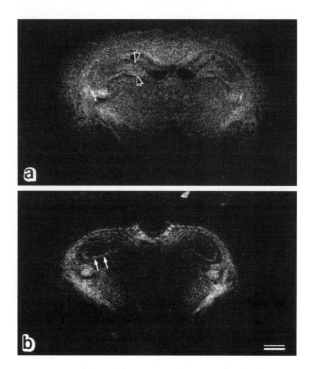

FIGURE 10 Macroradioautographs of the mouse brain incubated with ^{125}I-insulin in the absence (a) or presence (b) of excess unlabeled insulin (1 μM). The sections mounted on glass slide were apposed to film (Konica macroradioautograph film) for 5 days. Arrow heads indicate dentate gyrus and CA3 sector of the pyramidal cell layer of the hippocampus. Arrows indicate nonspecific binding sites of ^{125}I-insulin in CA3b and CA3c. Bar = 500 μm. (From Shimada, M., Hirose, Y., Nakanishi, M., Watanabe, H., Kawamoto, S., and Watanabe, M., *Recent Advances in Cellular and Molecular Biology*, Vol. 4, Peeters Press, Leuven, Belgium, 1992, 271. With permission.)

C. ANALYSIS OF THE MACRORADIOAUTOGRAPHS

The density of the radioautographic images in the films is determined by computer-assisted densitometry. As a standard for quantification of radioautographs, ^{125}I and ^{3}H microscales (Amersham International plc., Buckinghamshire, U.K.) are used. The sensitivity of Konica Macroradioautograph film (^{3}H type, Konica Co., Tokyo, Japan) to a 5-day exposure of ^{125}I-plastic standards ([I-125]microscales, RPA522, Amersham) is shown in Figure 11. The film begins to saturate above optical density 3.0. Instead of the microscales, brain pastes, prepared by mixing brain homogenate

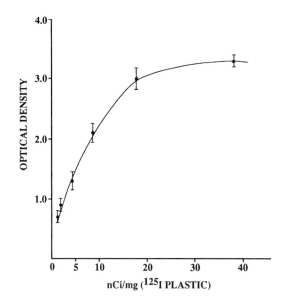

FIGURE 11 Response of Konica macroradioautograph film to a 5-day exposure of ^{125}I-plastic standard (mean ± SEM of 3 standards). (From Watanabe, M., Hirose, Y., Sugimoto, M., Nakanishi, M., Watanabe, H., and Shimada, M., *J. Recept. Res.*, 12, 13, 1992. With permission.)

and a known amount of ^{125}I or ^{3}H have been used as standards, but this results in a variance of outcomes among researchers. Saavedra and colleagues showed that 1 dpm/mg of the ^{125}I-microscales corresponds to 7.34 ± 0.22 dpm/mg of protein prepared from ^{125}I-brain paste using rat cerebral gray matter.[21]

Quantitative values are usually expressed as the amount of the ligand bound to a milligram of tissue protein. For more precise quantification, tissue sections, which are used for binding experiments, are stained by Commassie blue after exposure to Hyperfilm-[3]H®.[22] The staining is done with 2.5% (w/v) Commassie brilliant blue (B-1131, Sigma) in 10 mM Tris-HCl buffer (154 mM NaCl, pH 7.5) at room temperature after immersion in 10 mM Tris-HCl buffer (154 mM NaCl, pH 7.5) for 10 min. After being air dried, protein concentrations of the regions corresponding to the radiolabeled ligand binding sites are calculated by measuring the staining intensity with a microdensitometer.

Especially in the case of ^{3}H-ligands, there are large differences in the lipid content in the tissue, and error could occur between actual radioactivity from binding radiolabeled ligands and the film blackings (OD values), even after correction by protein concentrations. This results from the self-absorption of β energy. Careful attention must be paid, especially when one measures the receptor density in the brain's gray and white matter. This type of quenching effect could be estimated through a comparison of radioactivity before and after treatment with chloroform to remove lipids in the brain sections.

To obtain appropriate OD values of the radioautographs, it is necessary to know the relationship between isotope concentrations and film blackings at various exposure times. Therefore, in the binding experiments of radiolabeled ligands, with or without unlabeled ligand using serial cryosections, appropriate OD values for each section should be obtained by changing the exposure time for the different amount of binding.

Relevant to the quantitative method, instead of Hyperfilm-^{3}H, emulsion radioautography could be used. We used this technique to identify the presence of receptors at the light microscopic level because it is possible to quantify them with a high degree of accuracy by means of image analysis.

1. Progress of Quantity

There is a drawback to the use of photographic films in quantification, that is, the dynamic range is narrow. Recently, an imaging plate system (Bio-imaging Analyzer BAS 2000®, BAS 3000®, Fuji Photo Co., Tokyo, Japan) has been available for radioautography.[23-25] This system consists of an imaging plate and an image reading unit. The imaging plate is a two-dimensional sensor for the radiation energy. A polyester support is coated with microcrystals of BaFBr:Eu^{+2}. When the radiation energy is absorbed by the imaging plate, the excited states of very long lives are generated in the microcrystals. After an appropriate exposure, the imaging plate is inserted into an image reading unit and then scanned with a fine laser beam. The irradiation of the laser beam stimulates the emission of visible light from the excited states in BaFBr:Eu^{+2} (photostimulated luminescence). The image data are recorded as digital values in an analyzing unit for further analysis. The resolving powers of the BAS are 100 μm (BAS 2000) and 50 μm (BAS 3000), respectively. These are inferior to those of conventional photographic film. However, the imaging plate system is easy to manipulate and has several advantageous properties for detecting the radiation over photographic films. It has a wide dynamic range and is suitable for quantification. Moreover, it has a high sensitivity to the radiation and exposure time can be reduced markedly.

IV. MICRORADIOAUTOGRAPHY FOR TISSUE RECEPTOR DISTRIBUTION

To answer the question of where specific ligand binding sites are localized at the cellular level, radioautographic techniques with high spatial resolution are required. For this purpose, microradioautography at the light microscopic level is used. Similar to whole-body radioautography, microradioautography is also divided into two categories, *in vivo* and *in vitro*. *In vitro* microradioautography is especially recommended for the study of receptors in tissues and organs having a barrier function to a given ligand, e.g., brain. For other tissues and organs, *in vivo* microradioautography could be useful, though under *in vivo* conditions the cost of radiolabeled ligands is more expensive.

A. ROUTINE PROCEDURE FOR MICRORADIOAUTOGRAPHY[26,27]

The paraffin- and resin-embedded sections or the frozen sections of tissues obtained from the animals injected with radioactive compound are used for light microscopic radioautography. Nonradioactive tissue sections are incubated with radioactive compound. These sections are mounted on microscope slides and nuclear emulsion is applied. After appropriate exposure in a cool dark room, the emulsion is developed and fixed. Then the sections are stained and mounted per usual histologic procedure. Although there are certain ways of applying emulsion to a section, the dipping technique is commonly used.

1. Choice of Emulsion

Several types of nuclear emulsions suitable for microradioautography are commercially available (Table 1). Each emulsion has special characteristics. For example, Konica Co. (Tokyo, Japan) manufactures two different radioautographic emulsions: NR-M2 and NR-H2. NR-M2 is designed for use with light microscopes; NR-H2 is designed for use with electron microscopes.[28] NR-M2 has larger-diameter silver crystals to make observation of developed silver grains easy under low-power magnification. The sensitivity of NR-M2 emulsion is approximately one third when compared with that of NTB3, in our experience. For the microradioautographic study of receptors at the light microscopic level with ^{3}H and ^{125}I, Kodak NTB2, Ilford K2, and Konica NR-M2 are excellent. The emulsions should never be frozen and temperature exceeding 45°C must be avoided.

TABLE 1   Emulsions in General Use in Microradioautography

Manufacturer	Type	Crystal Diameter	Sensitivity	Characteristics
Ilford	G.5.	0.27 µm	5.0	For ^{14}C
Ilford	K.2.	0.20 µm	2.0	For tritium and ^{125}I
Eastman-Kodak	NTB-3	0.34 µm	5.0	For γ emitters
Eastman-Kodak	NTB-2	0.26 µm	3.0	For β emitters
Konica	NR-M2	0.14 µm	1.3	For light microscopic radioautography
Konica	NR-H2	0.08 µm	1.0	For electron microscopic radioautography

2. The Dipping Technique

This technique is based on brief dipping of the section-slide into liquefied emulsion and then drying the emulsion. The emulsions are in solid form at room temperature. Under safelighting, transfer the appropriate amount of emulsion into a glass cylinder from the bottle and place it in a water bath at 43°C for 10 min to melt the emulsion. The volume should then be increased by about 50% with distilled water depending on the emulsion used and thickness of emulsion layer desired. The emulsion must not be agitated too much to avoid the formation of bubbles. Pour the molten emulsion into a dipping jar designed to save the emulsion (Figure 12). Dip a slide in the emulsion for several seconds and withdraw it slowly. Wipe the back of the slide with a paper, then cool the slide quickly by laying it on the metal plate cooled with ice to prevent redistribution of silver grains. After about 10 min on the plate, the slide is laid flat on the bench for 1 h, transferred into a slide box, and then the box kept overnight in a desiccator with silica gel. An electric fan should not be used to dry the emulsion because quick drying may produce silver grains by pressure from the gelatin. The same phenomenon may occur when the emulsion is dried too much. Addition of 1% glycerol to the diluted emulsion reduces these phenomena

FIGURE 12   Emulsion application by dipping into liquid emulsion in a dipping jar.

considerably. Once dried overnight, the slide can be kept in the dark without silica gel at 4°C for a week to a month.

3. Controls for Radioautography

The nuclear emulsions are very sensitive to β particles and γ rays. However, the latent image is also produced by light, pressure, and certain chemicals. In addition, the latent image is lost by chemicals in the specimen. The interference with the photographic process by chemicals in the specimen is called chemography. There are two kinds of chemography, positive and negative. In positive chemography, latent images are produced during radioautographic exposure by reactive groups in the specimen other than the radioisotopes.[26] Conversely, the loss of latent image by reactive groups is called negative chemography (Figure 13). Chemography presents more problems in microradioautography than in macroradioautography because specimen contact with the nuclear emulsion is closer in microradioautography than in macroradioautography in which films are usually used to detect radioactivity. The creation of grains over the specimen by positive chemography is readily recognizable. The loss of latent image by negative chemography, however, may well pass unrecognized unless appropriate control slides are prepared and examined.[29] Certain fixatives such as osmium tetroxide, Shaudin, Zenker and potassium dichromate and many stains such as toluidine blue, methylene blue, May-Grünwald, Giemsa, and

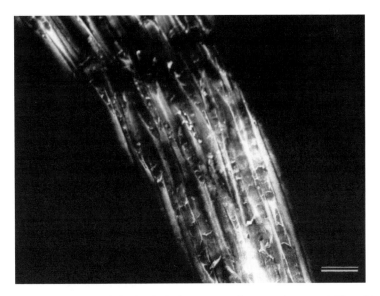

FIGURE 13 Image of mouse skeletal muscle for negative chemography control. The white areas correspond to the sites of negative chemography. The muscle was fixed in glutaraldehyde, dehydrated through a graded ethanol series, and embedded in paraffin. The deparaffinized 5-μm sections were coated with Konica NR-M2 emulsion, which was fogged by light before radioautographic exposure in the dark. Bar = 100 μm. (From Watanabe, M., Koide, T., Hayasaki, H., Kanbara, K., Jo, N., and Shimada, M., *Bull. Osaka Med. Coll*, 40, 17, 1994. With permission.)

Thionin nach Ehrlich cause severe chemographic effects. In receptor radioautography, however, low concentrations of glutaraldehyde and paraformaldehyde, which have no chemographic effects, may be used as fixatives, and staining is performed after development of nuclear emulsions. Therefore, one does not need to worry about the chemography, but it should be borne in mind that certain tissues have chemographic effects. Positive chemography was observed in the adult rat brain (the mesencephalic nucleus of the trigeminal nerve, the locus coeruleus, neuroglia of the corticospinal tract, cells of the cerebellar white matter, some cells near the central canal or ventricular system)[30] and human bone.[26] Negative chemography was associated with the mouse skeletal muscle,[31] human erythrocytes,[32] and submandibular gland and pancreas of rat.[33] If chemography is found, several strategies are now available to remove it:

1. Ensure that a solvent developer is used.[33]
2. Interpose a barrier layer between the specimen and emulsion, either of evaporated carbon[34] or of plastic.[35]
3. Chemically treat the specimen before coating with emulsion.[29]
4. Change types of emulsion. The relative sensitivities of radioautographic emulsions to chemography are different in various types of nuclear emulsions.[36]

To detect positive chemography, identical specimens that are not radioactive are exposed to photographic emulsions. To look for negative chemography, a few experimental whole-body sections adhering to the adhesive tapes or glass slides are exposed to fogged films by exposing to the light. In microradioautography, a few experimental slides from each batch of radioautographs are exposed to light. These controls are exposed and developed under identical conditions.

B. MICRORADIOAUTOGRAPHY OF INSULIN RECEPTOR AT THE LIGHT MICROSCOPIIC LEVEL

1. *In Vivo* Microradioautography

Mice were injected intravenously (tail vein) with ^{125}I-insulin. At the appropriate time after injection, the animals were perfused through the left ventricle with Ringer's solution to wash out unbound hormone and then perfused again with 2.5% glutaraldehyde in a 0.1 M phosphate buffer (pH 7.2) at room temperature for 10 min. At 3 min before the perfusion, the animals were anesthetized by intraperitoneal injection of sodium pentobarbital. Control animals received the same quantity of the radiolabeled ligand with excess unlabeled ligand. The tissues and organs were removed, immersed in the same fixative for 2 h, and embedded in paraffin after dehydration in graded concentrations of alcohol. Sections 5 μm thick were coated with a nuclear emulsion according to the dipping technique of Rogers.[26] After appropriate exposure, the emulsions were developed in Kodak D-19® diluted with the same volume of H_2O at 20°C for 7 min, fixed in a 30% sodium thiosulfate solution at 20°C for 8 min, and then stained with hematoxylin. Radioautographs thus obtained were observed by laser scanning microscope (Carl Zeiss, Germany) equipped with a Color Image Recorder® (CIR-310, Nippon Avionics Co. Ltd., Japan). Figure 14 is a microradioautograph showing the distribution of insulin receptors in the mouse liver lobules.[37]

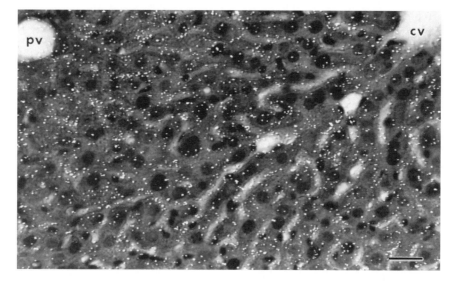

FIGURE 14 Microradioautograph of mouse liver 3 min after intravenous injection of 185 kBq ^{125}I-insulin without excess unlabeled insulin. The radioautograph was observed with a confocal laser microscope (see text, Section IV.C); pv, portal vein; cv, central vein. Bar = 50 μm. (From Nakanishi, M., Watanabe, M., and Shimada, M., *Cell. Mol. Biol.*, 41, 137, 1995. With permission.)

2. *In Vitro* Microradioautography

In vitro microradioautography visualizes specific binding sites or receptors through exposure of the thaw-mounted tissue sections on glass slides that are incubated with radiolabeled ligand to nuclear emulsions. This method is nearly the same as that of in vitro organ radioautography. By using this technique, we studied the distribution of insulin receptors in the mouse brain.[20]

Male mice were fasted overnight, anesthetized by intraperitoneal injection with sodium pentobarbital, and perfused through the left ventricle with Ringer's solution. The brains were rapidly removed on ice and frozen in isopentane cooled by liquid N_2. Coronal sections 25 μm thick were cut on a cryotome at –15°C, thaw-mounted on cold gelatin-coated slides, and placed on a plate cooled by ice for 1 h. Dried sections were preincubated with 30 mM KCl for 15 min twice at 20°C, incubated with 100 mM HEPES buffer (pH 7.8) containing 120 mM NaCl, 1.2 mM $MgSO_4$, 2.5 mM KCl, 15 mM Na acetate, 10 mM glucose, 1 mg/ml BSA, and 0.4 nM ^{125}I-insulin at 15°C for 100 min. To detect the nonspecific binding, 1 mM unlabeled insulin was added to the incubation medium. Sections were washed with 10 mM PBS (pH 7.4) to remove unbound ligand, and then the labeled sections were exposed to a liquid nuclear emulsion (Konica NR-M2). Figure 15 is an *in vitro* microradioautograph thus obtained.

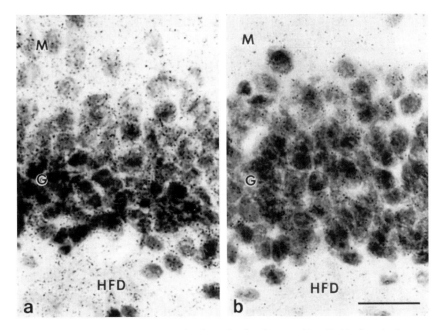

FIGURE 15 *In vitro* microradioautographs show the distribution of insulin bindings in the mouse dentate gyrus in the absence (a) and presence (b) of 1 μM unlabeled insulin; G, granular layer; M, stratum moleculare; HFD, hilus fascia dentate. Bar = 5 μm. (From Shimada, M., Hirose, Y., Nakanishi, M., Watanabe, H., Kawamoto, S., and Watanabe, M., *Recent Advances in Cellular and Molecular Biology*, Vol. 4, Peeters Press, Leuven, Belgium, 1992, 271. With permission.)

In *in vitro* microradioautography, sections are usually not fixed after incubation with radiolabeled ligand, and processing after the incubation must be done as soon as possible. Long-time fixation and washing of incubated sections may result in loss of radiolabeled ligand from binding sites because of the manner in which ligands bind to their receptors. Most ligands bind to their specific receptors reversibly but not irreversibly. As described above, we dipped the incubated sections into molten emulsion to detect radioactivity from radiolabeled ligand binding to the receptor. It has been assumed that dipping the slide-mounted tissue sections incubated with radiolabeled ligand into molten emulsion may result in loss of and rapid diffusion of radiolabeled ligand from receptor sites. In our case, however, it seems that excess washing of the sections after incubation with radiolabeled ligand is a more critical factor in dissociation of ligand from the binding sites than dipping the sections into liquid nuclear emulsion. Dipping requires only a few seconds, and the emulsion-coated glass slides are placed on a cool surface to allow the molten emulsion to gel as soon as possible after dipping.

As an alternative way of *in vitro* receptor radioautography, tissue blocks can be used instead of tissue sections. Incubate tissue blocks with radiolabeled ligands, and then section these blocks and localize the receptors by radioautography. In this technique, however, one should be careful about sectioning the incubated tissue blocks because the ligand does not penetrate throughout the block.[8]

3. Hinged Cover Glass Apposition Radioautography

This term for radioautography was first used by Stumpf and Duncan.[4] According to them, this technique was introduced by Hoecker and Roofe in 1949, the procedure was then modified by Roth and co-workers[38] and Young and Kuhar.[39] It was further modified by Peters and Barrack.[40] They successfully demonstrated the distribution of the steroid receptor in the human prostate by this technique.[41] Slide-mounted frozen sections are incubated with the radioligand, unbound ligand is washed out, and then the sections are apposed to dry, emulsion-coated cover glasses. These cover glasses are made by dipping the glasses into nuclear emulsion and air drying. The cover glass is glued on one end to create a hinge and to ensure the stability of the tissue (Figure 16). We use vinyl chloride resin glue (Scotch 3M®, Cat. No. 6425, Sumitomo 3M, Tokyo, Japan) or silicone resin glue (Bath-cork®, Cemedine Co. Ltd., Tokyo, Japan) as adhesive agents. These glues adhere well to glass and keep plasticity after the glue sets. The slide cover glass pairs are clamped with a binder clip and exposed in the dark box. In our laboratory, a vacuum sealer is used to keep a contact between the glass slide and the cover (see Figure 4). Before development, a razor's edge is inserted between the cover glass and the slide to separate them from each other. After photographic processing and staining of sections, the hinged cover glass is reattached to the slide with mounting medium. Instead of this technique, emulsion-coated slides can be used in *in vivo* microradioautography.[7] However, this procedure requires a good deal of technical skill to mount thin sections on emulsion-coated slides under safelight illumination. So, Tetsuji Nagata has developed dry-mounting radioautography by using dry emulsion film that is made inside a wire loop.[6] To produce the wire loop, the loop is dipped in melted emulsion containing 0.04% dioctyl sodium sulfosuccinate and dried for 1 min to gel. The sections mounted on the glass slide are covered with the gelled emulsion film.

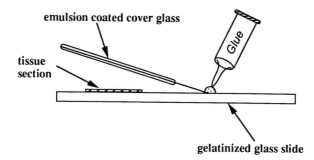

emulsion coated cover glass

tissue
section

gelatinized glass slide

FIGURE 16 Illustration shows the method of hinged cover glass apposition radioautography.

C. OBSERVATION OF MICRORADIOAUTOGRAPHS USING A CONFOCAL LASER SCANNING MICROSCOPE

Radioautographs prepared by microradioautography are usually observed through light microscopes. Recently, to improve resolving power of an radioautograph, radioautographs prepared with a thin layer of emulsion and a 1-μm-thick plastic section are frequently used. In these radioautographs, because silver grains are closely located to the tissue sections, it is not only relatively easy to correspond silver grains and tissue, but also a micrograph can be taken with both the silver grains and the tissue section in one focus. However, with thick sections made from paraffin and frozen blocks, using the ordinary dipping method, the resulting distances between the silver grains and tissue might be variable. There are many problems with this stained radioautographs.

1. It is difficult to correlate the silver grains with the tissue.
2. It is difficult to observe the silver grains on darkly stained tissue.
3. It is impossible to take a photograph within which both the silver grains and tissue are in one focus.
4. It is difficult to count the number of silver grains for quantitative analysis.
5. Resolving power tends to worsen with emulsion thickness.

These problems might be resolved by using a laser microscope.[42-44]

1. Laser Scanning Microscope

The laser scanning microscope (LSM) focuses laser light with the objective lens on a spot on the focal plane within the specimen, scans the specimen with the beam, captures the light from the specimen, changes the light into electronic signals, and then displays the image on the monitor (Figure 17). Therefore, we could not observe the images through the eyepiece like conventional light microscopes. The LSM has several features described below.

1. Images can be obtained with good contrast without the halo effect, which can be produced by optical systems or from objective specimen preparation.
2. High resolving power can be obtained by improving the detection system or through wavelength selection.

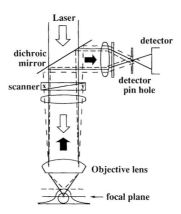

FIGURE 17   Principle of the confocal laser scanning microscope.

3. Images in the direction of the Z axis can be easily obtained.
4. The physical phenomena produced by spot illumination of a laser beam are easily visualized.

Currently, the LSM is mainly used to observe fluorescence in the fields of medicine and biology. However, because metals such as gold and silver also reflect lasers, it is possible to observe radioautographs with the LSM.

2.   Observation of Specimen by Differential Interference Contrast Mode

Because the LSM uses single-wavelength light such as argon 488 nm and 514 nm, differential interference contrast (DIC) images with high contrast are obtainable. A DIC image focused on a radioautograph specimen is shown in Figure 18.

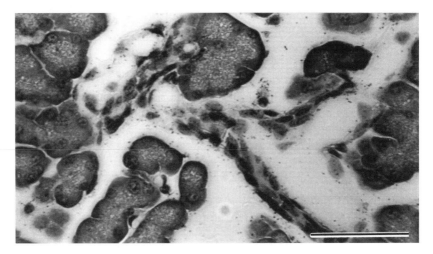

FIGURE 18 A differential interference contrast image of an radioautograph of mouse pancreas injected with 185 kBq ^{125}I-epidermal growth factor without excess unlabeled epidermal growth factor. The section was coated with NR-M2 emulsion and stained with hematoxylin. Bar = 50 μm.

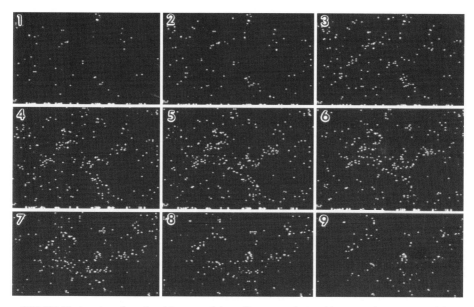

FIGURE 19 A series of confocal images of the reflectance obtained by optical sectioning of the radioautograph shown in Figure 18.

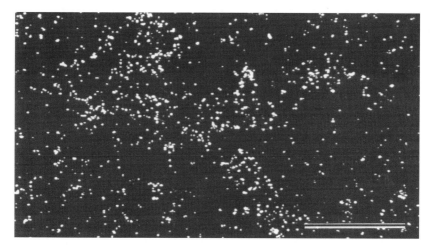

FIGURE 20 An overlay image of the series of confocal images shown in Figure 19. Bar = 50 μm.

3. Optical Sectioning

Images of silver grains at several depths of focus in the emulsion can be stored in the computer. This process is designated as optical sectioning because there is no actual physical sectioning. A series of images obtained by optical sectioning is shown in Figure 19, and an overlay image of this series is shown in Figure 20.

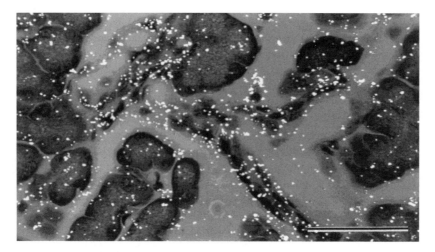

FIGURE 21 An overlay image of the differential interference contrast image shown in Figure 18 and the combined image of reflectance from developed silver grains shown in Figure 20. Bar = 50 µm.

4. Combining DIC Imaging and Optical Sectioning

Figures 18, 19, and 20 are all images obtained from the same radioautograph. An overlay of the DIC image (Figure 18), and the combined optical sectioning image (Figure 20), results in a new image (Figure 21), which displays all the silver grains dispersed in the emulsion over the specimen.

5. Further Advantages

In Figure 21, because all silver grains in the emulsion are visible over the specimen, easy matching of silver grains and specimen tissue is possible. Furthermore, the image is satisfactory for quantitative analysis of the silver grains. In contrast, the overlain image of the optical sectioning worsens resolution. With the LSM, if one takes an image along a line passing through the emulsion located close to the tissue section and overlays it with the DIC image of the tissue, the radioautographic image should correspond with that obtained with a very thin emulsion layer. Additionally, because DIC images can be obtained from unstained tissue sections, chemography from many prestainings can be avoided. Generally, optical microscopes have needed to use an objective lens with resolving power of at least 40× to take a microphotograph of small silver grains, but with this method, even a lens with resolving power of 10× can be used to observe the reflection from the silver grains. Thus, it is possible to take a microphotograph of an radioautograph at low magnification.

D. ANALYSIS OF THE MICRORADIOAUTOGRAPHS
1. Grain Counting

Microradioautographs are quantified by grain counting. The counting can be carried out on microphotographs taken by the recorder fitted in the laser scanning microscope. Conventional grain counting can be done under oil immersion at a magnification of 1000× using a light microscope equipped with an eyepiece-mounted circular mask that covers the area of interest on the tissue. The number of grains found in the cells of interest are scored, and the average number of grains per a certain square micrometer is determined.

2. Statistical Analysis of the Silver Grains

To estimate the specific ligand binding among the tissues and cells, statistical analysis is performed, usually with the *t*-test. In receptor radioautography, however, two different kinds of microradioautographs should be made to identify specific ligand binding. Specific binding of ligand in a given tissue is obtained from the differences between total and nonspecific binding. The total binding is obtained from radioautographs made from the tissue injected with radiolabeled ligand. The nonspecific binding is obtained from those made from the same radiolabeled ligand injected into tissue plus excess unlabeled ligand. Radioautographs showing total binding of a ligand and those showing the nonspecific binding are obtained from different specimens in receptor radioautography. Especially in microradioautography using nuclear emulsions, grains in a great number of microradioautographs are counted, and therefore it is difficult to correlate them to each other. For this purpose, the following statistical method has been developed.[45]

From two regions of tissues, we prepare several sections in which we count the numbers of total ligand bindings. We also prepare several sections for counting the numbers of nonspecific ligand bindings.

Let us use the symbol $i(i = 1,2)$ to distinguish between the first and the second regions. We are interested first in the mean, denoted by μ_i, of the population of the numbers of specific bindings of the *i*-th region. Our second interest is whether there is a significant difference between two populations of the specific ligand bindings, that is, the test of the hypothesis $\mu_1 = \mu_2$.

Let $x_{i1}, \ldots, x_{im_i}$ be the numbers of total ligand bindings from m_i sections of the *i*-th region. Similarly, let $y_{i1}, \ldots, y_{in_i}$ be the numbers of total ligand bindings from n_i sections of the *i*-th region. Then, for each i, the mean $\bar{d}$ and the variance u_i^2 of m_i numbers of total ligand bindings and the mean $\bar{y}_i$ and the variance v_i^2 of n_i numbers of nonspecific ligand bindings are given by

$$\bar{x}_i = \frac{\sum_{j=1}^{m_i} x_{ij}}{m_i}, \; \bar{y}_i = \frac{\sum_{k=1}^{n_i} y_{ik}}{n_i},$$

$$u_i^2 = \frac{\sum_{j=1}^{m_j} \left(x_{ij} - \bar{x}_i\right)^2}{m_i - 1}, \; v_i^2 = \frac{\sum_{k=1}^{n_i} \left(y_{ik} - \bar{y}_i\right)^2}{n_i - 1}.$$

For the estimate of μ_i for each i, by the usual Welch estimation, the approximate $100\,(1-2\alpha)$ % confidence interval of μ_i is given by $(\bar{x}_i - \bar{y}_i) \pm tSE_i$, where

$$SE_i = \sqrt{\frac{u_i^2}{m_i} + \frac{v_i^2}{n_i}}, \quad \tilde{D}_i = \frac{SE_i^4}{\dfrac{u_i^4}{m_i^2(m_i - 1)} + \dfrac{v_i^4}{n_i^2(n_i - 1)}},$$

D_i is the closest integer to $\tilde{D}_i$, and t denotes the upper 100α percentile point of t distribution with D_i degree of freedom.

To describe the method of testing of the hypothesis $\mu_1 = \mu_2$, we add some notation to the above. Let

$$u^2 = \frac{(m_1 - 1)u_1^2 + (m_2 - 1)u_2^2}{m_1 + m_2 - 2}, \quad v^2 = \frac{(n_1 - 1)v_1^2 + (n_2 - 1)v_2^2}{n_1 + n_2 - 2},$$

$$R^2 = u^2\left(\frac{1}{m_1} + \frac{1}{m_2}\right), \quad S^2 = v^2\left(\frac{1}{n_1} + \frac{1}{n_2}\right),$$

$$\tilde{d} = \frac{\left(R^2 + S^2\right)^2}{\dfrac{R^4}{m_1 + m_2 - 2} + \dfrac{S^4}{n_1 + n_2 - 2}}$$

and let d be the closest integer to $\tilde{d}$. Then, the test statistic

$$\tilde{T} = \frac{\left(\bar{x}_1 - \bar{y}_1\right) - \left(\bar{x}_2 - \bar{y}_2\right)}{\sqrt{R^2 + S^2}}$$

follows approximately the t distribution with d degree of freedom. So the p value of the two-sided test is equal to twice the probability of the upper tail right of the value $|\tilde{T}|$.

REFERENCES

1. Cuello, A. C., *Brain Microdissection Techniques*, John Wiley & Sons, Chichester, 1983.
2. MacLusky, N. J., Brown, T. J., Jones, E., Leranth, C., and Hochberg, R. B., Autoradiographic and microchemical methods for quantification of steroid receptors, in *Quantitative and Qualitative Microscopy*, Conn, P. M., Ed., Academic Press, New York, 1990, 3.
3. Bogoroch, R., Studies on the intracellular localization of tritiated steroids, in *Autoradiography of Diffusible Substances*, Roth, L. J. and Stumpf, W. E., Eds., Academic Press, New York, 1969, 99.
4. Stumpf, W. E. and Duncan, G. E., High-resolution autoradiographic mapping of drug and hormone receptors, in *Quantitative And Qualitative Microscopy*, Conn, P. M., Ed., Academic Press, San Diego, 1990, 35.
5. Roth, L. J. and Stumpf, W. E., *Autoradiography of Diffusible Substances*, Academic Press, New York, 1969.
6. Nagata, T., Radiolabeling of soluble and insoluble compounds as demonstrated by light and electron microscopy, in *Recent Advances in Cellular and Molecular Biology*, Vol. 6, Wegmann, R. J. and Wegmann, M. A., Eds., Peeters Press, Leuven, Belgium, 1992, 9.

7. Duncan, G. E. and Stumpf, W. E., High-resolution autoradiographic imaging of brain activity patterns with radiolabeled 2-deoxyglucose and glucose, in *Quantitative and Qualitative Microscopy*, Conn, P. M., Ed., Academic Press, San Diego, 1990, 50.
8. Kuhar, M. J., Perspectives on receptor autoradiography in human brain, in *Receptors in the Human Nervous System*, Mendelsohn, F. A. D. and Paxinos, G., Eds., Academic Press, San Diego, 1991, chap. 1.
9. Shimada, M. and Watanabe, M., The techniques and applications of whole-body radioautography, in *Radioautography in Medicine*, Nagata, T., Ed., Shinshu University Press, Matsumoto, Japan, 1993, 33.
10. Shimada, M. and Watanabe, M., Recent progress in whole-body radioautography, *Cell. Mol. Biol.*, 41, 39, 1995.
11. Curtis, C. G., Cross, S. A. M., McCulloch, R. J., and Powell, G. M., *Whole-Body Autoradiography*, Academic Press, London, 1981.
12. Ullberg, S., Studies on the distribution and fate of S35-labeled benzylpenicillin in the body, *Acta Radiol. Suppl.*, 118, 1, 1954.
13. Segu, L., Rage, P., Pinard, R., and Boulenguez, P., Computer-assisted quantitative receptor autoradiography, in *Computers and Computations in the Neurosciences*, Conn, P. M., Ed., Academic Press, San Diego, 1992, 129.
14. Watanabe, M., Hirose, Y., and Shimada, M., A new technique for autoradiography of whole-body sections mounted on glass slides, *Acta Histochem. Cytochem.*, 24, 167, 1991.
15. Ullberg, S., The technique of whole-body autoradiography, *Sci. Tools*, special issue, 1, 1977.
16. Watanabe, M., Shimada, M., and Kurimoto, K., Staining method for whole-body autoradiography, *Stain Technol.*, 50, 239, 1975.
17. Watanabe, M., Kihara, T., Shimada, M., and Kurimoto, K., Preparation and staining of whole-body sections, *Cell. Mol. Biol.*, 23, 311, 1978.
18. Shimono, R., Watanabe, M., and Goto, H., A new method of preparation of whole-body sections for autoradiography and staining, *Acta Anat. Nippon*, 57, 15, 1982.
19. Kuno, N., Watanabe, M., Akagi, N., Sugimoto, M., and Kihara, T., A technique of autoradiography for tubular organs, *Acta Histochem. Cytochem.*, 20, 547, 1987.
20. Shimada, M., Hirose, Y., Nakanishi, M., Watanabe, H., Kawamoto, S., and Watanabe, M., Autoradiographic studies on insulin receptor of the mouse hippocampus, in *Recent Advances in Cellular and Molecular Biology*, Vol. 4, Wegmann, R. J. and Wegmann, M. A., Eds., Peeters Press, Leuven, Belgium, 1992, 271.
21. Saavedra, J. M., Correa, F. M. A., Plunkett, L. M., Israel, A., Kurihara, M., and Shigematsu, K., Binding of angiotensin and atrial natriuretic peptide in brain of hypertensive rats, *Nature*, 320, 758, 1986.
22. Miller, J. A., Curella, P., and Zahniser, N. R., A new densitometric procedure to measure protein levels in tissue slices used in quantitative autoradiography, *Brain Res.*, 447, 60 1988.
23. Sato, T., Niwa, M., Himeno, A., Tsutsumi, K., and Amemiya, T., Quantitative receptor autoradiographic analysis for angiotensin II receptors in bovine retinal microvessels. Quantification with radioluminography, *Cell Mol. Neurobiol.*, 13, 233, 1993.
24. Amemiya, Y. and Miyahara, J., Imaging plate illuminates many fields, *Nature*, 336, 89, 1988.
25. Yanai, K., Ryu, J. H., Watanabe, T., Iwata, R., and Ido, T., Receptor autoradiography with ^{11}C and [3H]-labeled ligands visualized by imaging plates, *Neuroreport*, 13, 961, 1993.
26. Rogers, A. W., *Techniques of Autoradiography*, 3rd ed., Elsevier North-Holland, Amsterdam, 1979.
27. Rogers, A. W., *Practical Autoradiography*, Review No. 20, Amersham International Limited, Amersham, U.K., 1979.
28. Ono, K., Characteristics of Konica autoradiographic emulsion: NR-M2 and NR-H2, in *Radioautography in Medicine*, Nagata, T., Ed., Shinshu University Press, Matsumoto, Japan, 1993, 23.
29. Watanabe, M. and Rogers, A. W., Prevention of negative chemography in phenylenediamine stained autoradiographs of plastic semithin sections, *Stain Technol.*, 62, 173, 1987.
30. Daniels, J. S., Chemography associated with specific anatomic areas in autoradiographs of brain stems from adult rats, *Brain Res.*, 98, 343, 1975.
31. Watanabe, M., Koide, T., Hayasaki, H., Kanbara, K., Jo, N., and Shimada, M., Negative chemography with mouse skeletal muscle, *Bull. Osaka Med. Coll.*, 40, 17, 1994.

32. Dunham, P. B., Kerr, M. S., and Horowitz, M. C., Heterogeneous negative chemography with human erythrocytes, *Stain Technol.*, 53, 229, 1978.

33. Rogers, A. W. and John, P. N., Latent image stability in autoradiographs of diffusible substances, in *Autoradiography of Diffusible Substances*, Roth, L. J. and Stumpf, W. E., Eds., Academic Press, New York, 1969, 51.

34. Sechrist, J. W. and Upson, R. H., Prevention of chemography in prestained epoxy autoradiographs by an intervening carbon film, *Stain Technol.*, 49, 297, 1974.

35. Keyser, A. and Wijiffels, C., The preparation and use of polyvinylidenechloride protective films in autoradiography, *Acta Histochem.*, Suppl. 8, 359, 1968.

36. Rogers, A. W. and Watanabe, M., The relative sensitivities of autoradiographic emulsions to negative chemography, *J. Microsc.*, 149, 193, 1988.

37. Nakanishi, M., Watanabe, M., and Shimada, M., Heterogeneous distribution of ^{125}I-insulin binding sites in the liver of fed and fasted mice, *Cell. Mol. Biol.*, 41, 137, 1995.

38. Roth, L. J., Diab, I. M., Watanabe, M., and Dinerstein, R. J., A correlative autoradiographic, fluorescent, and histochemical technique for cytopharmacology, *Mol. Pharmacol.*, 10, 986, 1974.

39. Young, W. S., III. and Kuhar, M. J., A new method for receptor autoradiography: [^{3}H]opioid receptors in rat brain, *Brain Res.*, 179, 255, 1979.

40. Peters, C. A. and Barrack, E. R., A new method for the labeling and autoradiographic localization of androgen receptors, *J. Histochem. Cytochem.*, 35, 755, 1987.

41. Peters, C. A. and Barrack, E. R., Androgen receptor localization in the human prostate: demonstration of heterogeneity using a new method of steroid receptor autoradiography, *J. Steroid Biochem.*, 27, 533, 1987.

42. Watanabe, M., Kawamoto, S., Nakatsuka, Y., Koide, T., and Shimada, M., An application of the laser scanning microscope to microautoradiography, *Acta Histochem. Cytochem.*, 26, 93, 1993.

43. Watanabe, M., Koide, T., Konishi, M., Kanbara, K., and Shimada, M., Use of confocal laser scanning microscopy in radioautographic study, *Cell. Mol. Biol.*, 41, 131, 1995.

44. Watanabe, M., Koide, T., Nakatsuka, Y., and Shimada, M., Visualization of microautoradiography by confocal laser scanning microscope, in *Autoradiography in Medicine*, Nagata, T,. Ed., Shinshu University Press, Matsumoto, Japan, 1993, 215.

45. Watanabe, M., Nishimura, Y., Jo, N., Takafuchi, M., Kiyokane, K., and Shimada, M., Statistical analysis for receptor autoradiography, *Acta Histochem. Cytochem.*, 29, 135, 1996.

Chapter **6**

IDENTIFICATION BY RETROGRADE TRACING OR IMMUNOHISTOCHEMISTRY OF NEURONES LABELED BY RADIOLIGAND RECEPTOR HISTORADIOAUTOGRAPHY

Marie-Elisabeth Stoeckel
Marie-José Freund-Mercier
Elisabeth Waltisperger
Marie Jeanne Klein

CONTENTS

I. INTRODUCTION

The historadioautographic technique makes possible the detection of radioligand binding sites by light microscopy at a cellular scale. In neurobiological studies, it represents a valuable tool for localization of receptors on individual cells, especially on neurones. In such studies, further characterization of radioligand-labeled neurones is most often essential, for example, as regards the identification of the neurotransmitter or neuropeptides the labeled neurone synthesizes or the localization of its terminal field. However, processing of tissue sections for historadioautography impedes further application of most conventional morphofunctional techniques such as axonal tracer localization and immunohistochemistry. In the present chapter we report technical adaptations which permit simultaneous observation of a retrograde axonal tracer (fast blue) and radioautographic radioligand labeling. On the other hand, application of the recently developed microwave irradiation technique of histological sections made possible immunohistochemical staining of sections previously labeled by radioligand binding radioautography.

II. RADIOLIGANDS

The historadioautographic technique requires pharmacologically well-characterized, high-affinity ligands coupled to a radioisotope with high specific activity and low energy. [125]I, which emits Auger electrons, fulfills these condtions and is thus the most suitable for high resolution radioautography. It provides radioligands with high specific activity but with a consequently short utilization delay (half life of [125]I = 60 days). On the other hand, the ligands imperatively need to include a free amino group (NH_2) which permits cross-linking by covalent bonds to the binding site by means of aldehydes (e.g., formaldehyde or glutaraldehyde), which avoids leakage during application of the radioautographic procedure.[1]

Radioiodinated ligands are commercially available; a preparation technique is described in Protocols 1 and 2 of Chapter 4.

III. TISSUE PREPARATION

A. TISSUE FREEZING

Frozen sections from unfixed specimens provide optimal radioligand binding with minimal background. They are thus most suitable, despite the relatively poor preservation of cellular structures.

The specimens are frozen by immersion for 1 to 2 min in isopentane cooled between –40°C and –45°C with dry ice. Small specimens are embedded in adapted medium (e.g., Tissue-Teck, Miles Laboratories). The frozen samples can be stored at –20°C for several days before sectioning (avoid desiccation by wrapping them in aluminum foil or parafilm and store in stoppered vials).

B. FROZEN SECTIONS

Sections 10 to 20 μm thick are cut on a cryostat microtome and thaw-mounted on gelatin-coated slides (see Appendix). Sections from specimens containing fast-blue-labeled cells are immediately placed on a hot plate (45°C) for 10 to 15 min. Sections are then stored for at least 24 h at –20°C until application of the binding procedure, to ensure their adherence to the slide.

C. PRESERVATION OF RETROGRADE TRACER IN THE SECTIONS

Before application of the binding procedure the sections from samples containing fast-blue-labeled cells are thawed and immersed for 20 min in 4% paraformaldehyde at 4°C, and washed 3 times in PBS (see Appendix). This treatment ensures acceptable preservation of both the binding sites and the fast-blue fluorescence.

IV. BINDING PROCEDURE

The incubation conditions are those generally used for film radioautography on unfixed tissue sections (see Protocol 1):

- The concentration of the radioligand, close to the dissociation constant (K_D), is determined on the basis of its specific activity. The persistence or not of the biological activity of the radioligand which results from the decay of ^{125}I must also be taken into account.

- If the tellurium-labeled compound is biologically inactive, the specific activity of the ligand is constant and the cpm/ml determined at day 0 are used whatever the age of the ligand.
- If the tellurium-labeled compound has the same biological activity as the radioiodine-labeled compound (which is the case for the oxytocin antagonist), the concentration of the radioligand is calculated according to the decay rate of ^{125}I as given on tables by the manufacturer.

- The endogenous ligands are normally eliminated by the preincubation step. Their dissociation can be favored by guanyl nucleotides.[2] Adjunction of GTP to the incubation medium, however, did not notably affect labeling intensity of the oxytocin binding sites.
- Low temperature of the incubation medium improves the signal/background ratio. Duration of incubation must last until the binding equilibrium is reached. For oxytocin and vasopressin antagonists, optimal and reproducible labeling is obtained by incubating sections at 4°C for 24 h.
- Protection against peptidases of the radioligand by adding bacitracin (5 μg/ml) or aprotinin (100 kIU/ml) to the incubation medium has been recommended by some authors.[2] In the case of tritiated oxytocin and vasopressin, these agents did not appreciably modify labeling intensity.

PROTOCOL 1: INCUBATION OF SECTIONS IN THE RADIOLIGAND

Place the slides back-to-back in a Coplin jar, the volume of which can be reduced with dental wax in order to reduce the amount of radioligand used.

1. Preincubate for 20 min at room temperature in Tris-HCl buffer (see Appendix).
2. Incubate for 24 h at 4°C in Tris-HCl buffer added with the radioligand.
3. Wash 2 × 5 min in Tris-HCl buffer at 4°C.
4. Rinse rapidly with distilled water.
5. Dry in a cold airstream for 12 h.

Incubation of the sections can also be performed by covering the slides with 300 to 500 μl of the radioligand solution, coverslipped to avoid desiccation.

V. FILM RADIOAUTOGRAPHY

This step is facultative but is nevertheless very useful to check radioligand binding site distribution and intensity in the sections at a macroscopic level.

Film radioautographs from the sections are generated by apposing the slides against tritium-sensitive film (Hyperfilm, Amersham) in X-ray cassettes. After 4 to 6 days of exposure at 4°C the films are developed in D19 for 5 min at 18°C.

It is important to keep in mind that very small radiolabeled structures, such as isolated cells or tiny cell groups, may not be detectable on the films because of their poor resolution.

VI. HISTORADIOAUTOGRAPHY

A. PREPARATION OF SECTIONS FOR HISTORADIOAUTOGRAPHY

In view of the reversibility of its interaction with the receptor, the ligand needs to be attached by covalent links to the binding site by aldehyde fixatives. As already mentioned by Herkenham and Pert,[3] fixation by postincubation immersion of sections into liquid fixatives appears by far less effective than exposure to paraformaldehyde vapors at 80°C (see Protocol 2). This treatment does not notably affect the fast-blue fluorescence in retrogradely labeled neurones.

PROTOCOL 2: FIXATION OF THE SECTIONS WITH PARAFORMALDEHYDE VAPORS

1. Place 10 to 30 g paraformaldehyde* in a Petri dish at the bottom of a desiccator jar.
2. Equilibrate paraformaldehyde to room humidity (30 to 60%) by leaving exposed to room air for at least 1 day.
3. Place the thoroughly dried sections on racks into the desiccator jar.
4. Reduce pressure in the jar with a vacuum pump, sufficient to maintain a seal between the lid and the jar.
5. Heat the jar in an oven at 80°C for 3 h.
6. Open the jar under a hood and immediately remove the racks to prevent the cooling vapors from condensing on the slides.
7. Leave the sections overnight in a drafted hood to rid them of residual vapors.

To ensure adherence of the nuclear emulsion to the sections, to improve staining capacity, and to remove any unfixed radioligand, the sections are demyelinated (defatted) according to the method of Herkenham and Pert.[3] The sections are successively immersed in:

- Increasing concentrations of ethanol: 70°, 95°, 2 × 100° (5 min each),
- Xylene 2 × 15 min each,
- Decreasing concentrations of ethanol: 100°, 95°, 70° (5 min each),
- Distilled water (rapid rinse) and air dried.

---

* Preferably use small granules of paraformaldehyde (e.g., Kodak); finely powdered paraformaldehyde provoked deposits on the sections.

B. HISTORADIOAUTOGRAPHIC PROCEDURE
1. General Considerations

The radioligand binding sites are detected by covering the sections with nuclear emulsion which consists in a suspension of silver halide crystals in gelatin. The Auger electrons emitted by ^{125}I (or the β particles emitted by ^{3}H) induce formation of a latent image in overlying crystals. The latent image catalyzes reduction of the silver halide of the crystal to metallic silver grains by the developer, and subsequent fixation eliminates the nonimpressed crystals.

Emulsion coating is performed by the conventional "dipping" method (see Protocol 3). Other methods, e.g., the loop technique or stripping film, can also been used.

PROTOCOL 3: EMULSION COATING

Materials and reagents:

In a darkroom equipped with a safelight adapted to the nuclear emulsion used, the following materials and reagents are required:

- Nuclear emulsion (Amersham LM-1, ready for use when melted or Ilford K5 to be diluted 1:1).
- Water bath at 43 to 44°C.
- Jar to melt (or dilute) the emulsion.
- Jolly tube (a specially manufactured tube with minimal volume reduces the amount of emulsion for a small series of slides).
- Glass stirring rod.
- Five clean histological slides.
- Light-tight slide boxes containing drying agent in the lid or wrapped in paper tissue.

Procedure:

1. Place the emulsion jar in the water bath until emulsion melts (which takes approximately 1 h).
2. Dip clean glass slides to remove the bubbles.
3. Allow to dry in a position close to vertical for 6 h (or overnight).
4. Place the slides in the light-tight boxes.
5. Store at 4°C until development.

2. Exposure and Development

Exposure time is approximately 3 to 4 times that required for film radioautographs. Development of test slides at regular intervals permits the determination of optimal exposure times.

For development (see Protocol 4), the use of fresh reagents and strict respect for the development time and temperature are prerequisites for optimal labeling.

PROTOCOL 4: DEVELOPMENT OF HISTORADIOAUTOGRAPHS

Operate in a darkroom equipped with a safety light.

Reagents:

Developer D19 Kodak (filtered if necessary).
Distilled water.
Fixer: 30% sodium thiosulfate (commercial fixers, e.g., Hypam, Ilford can also been used).

Procedure:

1. Remove the boxes containing the sections from the refrigerator at least 1 h before development to reach room temperature.
2. Immerse the sections for exactly 5 min in D19 at 18°C.
3. Rinse in distilled water for a few seconds.
4. Immerse in fixer.
5. Wash in running water for at least 30 min.
6. Rinse rapidly 3 times in distilled water.
7. Allow to dry on racks in a dust-free atmosphere.

VII. SIMULTANEOUS OBSERVATION OF RADIO-LIGAND LABELING AND RETROGRADE TRACING

Observations are performed on either unstained sections or after slight thionine staining (see Protocol 5). In both cases dry sections are coverslipped in a drop of Permount or Eukitt. Mounting the sections after ethanol dehydration and clearing in xylene, as done in conventional histology, is avoided because it frequently results in disappearance of the silver grains (by a not yet understood oxidation process) after a variable time lag.

Unstained coverslipped sections are examined under a light microscope equipped with both ultraviolet epifluorescence (e.g., Leitz ploem A) for fast-blue detection, and transmission darkfield illumination which reveals the silver grains of the nuclear emulsion. So, retrograde labeling with fast-blue (blue fluorescence) and radioligand binding (bright silver grains) can be simultaneously observed.

On thionine-stained sections (according to Protocol 5), fluorescence intensity of fast-blue-labeled cells is slightly decreased. However, observation with brightfield illumination of these preparations facilitates identification of structures labeled by either technique.

PROTOCOL 5: STAINING SECTIONS WITH THIONINE

Staining solution:

95 ml Distilled water

4 ml Acetic acid 0.2 M (1.11% v/v)
1 ml Sodium acetate 0.2 M (2.72 g%, w/v $CH_3COONa \cdot 3H_2O$)
0.1 g Thionine

Procedure:

1. Bring frozen sections to water.
2. Rinse rapidly with distilled water.
3. Immerse for 5 s in 70° alcohol.
4. Rinse rapidly with distilled water.
5. Immerse for 10 to 15 s in the thionine solution.
6. Rinse in distilled water.

VIII. IMMUNOHISTOCHEMICAL IDENTIFICATION OF RADIOLABELED CELLS

A. GENERAL CONSIDERATIONS

The immunological identification of radioligand binding cells results from the comparison of labelings obtained by successive treatments of the section with histo-radioautographic and immunohistochemical procedures. Presence of radioautographically labeled radioligand binding sites on immunostained neurones has been demonstrated on double-labeled cell cultures.[4] The technique consists of successive radioligand binding, aldehyde fixation, immunoperoxidase staining, and finally, application of the radioautographic procedure. In our hands, application of the same procedure to brain sections did not result in clear-cut radioautographic labeling of radioligand binding sites. Another approach consists of immunoperoxidase identification of radioligand-labeled neurones on paraformaldehyde-fixed adjacent sections.[5]

The recent introduction of microwave irradiation of sections makes possible immunohistochemical labeling of preparations on which application of this technique has so far been ineffective. Microwave treatment thus permitted immunolabeling of the sections already processed for radioautography after elimination of the nuclear emulsion. For a discussion of the action mechanisms of microwave irradiation see Reference 6.

In our preparations, this treatment was very effective for the immunolabeling of proteins or large peptides (e.g., neurophysins, glial fibrillary acidic protein), but not for small peptides such as oxytocin or vasopressin which are possibly not preserved in these preparations.

B. OBSERVATION OF THE HISTORADIOAUTOGRAPHS

The sections, processed for historadioautography as described in Sections III to IV, are stained with thionine (see Protocol 5), allowed to dry, and coverslipped with a drop of Permount or Eukitt.

At this stage, the sections are carefully examined under the brightfield and darkfield light microscope and the zones of interest photographed (noting the graduations of the stage helps to find a cell again after immunolabeling).

C. PRELIMINARY TREATMENTS FOR IMMUNOHISTOCHEMISTRY

1. Enzymatic Digestion of the Nuclear Emulsion

Immunolabeling of sections already processed for historadioautography requires elimination of the nuclear emulsion, which is done by enzymatic digestion (see Protocol 6). Such an approach has already been described by Bugnon et al.[7] but revealed to be ineffective in our hands, possibly because of changes in the composition of the emulsion.

PROTOCOL 6: ENZYMATIC DIGESTION OF THE NUCLEAR EMULSION

1. Place the coverslipped slides in xylene until the coverslip detaches (this can be hours or days, depending on the age of the sections).
2. Rehydrate the slides by immersion in decreasing ethanol concentrations.
3. Immerse in PBS (see Appendix).
4. Place the slides on a hot plate (approximately 37°C), without allowing them to dry, and cover the sections with a mixture of 1.6 mg trypsin, 1.6 mg collagenase, and 3 mg proteinase K dissolved in 1 ml PBS (see Appendix). Control under a light microscope the mobilization of the silver grains, which progressively form coarse aggregates. Detachment of the silver grains takes 1 to 5 min, depending on the emulsion.
5. Agitate the slide in PBS at 40°C for 1 min.
6. Check elimination of silver grains under the light microscope. A few grains may unavoidably remain encrusted in the tissue.
7. Place the slides in distilled water.

2. Paraformaldehyde Postfixation

The sections previously processed for radioautography and free of the nuclear emulsion are immersed in 4% paraformaldehyde in 0.1 M phosphate buffer pH 7.4 for 20 min and rinsed in PBS (see Appendix). The same treatment is applied to adjacent, previously unlabeled sections.

3. Microwave Irradiation

The sections are treated according to the technique we routinely use for detection of various antigens in frozen or paraffinized sections (see Protocol 7).

PROTOCOL 7: MICROWAVE IRRADIATION

A conventional domestic microwave oven can be used.

1. Immerse the slides in citrate buffer (see Appendix) in a heat-resistant Coplin jar.
2. Place besides a beaker containing 500 ml tap water.

3. Irradiate at 750 W for 3 × 5 min (the solution normally reaches ebullition), readjust the volume of the buffer after each cycle.
4. Remove the slides from the oven and, remaining in the citrate buffer, let them return to room temperature for 20 min.
5. Place the slides in PBS (see Appendix).

D. IMMUNOHISTOCHEMISTRY

The antigenic sites are detected using any immunoperoxidase technique (see Protocol 8). The avidin biotin peroxidase complex technique appeared most suitable for this purpose.

PROTOCOL 8: ABC IMMUNOPEROXIDASE LABELING

Reagents:

Primary antibody.
Biotinylated antibody against the IgG of the species producing the primary antibody.
Avidin-biotin complex kit (e.g., ABC Vector Elite).
DAB substrate kit for peroxidase (Vector).

Procedure:
In a humid chamber treat the sections with:

1. PBS (see Appendix) containing 0.5% Triton X100 (PBST) and 20 mg/ml defatted dry milk to block nonspecific sites for 10 min.
2. Primary antibody diluted in PBST, added with 0.02% sodium merthiolate, overnight at room temperature.
3. PBS, 3 × 10 min.
4. Biotinylated IgG against the IgG of the species producing the primary antibody diluted in PBST for 1 h.
5. PBS, 3 × 10 min.
6. ABC complex prepared according to the manufacturer's instructions for 1 h.
7. PBS, 3 × 10 min.
8. DAB, according to instructions of the manufacturer. When labeling/background ratio appears optimal, stop the reaction by a PBS rinse.
9. Tap water.
10. Dehydrate in graded alcohols, clear in xylene, and coverslip with a drop of Eukitt or Permount.

Note: faint immunolabeling can be reinforced by treating the sections for 1 min in 0.4% osmium tetroxide after step 9.
Slight counterstaining with thionine facilitates the localization of labeled cells.

IX. RESULTS

A. RADIOLIGAND LABELING AND RETROGRADE TRACING

Combined radioautographic radioligand detection and retrograde labeling were applied to detect oxytocin binding sites on sympathetic preganglionic neurones of the intermediolateral column in the rat spinal cord (Figure 1).

Fast-blue-labeled neurones could clearly be detected under the fluorescence microscope, although less clearly outlined than on sections from conventional perfusion-fixed samples. In the same preparations, radioautographic labeling of oxytocin binding sites was similar to that previously observed,[8] i.e., concentrated on cell bodies of the intermediolateral cell column (Figures 1 A and B). Nonspecific labeling, however, was slightly higher in such preparations, probably due to the paraformaldehyde fixation just before incubation in the radioligand, a prerequisite for preservation of the retrograde tracer.

Observation of the preparations by simultaneous epifluorescence and darkfield transmission illumination revealed cells exhibiting blue fluorescence and high concentrations of bright silver grains, indicating the presence of oxytocin receptors on retrogradely labeled neurones (not shown). Thionine staining of the sections, although decreasing the intensity of fast-blue labeling, permitted conventional brightfield observation of the radioautographic label, which was useful to precisely localize the labeled cells (Figure 1s C-F).

B. RADIOLIGAND LABELING AND IMMUNOHISTOCHEMISTRY

Historadioautographic detection of oxytocin binding sites and immunoperoxidase staining of oxytocin-neurophysin and vasopressin-neurophysin were performed on sections of lactating rats having previously received an intracerebroventricular injection of oxytocin antagonist.[9] In the hypothalamic magnocellular nuclei, where the oxytocin- and vasopressin-producing neurones are concentrated, adjacent sections revealed distribution of oxytocin binding sites superimposable to those of oxytocinergic neurones, but clearly distinct from those of vasopressinergic cells (Figure 2). Successive application of radioautographic and immunohistochemical procedures on the same sections clearly indicated concentrations of oxytocin binding sites on neurones immunoreactive for oxytocin-neurophysin (Figure 3). This colocalization was also obvious on isolated neurones scattered in the periventricular area (Figure 4). In view of previous observations, strongly suggesting that oxytocin binding sites detected by the radioligand binding technique presently used are functional receptors, these data provided first evidence for peptide autoreceptors on neuronal somata.

Whether such autoreceptors concern all oxytocin-producing cells of the areas examined remains to be established. Indeed, almost all cells obviously exhibiting oxytocin binding sites were also immunoreactive for oxytocin-neurophysin, although with variable intensities. In contrast, several immunoreactive cells were not labeled for binding sites. Absence of radioautographic labeling of immunoreactive cells may either be due to the absence of binding sites or to their localization within the section.

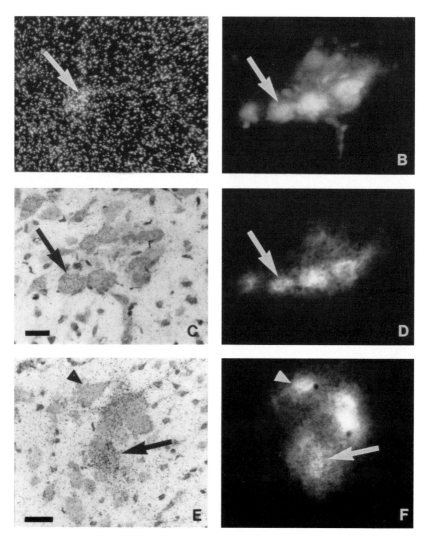

FIGURE 1   Rat spinal cord intermediolateral cell column. A, B: photomicrographs of an unstained section in darkfield illumination (A), revealing radioautographic labeling of oxytocin binding sites and in epifluorescence illumination (B), showing neurones retrogradely labeled by fast-blue. C, D: same section, slightly stained with thionine (C), shows concentration of silver grains on a neurone in brightfield illumination, whereas fast-blue fluorescence (D), although decreased, is still detectable. A double-labeled neurone is indicated by arrows. E, F: similar preparation: brightfield detection of oxytocin binding sites (E), epifluorescence labeling of fast-blue retrograde labeling (F). The arrows indicate a double-labeled neurone, the arrowheads a retrogradely labeled neurone deprived of oxytocin binding sites. Calibration bars = 25 μm.

In the relatively thick (16 to 20 μm) sections used, the low-energy Auger electrons emitted by the ^{125}I-radiolabeled structures deep within or at the bottom of the section possibly can not reach the emulsion. In contrast, the immunohistochemical proce-dure, applied in the presence of detergent labels the structures, whatever their position in the section. More precise correlation between radioisotope labeling and

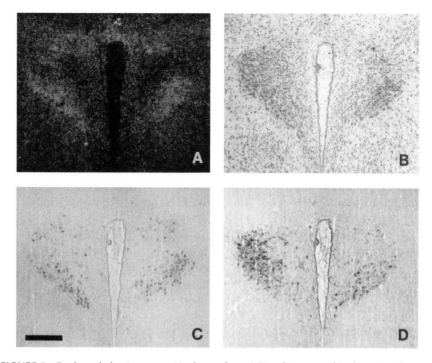

FIGURE 2   Rat hypothalamic paraventricular nucleus. A,B: radioautographic detection of oxyto-cin binding sites in darkfield (A) and brightfield (B) illumination. C,D: adjacent sections immu-noperoxidase-labeled for oxytocin-neurophysin (C), and vasopressin-neurophysin (D). The overall distribution of the oxytocin binding sites corresponds to that of the oxytocin-neurophysin immu-noreactive neurons. Calibration bar = 500 μm. (From Feund-Mercier et al., *J. Physiol.*, 480, 158, 1994. With permission.)

immunolabeling would need thinner sections, which would, however, result in wors-ened morphological preservation.

X. CONCLUSION

The historadioautographic technique which makes possible cellular localizations of radioligand binding sites can be combined with other morphofunctional tech-niques, which permits further identification of the radiolabeled cells as regards their projection sites or their neurochemical characteristics. The resolution of these tech-niques, however, is limited by the relatively poor histological preservation of the tissues, which is partly due to the absence of chemical fixation preceding preparation of the frozen sections. In the combinations reported here, preference was given to radioligand labeling with regard to labelings with the other techniques, thus exclud-ing any modification detrimental to the intensity of binding site detection or increas-ing the background. These techniques could certainly be improved in accordance with other specific fields of interest.

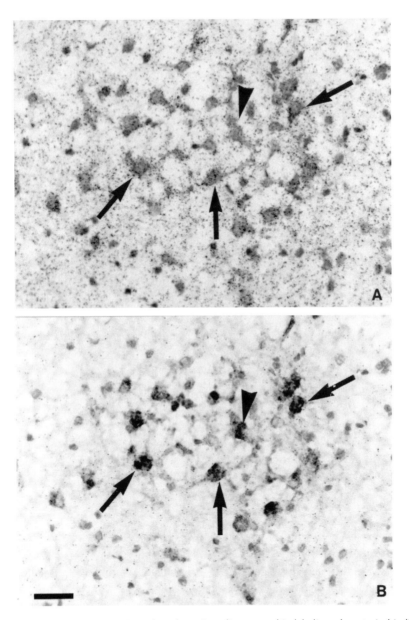

FIGURE 3   Rat anterior commissural nucleus. A: radioautographic labeling of oxytocin binding sites; B: same area of the same section after digestion of the nuclear emulsion, immunoperoxidase-labeled with an antibody against oxytocin-neurophysin. Arrows indicate double-labeled cells and arrowheads an oxytocinergic cell unlabeled with the ^{125}I-oxytocin antagonist. Calibration bar = 50 μm. (From Freund-Mercier and Stoeckel, in *Oxytocin,* Ivell and Russel, Eds., Plenum Press, New York, 1995, p. 185. With permission.)

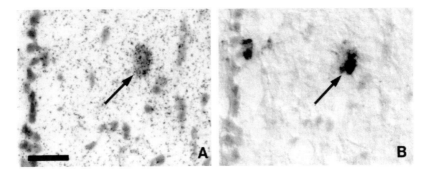

FIGURE 4 Isolated neurones in the subependymal area. The arrows indicate a neurone bearing oxytocin binding sites (A) and intensely labeled with the antibody against oxytocin-neurophysin (B); a smaller, subependymal neurone (B, left) is solely immunoreactive. Calibration bar = 25 μm. (From Freund-Mercier and Stoeckel, in *Oxytocin,* Ivell and Russel, Eds., Plenum Press, New York, 1995, p. 185. With permission.)

ACKNOWLEDGMENTS

We are indebted to Dr. M. Manning for generously providing the oxytocin antagonist and to Drs. H. Gainer and R. Acher for gifts of antibodies. We thank J.M. Gachon for skillful photographic work.

APPENDIX

Citrate Buffer

   A: Citric acid 0.1 M, ($C_6H_8O_7 \cdot H_2O$, 21.01 g/l).
   B: Sodium citrate 0.1 M ($C_6H_5O_7Na_3 \cdot 2H_2O$, 29.4 g/l).
   1. Mix 9 ml of A with 41 ml of B.
   2. Add distilled water up to 500 ml.

Gelatin-Coated Slides

   1. In 250 ml distilled water.
   2. Dissolve by heating moderately (below 45°C) 1.25 g powdered gelatin and 0.125 g chrome alum.
   3. Filter.
   4. Dip thoroughly cleaned sections in the solution.
   5. Let dry.
   6. Heating the gelatin-coated slides in an oven at 120°C for 3 h improves adherence of frozen sections.

Phosphate Buffer 0.1 M, pH 7.35

Disodium phosphate ($Na_2HPO_4 \cdot 12H_2O$) 2.9 g
Monosodium phosphate ($NaH_2PO_4 \cdot H_2O$) 0.264 g
Distilled water to 100 ml.

Phosphate-Buffered Saline (PBS)

NaCl 9 g
Disodium phosphate ($Na_2HPO_4 \cdot 12H_2O$) 4.7 g
Monosodium phosphate ($NaH_2PO_4 \cdot H_2O$) 0.346 g
Distilled water to 1 l.

Tris-HCl Buffer, pH 7.4
85 mM Tris-HCl:

- Solution A: Tris base (MW 121.14) 170 mM: 20.787 g/l.
- Solution B: hydrochloric acid 170 mM: HCl N 170 ml/l.
- Mix equal parts of solutions A and B.
- Adjust pH at 7.4 with A if necessary.
- Add 5 mM magnesium chloride ($MgCl_2 \cdot 6H_2O$): 1.016 g/l; and 0.1% bovine serum albumin: 1 g/l

REFERENCES

1. Hamel, E. and Beaudet, A., Localization of opioid binding sites in rat brain by electron microscopic autoradiography. *J. Electron. Microsc. Technol.*, 1, 317, 1984.
2. Leroux, P., Gonzalez, B. J., Laquerriere, A., Bodenant, C., and Vaudry, H., Autoradiographic study of somatostatin receptors in the rat hypothalamus: validation of a GTP-induced desaturation procedure. *Neuroendocrinology*, 47, 533, 1988.
3. Herkenham, M. and Pert, C. B., Light microscopic localization of brain opiate receptors. A general autoradiographic method which preserves tissue quality, *J. Neurosci.*, 2, 1129, 1982.
4. Hösli, E., Stauffer, S., and Hösli, L., Autoradiographic and electrophysiological evidence for the existence of neurotensin receptors on cultured astrocytes. *Neuroscience*, 66, 627, 1995.
5. Szigethy, E. and Beaudet, A., Correspondence between high affinity ^{125}I-neurotensin binding sites and dopaminergic neurones in the rat substantia nigra and ventral tegmental area: a combined autoradiographic and immunohistochemical light microscopic study. *J. Comp. Neurol.*, 279, 128, 1989.
6. Boon, M. E. and Kok, L. P., Microwaves for immunohistochemistry, *Micron*, 25, 151, 1994.
7. Bugnon, C., Bahjaoui, M., and Fellmann, D., A simple method for coupling *in situ* hybridization and immunocytochemistry: application to the study of peptidergic neurons. *J. Histochem. Cytochem.*, 39, 859, 1991.
8. Reiter, M. K., Kremarik, P., Freund-Mercier, M. J., Stoeckel, M. E., Desaulles, E., and Feltz, P., Localization of oxytocin-binding sites in the thoracic and upper lumbar spinal cord in the adult and postnatal rat: a histoautoradiographic study. *Eur. J. Neurosci.*, 6, 98, 1994.
9. Freund-Mercier, M. J., Stoeckel, M. E., and Klein, M. J., Oxytocin receptors on oxytocin neurones: histoautoradiographic detection in the lactating rat. *J. Physiol. (London)*, 480, 155, 1994.

Chapter 7

ULTRASTRUCTURAL VISUALIZATION OF RADIOLABELED LIGANDS

Hichem C. Mertani
Gérard Morel

CONTENTS

0-8493-2644-3/97/$0.00+$.50

I. INTRODUCTION

Electron microscopic (EM) radioautography has been frequently used as a tool to follow a biological activity (e.g., cellular uptake or incorporation of a molecule). It is utilized by several laboratories to visualize the specific binding sites (receptors) of radioactive tracer molecules and to determine their kinetics of association with subcellular compartments. Table 1 lists the binding sites for some hormones, growth factors, neurotransmitters, vitamins, and neurotoxins which have been successfully demonstrated by using EM radioautography. A historical viewpoint on EM radio-autography has been given.[61] Although simple in practice, EM radioautography requires rigorous processing controls to achieve an adequate interpretation. Technically, demonstration of receptors by EM radioautography requires labeling the ligand with an appropriate isotope, administration of the labeled ligand, preparation of the tissue, preparation of the radioautographs, observation, and quantification. Controls are absolutely necessary to validate the specificity of the cellular ligand uptake and are the basis for statistical quantitative analysis.

II. RADIOAUTOGRAPHIC CONSIDERATIONS

A. GENERAL

The radioautographic process is based upon the detection of a radioactive source in a nuclear emulsion. Particle disintegration causes the formation of a "latent image" in some emulsion crystals which appears after development as silver grains (filamentous silver bodies or silver particles) at the EM level. Thus, the maximum particle energy and the half-life of the radioisotopes must be considered. Only the emissions of α particles, electrons (β particles), Auger electrons from electron capture reactions (iodine-125 decay), and positrons are effective for the production of radioautographs; γ and X-rays are not. The most common isotopes used are iodine-125 (^{125}I), sulfur-35 (^{35}S), and tritium (^{3}H). The particle energy determines the EM radioautography efficiency, which is defined as the number of silver grains generated for each radioactive disintegration. The spatial resolution of the radioautograph is inversely related to the maximum energy of the particle resulting from its decay. In thin nuclear emulsion as used in EM radio-autography, the higher the β particle, the lower the efficiency of the process. Particle energies above 0.5 MeV are usually very difficult to detect.

The half-life of the radioisotope is also important to consider as it determines the specific activity (amount of radioactivity, in Becquerels [Bq] for one mole of labeled molecule) which can be achieved for labeled molecules, and directly affects the exposure time. The longer the half-life, the lower the number of nuclear transformations per unit time. Nevertheless, with too short a half-life an important

**TABLE 1 Binding Sites of Some
Ligands Visualized by
Ultrastructural
Radioautography**

Type of Ligand	Ref.
^{3}H-serotonin	1
^{125}I-opioid peptide agonist	2, 3
^{3}H-opioid antagonist	4
^{3}H-benzodiazepine agonist	5, 6
^{3}H-dopamine	7
^{125}I-α neurotoxins	8, 9
^{125}I-peptide YY	10
^{125}I-gonadoliberin/agonists	11–15
^{3}H-thyrotropin releasing hormone	16
^{125}I-corticotropin releasing factor	17
^{125}I-chorionic gonadotropin	18,19
^{125}I-human growth hormone	20,21
^{125}I-bovine growth hormone	22–24
^{125}I-prolactin	25, 26
^{125}I-triiodothyronin	27
^{125}I-parathyroid hormone	28
^{125}I-elcatonin	29
^{125}I-cholecystokinin	30
^{125}I-somatostatin	31–33
^{125}I-glucagon	34
^{125}I-insulin/agonists	35–43
^{125}I-atrial natriuretic peptide	44–48
^{125}I-epidermal growth factor	49–54
^{125}I-transforming growth factor β1	55
^{125}I-pregnancy zone protein[3]	56
^{3}H-œstradiol-17β	57
^{3}H-androgen binding protein	58–60

amount of radioactivity will disappear during EM processing. Only radioisotopes with a half-life greater than 1 month are suitable.

B. RADIOLABELED MOLECULES

Labeling of the binding molecules can be achieved by three means: (1) substitution of one or more atoms of the stable isotope by a radioactive one; (2) addition of a radioactive isotope to the molecule; or (3) incorporation of the radioactive isotope into the molecule during its synthesis. Purification steps are necessary to separate the labeled ligand from free radioisotopes and to verify its integrity. Finally, one must calculate the specific activity of the labeled ligand and its pharmacological activity.

III. INJECTION OF LABELED LIGAND AND TISSUE PREPARATION

The mode of administration of the labeled ligand may influence the radioautographic signal distribution in animal tissues.[62] Prolonged injection into the third cerebral ventricle is usually performed for examining ligand binding in brain, intra-peritoneal injection is performed for examining ligand binding in the gastrointestinal tract, and intraaortic or intracardiac injections are performed for examining binding sites in kidney, liver, and pituitary.[61]

The concentration of labeled ligand required for EM radioautography will determine the exposure time, which is usually 10 times greater than in light microscopic radioautography and 10 to 100 times greater than in scintillation counting.[63] Such amounts are necessary considering the very small mass of the ultrathin sections. Less than 1 µCi of labeled ligand per gram of body weight is usually used.[31] Effective dose determination is described in details in Reference 63.

Tissue processing must retain the specificity of the labeling and preserve the ultrastructural cell characteristics. Most common EM fixing methods can be used (e.g., paraformaldehyde, glutaraldehyde plus osmium tetroxide) (see Chapter 12). The fixation step extracts about 20 to 25% of the radioactivity, which could correspond to unbound or denatured radioactive ligand (e.g., nonspecific binding). This loss can be determined by counting the radioactivity following each step. The cellular radioactivity uptake as a function of the time after injection must be determined by counting the radioactivity in the tissue and in the plasma immediately after sacrifice. Epoxy resins are utilized for embedding tissue samples (see Chapter 12).

IV. PREPARATION OF THE RADIOAUTOGRAPHS

A. SEMITHIN SECTIONS EMBEDDED IN EPOXY RESINS
1. Semithin Sections Preparation

Semi-thin sections of 0.5 to 1 µm must be used to check the specificity of the cellular uptake and the density of the radioautographic signal (Figure 1). The results will allow one to gauge the exposure period required for developing the ultrathin sections. Semithin sections may be collected on coated or noncoated histological slides (gelatin, chrome alum, polylysine, aminopropyl triethoxysilane) (see Protocol 1).

PROTOCOL 1: COATED-SLIDES FOR SEMITHIN SECTIONS

Reagents

- Ethanol 95°
- HCl 10 N
- 3-Aminopropyl triethoxysilane
- Acetone

Method

1. Overnight washing in 95° ethanol containing 1% HCl.
2. Rinse in running water for 2 h.
3. Rinse in distilled water 2 times for 1 min.
4. Dry in an oven.

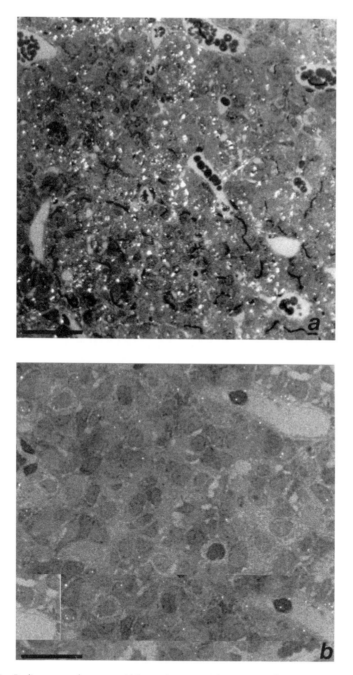

FIGURE 1   Radioautography on semithin sections. (a) Light micrographs from semi-thin sections of rat pituitary obtained 30 min after injection of ^{125}I-bGH showing the distribution of bright silver grains (under epipolarized light) over some cells; (b) the radioautographic specificity of the uptake is demonstrated by concomitant injection of the animal with 200-fold excess of unlabeled bGH, leading to a total abrogation of the signal density. Magnification bar equals 50 µm. (From Mertani, H.C., Waters, M.J., and Morel, G., *Neuroendocrinology,* 63, 257, 1996. With permission.)

5. Dip in acetone 15 s.
6. Dip in acetone containing 2% aminopropyl triethoxysilane for 10 s.
7. Rinse in acetone 2 times for 1 min.
8. Rinse in distilled water.
9. Dry in an oven and store at 4°C.

2. Radioautography on Semithin Sections

The radioautographic processing of semithin sections (Protocol 2) is quite similar to that performed for the radioautographic detection of ligand binding (see Chapter 3) or *in situ* hybridization (see Chapter 13) at the light microscopic level.

PROTOCOL 2: RADIOAUTOGRAPHS OF SEMITHIN SECTIONS

1. Cut the resin-embedded tissue at 0.5 to 1 μm thickness.
2. Expand the section with solvent vapor while it is in the water bath.
3. Transfer the section on the slide with a drop of distilled water.
4. Dry the section by warming the slides on a hot plate (40 to 50°C). At this step epoxy resin can be dissolved (see Protocol 3).
5. Apply the nuclear photographic emulsion (see Protocol 4).
6. Dry at room temperature.
7. Expose the emulsion-coated slides at 4°C in a light-tight box containing drying agent for 2 to 30 days.

PROTOCOL 3: REMOVAL OF EPOXY RESIN FROM SEMITHIN SECTIONS

Solution

Prepare and use immediately a 50% methanolic sodium methoxide solution: 2 g of sodium metal plus 110 ml methanol; after dissolution add 100 ml of benzene.

Method

1. Dip the slides into methanolic sodium methoxide solution for 3 min.
2. Wash in methanol-benzene (vol/vol) for 5 min.
3. Wash in acetone for 5 min.
4. Air dry.

PROTOCOL 4: PREPARATION OF LIQUID NUCLEAR EMULSION

This part requires important care. Precaution in the storage and use of the emulsion must be observed and consists of protection from light, radioactive sources, and chemicals. It is stored at 4°C.

Materials

Darkroom checklist:

- Suitable safelighting recommended by the supplier.
- Temperature: 18 to 20°C.
- Humidity: 40 to 50%.

Equipment checklist:

- Water bath at 45°C.
- Dipping glass beaker.
- Rack to hold the drying slides.
- Light-tight box.
- Silica gel.

Method

Under safelight:

1. Place the glass beaker half full of distilled water in the water bath at 45°C for 30 min to equilibrate.
2. Allow the emulsion to equilibrate to room temperature and mix with distilled water (vol/vol).
3. Dip a test slide vertically into the diluted emulsion and shake it very gently to mix the emulsion. Avoid bubble formation.
4. Withdraw the slide at a steady rate, keep it vertical, and drain off the excess emulsion on absorbing paper. Then examine the thickness and the homogeneity of the solution.
5. Dip the slides with semithin sections in exactly the same manner as the test slide, and dry for 2 to 3 h.
6. Place the slides in the light-tight box plus silica gel and seal it with black tape.
7. Store at 4°C for 2 to 30 exposure days.

3. Development and Staining of Radioautographs

PROTOCOL 5: DEVELOPMENT PROCESSING

Reagents

- Developer: in agreement with the emulsion
- Fixative: 30% sodium thiosulfate

Method

1. Develop at 18 to 20°C for 2 to 3 min with D19 or Dektol (Kodak® developers) using pure or diluted distilled water vol/vol. Dektol generated bigger silver grains than D19.

2. Rinse briefly in water.
3. Fix with 30% sodium thiosulfate at room temperature for 1 to 3 min.
4. Rinse carefully in running water.
5. Stain the sections (see Protocol 6) and mount with coverslide.

PROTOCOL 6: STAINING OF THE SEMITHIN SECTIONS

Reagent

- Toluidine blue solution 1% in 1% sodium borax
- Ethanol 70%

Method

1. Apply a drop of 0.5% toluidine blue on the sections and warm on a hot-plate at 60°C for 30 s.
2. Differentiate the staining with 70% ethanol.
3. Air dry.
4. Mount with coverslide.
5. Observe under a light microscope under brightfield or epipolarized light.

These micrographs are used to control the specificity of the radioautographic signal and its cellular localization (Figure 1).

B. ULTRATHIN SECTIONS EMBEDDED IN EPOXY RESINS

Ultrathin sections must be processed for radioautography with absolute handling precautions. One must consider that the exposure period takes 2 to 8 months, but when well handled, it yields very satisfactory results. Provide a large number of grids for each tissue in order to follow the appearance of the radioautographic signal and to quantify it. The radioautographic processing of ultrathin frozen sections was previously described in details in Reference 61.

1. Ultrathin Sections and Photographic Emulsion Characteristics

The quality, surface, and thickness of the sections are important variables to consider. Avoid the knife marks, which will produce artifacts, and cut a large surface to achieve better quantification. Sections about 100 nm (gold interference color) will yield a silver grain density about 1.5 to 2 times that obtained with sections about 50 nm (silver color), depending on the isotope and the developer used.[63] Thicker sections will also shorten the time of exposure but they result in a noticeable loss in electron optical resolution, image quality, and radioautographic resolution. Ultrathin sections are collected on formvar- (see Protocol 7) or collodion- (see Chapter 12) coated grids.

Ultrastructural radioautography is poorly compatible with the use of liquid emulsion and stripping film techniques, as they produce emulsion layers of only a few microns thick over the tissue. The photographic emulsions used are from Ilford®

(L4) or Amersham® (EM1) and have to be prepared as a monolayer. Before apposing the emulsion over the ultra-thin sections check its uniformity on a test grid.[63] Currently two techniques have been described for the application of the monolayer over the grids: the loop method (see Protocol 8) and the dipping technique (see Chapter 4). We routinely used the loop method, which requires less materials and handling than the dipping technique previously described.[61,63]

PROTOCOL 7: PREPARATION OF FORMVAR-COATED GRIDS

Equipment

- Histological slide cleaned with dichloro-1 2-ethane
- Glass rod
- Tall beaker
- Dropper
- Acetate film
- Formvar solution 0.5% in dichloro-1 2-ethane
- Nickel grids, 200 mesh

Method

1. Put some drops of the formvar solution at the end of the glass slide.
2. Spread the drops over the glass slide with the glass rod.
3. Remove the membrane by introducing the slide at a 45° angle into the water.
4. Let the detached formvar film float on the water surface.
5. Put the grids on the floating formvar film.
6. Carefully put the acetate film over the formvar film and grids to remove the whole, and air dry.

PROTOCOL 8: EMULSION OVERLAY BY THE LOOP METHOD

Material
Darkroom checklist:

- Small room, completely light-tight, suitable safelighting recommended by the supplier.
- Temperature of 35 to 45°C.
- Near 100% humidity.

Equipment checklist:

- Water bath at 45°C.
- Dipping glass beaker.

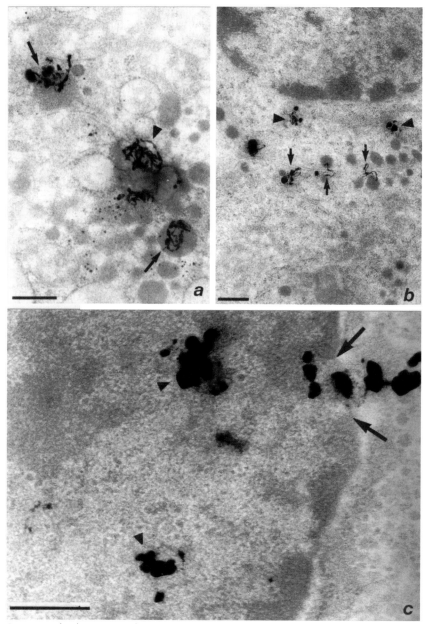

FIGURE 2   Ultrathin section of resin-embedded rat tissues processed for radioautographic detection of ^{125}I-bGH at different times after injection. (a) Silver grains localized 2 min after injection over the secretory granules (arrows) and the plasma membrane (arrowhead) of a pituitary cell. (b) Silver grains are present 15 min after injection over the plasma membrane (arrows) and the cytoplasmic matrix (arrowheads) of a pituitary cell. (c) Silver grains localized over the nuclear membrane particularly the nuclear pore delimited by arrows, and the nuclear matrix (arrowheads) of a hepatocyte. Bar equals 0.5 µm. (Radioautograph (c) reproduced from Lobie, P.E., Mertani, H.C., Morel, G., Morales-Bustos, O., Norstedt, G., and Waters, M.J., *J. Biol. Chem.,* 269, 21330, 1994; radioautograph (a) reproduced from Mertani, H.C., Waters, M.J., and Morel, G., *Neuroendocrinology,* 63, 257, 1996. With permission.)

Method

1. Incubate the grids on a drop of aqueous or ethanol uranyl acetate for 10 to 60 min at room temperature.
2. Contrast the grid on a drop of lead citrate for 3 to 10 min at room temperature.*
3. Rinse with distilled water.
4. Air dry.
5. Observe in the electron microscope.

The staining of the gelatin is not disturbing when the emulsion monolayer is well done. Otherwise, if the emulsion coating is too thick, the gelatin might be removed with 0.5 N acetic acid.

V. CONTROLS

EM radioautography imperatively needs to be validated by several controls before any conclusion should be drawn on the signal specificity. They include control of the reaction processing (background), control of the specificity of the cellular uptake, and control of the integrity of the radioactively injected ligand.

A. BACKGROUND NOISE

As in any revelation systems, some silver grains observed in radioautographs do not result from the ionizing particles emitted by the isotope. Rather, they are due to not enough care taken during emulsion handling and storage. Determination of this background level is achieved by taking several micrographs of the areas around the tissue section. The density of silver grains (number/unit area) must be determined by direct counting. If it exceeds 5% of the grain density determined over labeled cells, then it is suggested to abandon this grid.

The background noise may be generated only over the tissue sections by chemical effects (positive chemography) of tissue components, such as lysosomes.[63]

B. CONTROL OF THE CELLULAR UPTAKE

Specificity of the radioactive ligand binding is first determined at the tissue level by determining in a γ counter the ratio between the amount of radioactivity when the labeled ligand is injected alone and when it is injected with a large excess (40 to 200-fold) of unlabeled ligand, leading to a decrease of 60 to 90% of the radioactivity counts.[23,31]

At the cellular level, one might check the specificity on semithin sections by comparing the amounts of silver grains generated on tissue obtained when the labeled ligand is injected alone vs. with an excess of unlabeled ligand. Almost full abrogation of the radioautographic signal is expected (Figure 1). The use of an antagonist molecule in excess provides another specificity test of the ligand binding.[15,64]

---

* This step can be avoided.

C. CONTROL OF THE MOLECULAR INTEGRITY OF THE LIGAND

Most ligands are subjected to cellular catabolism after injection. It is thus preferable to inject the ligand via the jugular vein or intracardially to avoid its degradation by the hepatic portal circulation. In the tissue, it is necessary to ascertain whether the radioactivity present is due to intact molecule, free isotopes, free catabolite molecules containing the isotope, or newly synthesized molecules with incorporated isotope. Gel separation methods are mostly used to determine the molecular integrity of the injected ligand.[64]

D. OTHER CONTROLS

Experimental and physiological conditions regulate the quality and the quantity of the EM radioautographic signal. Kinetics of cellular ligand uptake are usually described in the literature.[61] Comparisons between *in vivo* and *in vitro* results might also cast light on the observed specificity of the uptake.[61]

VI. METHODS OF ANALYSIS AND QUANTIFICATION

The analysis of silver grain distribution may be quantitative as well, on the basis that the number of silver grains is proportional to the radioactivity content of the tissue. The main objective is to establish a correlation between the number of silver grains over a compartment to the relative proportion of radioactive ligand in this compartment. Methods to quantify the radioautographs are easy to apply when the concepts of efficiency and resolution are integrated.

A. EFFICIENCY

The efficiency is the percentage of disintegrations recorded. It can be determined empirically in EM radioautography by experiment. Overall, radioautographic efficiencies according to the type of isotope and to different emulsion-developer combinations have been described.[63] Assessment of the efficiency depends on various elements:

- On account of the presence of the emulsion layer over only one face of the section and the fact that disintegrations do not have any direction preference, the efficiency is less than 50%.
- For a β particle energy, the efficiency increases with the particle size.
- For bromide crystal diameters, the efficiency increases with their size.
- Due to the nature of the developer, all latent images are not transformed into silver grains.
- According to the thickness of the sections, self-absorption is in the range of 40 to 100 nm.
- During fixation with a heavy metal (OsO_4-uranyl salt), self-absorption is less than 10%.
- The size of the radiation dose is inversely proportional to the grain density.

B. RESOLUTION

For light microscopic radioautographs, the silver grain dispersion around the radioactive source is largely due to the multilayer emulsion, the thickness of the semithin section, the developers used which do not develop like latent images, and the size of the bromide crystal.[66]

In EM radioautography the silver grains dispersion is close to the radioactive source and correlated to the β particle energy. The geometric error is defined as the distance between the bromide crystal and the radioactive source, and it depends on the section thickness, the type of emulsion, and the β emission angle. The photographic error is defined as the distance between the developed silver grain and the center of the bromide crystal, and it depends on the bromide crystal size.[62,66-69] Detailed analyses from Salpeter et al.,[69,70] Williams,[63,66] and Nadler[71,72] are the basis for the interpretation and counting of silver grains. The distance from a point radioactive source within which half of the silver grain fall is the half radius (HR) value, and the distance from a radioactive line within which half of the silver grain fall is the half distance (HD) value. They are directly related by $HR = \sqrt{3} \times HD$. Both of these physical characteristics are the "point" and "line-resolution" values of the system. They are important to determine the extent to which the radioactivity can be localized. For example, HR values of the main isotopes used are 170 to 240 nm for ^{125}I, 250 to 310 nm for ^{3}H, and about 390 nm for ^{14}C and ^{35}S. Thus for the isotope ^{125}I, the HR value means that 50% of silver grains generated by the disintegration from a point radioactive source are within a circle of 170 to 240 nm in diameter located upon this source. This value is the basis of the probability circle method used to quantify the EM radioautographic signal. It is also essential to realize that some silver grains are generated outside the outline of the feature of origin and are said to be "cross-fired," their proportion depending on the size and shape of the radioactive feature.[66,68]

C. STATISTICAL ANALYSIS

Two methods of quantification are available: the probability circle and the cross-fire methods. However, results obtained with the cross-fire method are somewhat controversial and nowadays the probability circle method is mostly used.

The determination of cellular compartments which have specifically taken up the radioactive ligand is easily achieved by the probability circle method. A common procedure expresses data in terms of the percentage of total grains associated with a compartment. These data are transformed into a percentage of grain density by dividing the percentage of total grains in the compartment by the percentage of total area of the same compartment. If the distribution of silver grains is purely random, then the percentage of grain density should have a value of 1. All values above 1 indicate a signal level above the average, thus specific, and values below 1 indicate a nonspecific labeling.[69] A circle of a diameter related to its probability of containing the type of radioisotope is drawn around every grain center. A cellular compartment is considered to be associated with a grain if any part of the circle falls within that compartment.[71] If only one compartment is located within the circle then the grain is classified as exclusive. The grains shared between two or more compartments are assigned on the basis of probability to only one structure. The corrected number of silver grains for each compartment is the sum of exclusive and adjusted shared grains and is expressed as the percentage of the total corrected number of silver grains.[69,71] When the corrected number of silver grains is achieved, one might determine the concentration of signal by dividing the percentage of silver grains by the percentage of the area occupied by the cellular compartment.[69] If the density of silver grains is large enough over the tissue (≥ 400 for each experimental point), and if the background is low, then the correction proposed by Nadler[71] is not obligatory. In such a case the localization of a silver grain can be determined by its center.

VII. CONCLUSION

As shown in Table 1, *in vivo* ultrastructural radioautography has been used by several authors to study the localization of binding sites for a wide spectrum of molecules, including hormones, growth factors, peptides, toxins, neurotransmitters, and drugs.[73] This is a powerful technique to demonstrate the binding, uptake, and intracellular localization of a molecule which may account for a specific cellular effect. When it is combined with biochemical methods, this technique offers the unique opportunity to visualize with high resolution the association of a labeled ligand with an intracellular compartment.

ACKNOWLEDGMENT

H. C. Mertani has a fellowship and G. Morel a grant from ARC (Association de Recherche Contre le Cancer).

REFERENCES

1. **Beaudet, A.,** High resolution radioautography of central 5-hydroxytryptamine (5-HT) neurons, *J. Histochem. Cytochem.*, 30, 765, 1982.
2. **Hamel, E. and Beaudet, A.,** Electron microscopic radioautographic localization of opioid receptors in rat neostriatum, *Nature*, 312, 155, 1984.
3. **Jomary, C., Gairin, J.E., and Beaudet, A.,** Synaptic localization of k opioid receptors in guinea pig neostriatum, *Proc. Natl. Acad. Sci. U.S.A.*, 89, 564, 1992.
4. **Hanke, J., Jaros, P.P., and Willig, A.,** Radioautographic localization of opioid binding sites combined with immunogold detection of Leu-enkephalin, crustacean hyperglycaemic hormone and moult inhibiting hormone at the electron microscopic level in the sinus gland of the shore crab, *Carcinus maenas, Histochemistry*, 99, 405, 1993.
5. **Pinard, R., Segu, L., Cau, P., and Lanoir, J.,** Distribution of benzodiazepine receptors in the rat superior colliculus, a light and electron microscope quantitative radioautographic study, *Brain Res.*, 474, 48, 1988.
6. **Cahard, D., Canat, X., Carayon, P., Roque, C., Casselas, P., and Le Fur, G.,** Subcellular localization of peripheral benzodiazepine receptors on human leukocytes, *Lab. Invest.*, 70, 23, 1994.
7. **Morel, G. and Pelletier, G.,** Endorphinic neurons are contacting the tuberoinfundibular dopaminergic neurons in the rat brain, *Peptides*, 7, 1197, 1986.

8. **Miller, M.M., Billiar, R.B., and Beaudet, A.,** Ultrastructural distribution of alpha-bungarotoxin binding sites in the suprachiasmatic nucleus of the rat hypothalamus, *Cell Tissue Res.*, 250, 13, 1987.

9. **Boudier, J.L., Jover, E., and Cau, P.,** Radioautographic localization of voltage-dependent sodium channels on the mouse neuromuscular junction using ^{125}I-alpha scorpion toxin. I. Preferential labeling of glial cells on the presynaptic side, *J. Neurosci.*, 8, 1469, 1988.

10. **Sheikh, S.P., Roach, E., Fuhlendorff, J., and Williams, J. A.,** Localization of Y1 receptors for NPY and PYY on vascular smooth muscle cells in rat pancreas, *Am. J. Physiol.*, 260, G250,1991.

11. **Pelletier, G., Dubé, G., Guy, J., Séguin, C., and Lefebvre, F.A.,** Binding and internalization of a luteinizing hormone-releasing hormone agonist by rat gonadotrophic cells. A radioautographic study, *Endocrinology*, 111, 1068, 1982.

12. **Jennes, L., Stumpf, W.E., and Conn, P.M.,** Receptor-mediated binding and uptake of GnRH agonist and antagonist by pituitary cells, *Peptides*, 5, 215, 1984.

13. **Hazum, E., Koch, Y., Liscovitch, M., and Amsterdam, A.,** Intracellular pathways of receptor-bound GnRH agonist in pituitary gonadotropes, *Cell Tissue Res.*, 239, 3, 1985.

14. **Wynn, P.C., Suarez-Quian, C.A., Childs, G.V., and Catt, K.J.,** Pituitary binding and internalization of radiodinated gonadotropin-releasing hormone agonist and antagonist ligands *in vitro* and *in vivo*, *Endocrinology*, 119, 1852, 1986.

15. **Morel, G., Dihl, F., Aubert, M.L., and Dubois, P.M.,** Binding and internalization of native gonadoliberin (GnRH) by anterior pituitary gonadotrophs of the rat. A quantitative radioautographic study after cryoultramicrotomy, *Cell Tissue Res.*, 248, 541, 1987.

16. **Shioda, S. and Nakai, Y.,** Imunocytochemical localization of TRH and radioautographic determination of 3H-TRH binding sites in the arcuate nucleus-median eminence of the rat, *Cell Tissue Res.*, 228, 475, 1983.

17. **Leroux, P. and Pelletier, G.,** Radioautographic study of binding and internalization of corticotropin-releasing factor by rat anterior pituitary corticotrophs, *Endocrinology*, 114, 14, 1984.

18. **Han, S.S., Rajaniemi, H.J., Cho, M.I., Hirshfield, A.N., and Midgley, A.R., Jr.,** Gonadotropin receptors in rat ovarian tissue. II. Subcellular localization of LH binding sites by electron microscopic radioautography, *Endocrinology*, 96, 589, 1974.

19. **Hermo, L. and Lalli, M.,** Binding and internalization *in vivo* of (^{125}I)hCG in Leydig cells of the rat, *J. Androl.*, 9, 1, 1988.

20. **Barazzone, P., Lesniak, M.A., Gorden, P., Van Obberghen, E., Carpentier, J.L., and Orci, L.,** Binding, internalization, and lysosomal association of ^{125}I-human growth hormone in cultured human lymphocytes. A quantitative morphological and biochemical study, *J. Cell. Biol.*, 87, 360, 1980.

21. **Hizuka, N., Gorden, P., Lesniak, M.A., Carpentier, J.L., and Orci, L.,** Effect of pH and lysosomotropic agents on membrane-associated and internalized ^{125}I-iodinated human growth hormone in cultured human lymphocytes, a quantitative biochemical and electron microscopic study, *Endocrinology*, 111, 1576, 1982.

22. **Groves, W.E., Houts, G.E., and Bayse, G.S.,** Subcellular distribution of ^{125}I-labeled bovine growth hormone in rat liver and kidney, *Biochim. Biophys. Acta*, 264, 472, 1972.

23. **Lobie, P.E., Mertani, H.C., Morel, G., Morales-Bustos, O., Norstedt, G., and Waters, M.J.,** Receptor-mediated nuclear translocation of growth hormone, *J. Biol. Chem.*, 269, 21330, 1994.

24. **Mertani, H.C., Waters, M.J., and Morel G.,** Cellular trafficking of exogenous growth hormone in dwarf rat pituitary, *Neuroendocrinology*, 63, 257, 1996.

25. **Bergeron, J.J., Resch, L., Rachubinski, R., Patel, B.A., and Posner, B.I.,** Effect of colchicine on internalization of ovine prolactin in female rat liver, an *in vivo* radioautographic study, *J. Cell. Biol.*, 96, 875, 1983.

26. **Giss, B.J. and Walker, A.M.,** Mammotroph autoregulation, intracellular fate of internalized prolactin, *Mol. Cell. Endocrinol.*, 42, 259, 1985.

27. **Ardail, D., Lerme, F., Puymirat, J., and Morel, G.,** Evidence for the presence of alpha and beta-related T3 receptors in rat liver mitochondria, *Eur. J. Cell Biol.*, 62, 105, 1993.

28. **Rouleau, M.F., Mitchell, J., and Goltzman, D.,** *In vivo* distribution of parathyroid hormone receptors in bone, evidence that a predominant osseous target cell is not the mature osteoblast, *Endocrinology*, 123, 187, 1988.

29. **Ikegame, M., Ejiri, S., and Ozawa, H.,** Histochemical and radioautographic studies on elcatonin internalization and intracellular movement in osteoclasts, *J. Bone Min. Res.*, 9, 25, 1994.

30. **Rosenzweig, S.A., Miller, L.J., and Jamieson, J.D.,** Identification and localization of cholecystokinin-binding sites on rat pancreatic plasma membranes and acinar cells, a biochemical and radioautographic study, *J. Cell Biol.*, 96, 1288, 1983.

31. **Morel, G., Leroux, P., and Pelletier, G.,** Ultrastructural radioautographic localization of somatostatin-28 in the rat pituitary gland, *Endocrinology*, 116, 1615, 1985.

32. **Morel, G., Pelletier, G., and Heisler, S.,** Internalization and subcellular distribution of radiolabeled somatostatin-28 in the mouse anterior pituitary tumor cells, *Endocrinology*, 119, 1972, 1986.

33. **Morel, G. and Pelletier, G.,** Ultrastructural radioautographic localization of somatostatin-14 (SS-14) in the rat pituitary gland, *Exp. Clin. Endocrinol.*, 5, 31, 1986.

34. **Watanabe, J., Kanai, K., and Kanamura, S.,** Glucagon receptors in endothelial and Kupffer cells of mouse liver, *J. Histochem. Cytochem.*, 36, 1081, 1988.

35. **Gorden, P., Carpentier, J.L., Moule, M.L., Yip, C.C., and Orci, L.,** Direct demonstration of insulin receptor internalization. A quantitative electron microscopic study of covalently bound 125I-photoreactive insulin incubated with isolated hepatocytes, *Diabetes*, 31, 659, 1982.

36. **Fan., J.Y., Carpentier, J.L., Van-Obberghen, E., Blackett, N.M., Grunfeld, C., Gorden, P., and Orci, L.,** The interaction of ^{125}I-insulin with cultured 3T3-L1 adipocytes, quantitative analysis by the hypothetical grain method, *J. Histochem. Cytochem.*, 31, 859, 1983.

37. **Carpentier, J.L., Fehlmann, M., Van-Obberghen, E., Gorden, P., and Orci, L.,** Redistribution of ^{125}I-insulin on the surface of rat hepatocytes as a function of dissociation time, *Diabetes*, 34, 1002, 1985.

38. **Podlecki. D.A., Smith, R.M., Kao, M., Tsai, P., Huecksteadt, T., Branderburg, D., Lasher, R.S., Jarret, L., and Olefsky, J.M.,** Nuclear translocation of the insulin receptor. A possible mediator of insulin's long term effects, *J. Biol. Chem.*, 262, 3362, 1987.

39. **Goldfine, I.D., Jones, A.L., Hradek, G.T., and Wong, K.Y.,** Electron microscope radioautographic analysis of (^{125}I)iodoinsulin entry into adult rat hepatocytes *in vivo*, evidence for multiple sites of hormone localization, *Endocrinology*, 108, 1821, 1981.

40. **James, C.R. and Cotlier, E.,** Fate of insulin in the retina, an radioautographic study, *Br. J. Ophthalmol.*, 67, 80, 1983.

41. **Bar, R.S., DeRose, A., Sandra, A., Peacock, M.L., and Owen, W.G.,** Insulin binding to microvascular endothelium of intact heart, a kinetic and morphometric analysis, *Am. J. Physiol.*, 244, E447, 1983.

42. **Sakamoto, C., Williams, J.A., Roach, E., and Goldfine, I.D.,** *In vivo* localization of insulin binding to cells of the rat pancreas, *Proc. Soc. Exp. Biol. Med.*, 175, 497, 1984.

43. **Cruz, J., Posner, B.I., and Bergeron, J.J.,** Receptor-mediated endocytosis of (^{125}I)insulin into pancreatic acinar cells *in vivo*, *Endocrinology*, 115, 1996, 1984.

44. **Chabot, J.G., Morel, G., Kopelman, H., Belles-Isles, M., and Heisler, S.,** Atrial natriuretic factor and exocrine pancreas, radioautographic localization of binding sites and ultrastructural evidence for internalization of endogenous ANF, *Pancreas*, 2, 404, 1987.

45. **Chabot, J.G., Morel, G., Belles-Isles, M., Jeandel, L., and Heisler, S.,** ANF and exocrine pancreas, ultrastructural radioautographic localization in acinar cells, *Am. J. Physiol.*, 254, E301, 1988.

46. **Morel, G., Chabot, J.G., Belles-Isles, M., and Heisler, S.,** Synthesis and internalization of atrial natriuretic factor in anterior pituitary cells, *Mol. Cell. Endocrinol.*, 55, 219, 1988.

47. **Morel, G., Chabot, J.G., Garcia-Caballero, T., Gossard, F., Dihl, F., Belles-Isles, M., and Heisler, S.,** Synthesis, internalization, and localization of atrial natriuretic peptide in rat adrenal medulla, *Endocrinology*, 123, 149, 1988.

48. **Morel, G., Chabot, J.G., Gossard, F., and Heisler, S.,** Is atrial natriuretic peptide synthesized and internalized by gonadotrophs? *Endocrinology*, 124, 1703, 1989.

49. **Burwen, S.J., Barker, M.E., Goldman, I.S., Hradek, G.T., Raper, S.E., and Jones, A. L.,** Transport of epidermal growth factor by rat liver, evidence for a nonlysosomal pathway, *J. Cell Biol.*, 99, 1259 1984.

50. **Chabot, J.G., Walker, P., and Pelletier, G.,** Distribution of epidermal growth factor binding sites in the adult rat anterior pituitary gland, *Peptides*, 7, 45, 1986.

51. **Green, M.R., Mycock, C., Smith, C.G., and Couchman, J.R.,** Biochemical and ultrastructural processing of (^{125}I)epidermal growth factor in rat epidermis and hair follicles, accumulation of nuclear label, *J. Invest. Dermatol.*, 88, 259, 1987.

52. **Péchoux, C., Boumendil, J., Dolbeau, D., Souchier, C., and Frappart, L.,** Visualization and rapid quantification of radioautographic labeling in scanning electron microscopy applied to localization of receptor sites on the surface of whole cells, *Virchows Arch. B Cell Pathol.*, 62, 377, 1992.

53. **Frappart, L., Lefebvre, M.F., and Saez, S.,** The effect of 1,25-dihydroxyvitamin D3 on binding and internalization of epidermal growth factor in cultured cells, *Exp. Cell Res.*, 184, 329, 1989.

54. **Falette, N., Frappart, L., Lefebvre, M.F., and Saez, S.,** Increased epidermal growth factor receptor level in breast cancer cells treated by 1,25-dihydroxyvitamin D3, *Mol. Cell. Endocrinol.*, 63, 189, 1989.

55. **Dickson, K., Philip, A., Warshawsky, H., O'Connor-McCourt, M., and Bergeron, J.J.M.,** Specific binding of endocrine transforming growth factor-β1 to vascular endothelium, *J. Clin. Invest.*, 95, 2539, 1995.

56. **Moestrup, S.K., Christensen, E.I., Sottrup-Jensen, L., and Gliemann, J.,** Binding and receptor-mediated endocytosis of pregnancy zone protein-proteinase complex in rat macrophages, *Biochim. Biophys. Acta*, 930, 297, 1987.

57. **Cheng, C.Y., Rose, R.J., and Boettcher, B.,** The binding of oestradiol-17beta to human spermatozoa—an electron microscope radioautographic study, *Int. J. Androl.*, 4, 304, 1981.

58. **Gerard, A., Egloff, M., Gerard, H., el-Harate, A., Domingo, M., Gueant, J.L., Dang, C.D., and Degrelle, H.,** Internalization of human sex steroid-binding protein in the monkey epididymis, *J. Mol. Endocrinol.*, 5, 239, 1990.

59. **Gerard, H., Gerard, A., En-Nya, A., Felden, F., and Gueant, J.L.,** Spermatogenic cells do internalize sertoli androgen-binding protein, a transmission electron microscopy radioautographic study in the rat, *Endocrinology*, 134, 1515, 1994.

60. **Gueant, J.L., Fremont, S., Felden, F., Nicolas, J.P., Gerard, A., Leheup, B., Gerard, H., and Grignon, G.,** Evidence that androgen-binding protein endocytosis *in vitro* is receptor mediated in principal cells of the rat epididymis, *J. Mol. Endocrinol.*, 7, 113, 1991.

61. **Morel, G.,** Electron microscopic radioautographic techniques, in *Electron Microscopy in Biology. A Practical Approach,* Harris, R., Ed., IRL Press, Oxford, 1991, 83.

62. **Skougaard, M.R.K. and Stewart, P. A.,** Comparative effectiveness of intraperitoneal and intramuscular 3H-TDR injection routes in mice, *Exp. Cell. Res.*, 45, 158, 1967.

63. **Williams, M. A.,** radioautography and immunocytochemistry, in *Practical Methods in Electron Microscopy,* Glauert, A. M., Ed., North Holland, Amsterdam, 1985, Vol 6, part 1, 77.

64. **Morel, G. and Heisler, S.,** Internalization of endogenous and exogenous atrial natriuretic peptide by target tissues, *Electron Microsc. Rev.*, 1, 221, 1988.

65. **Salpeter, M.M. and McHenry, F.A.,** Electron microscope radioautography. Analysis of radioautograms, in *Advanced Techniques in Biological Electron Microscopy,* Koehler, J. K., Ed., Springer, New York, 1973, 113.

66. **Williams, M.A.,** Autoradiography, its methodology at the present time, *J. Microsc.*, 128, 79, 1982.

67. **Bachmann, L. and Salpeter, M.M.,** Absolute sensitivity of electron microscope radioautography, *J. Cell Biol.*, 33, 299, 1967.

68. **Caro, L.G.,** High-resolution radioautography. II. The problem of resolution, *J. Cell Biol.*, 15, 189, 1962.

69. **Salpeter, M.M., Bachmann, L., and Salpeter, E.E.,** Resolution in electron microscope radioautography, *J. Cell Biol.*, 41, 1, 1969.

70. **Salpeter, M.M., Fertuck, H.C., and Salpeter, E.E.,** Resolution in electron microscope radioautography. III. Iodine-125, the effect of heavy metal staining, and a reassessment of critical parameters, *J. Cell Biol.*, 72, 161, 1977.

71. **Nadler, N.J.,** The interpretation of grain counts in electron microscope radioautography, *J. Cell Biol.*, 49, 877, 1971.

72. **Nadler, N.J.,** Quantitation and resolution in electron microscope radioautography. *J. Histochem. Cytochem.,* 27, 1531, 1979.

73. **Bergeron, J.J.M. and Posner, B.I.,** *In vivo* studies on the initial localization and fate of polypeptide hormone receptors by the technique of quantitative radioautography, *J. Histochem. Cytochem.,* 27, 1512, 1979.

Chapter 8

USE OF PLASMON RESONANCE (BIAcore™) FOR THE ANALYSIS OF LIGAND-RECEPTOR INTERACTIONS

Hugues Lortat-Jacob
Sylvie Ricard-Blum

CONTENTS

0-8493-2644-3/97/$0.00+$.50
© 1997 by CRC Press LLC

I. INTRODUCTION

As reviewed recently, protein-protein interactions can be detected and investigated by several methods, including protein affinity chromatography, affinity blotting, and immunoprecipitation.[1,2] Binding constants can be determined by sedimentation through gradients, gel filtration assays for solubilized receptors, centrifugation binding assays (sedimentation equilibrium), fluorescence, and surface plasmon resonance methods.[2] A biosensor-based analytical system, based on the optical phenomenon of surface plasmon resonance, was introduced by Pharmacia Biosensor AB (Uppsala, Sweden) in 1990 to monitor and visualize molecular interactions in real time. It is one of the tools allowing the analysis of macromolecular interactions using immobilized ligands, but it is not limited to the study of protein-protein interactions.[3]

II. PRINCIPLE OF THE BIAcore™

The BIAcore™* (Biomolecular Interaction Analysis) technology has been designed to monitor biomolecular interactions in real time without labeling of the interactants in a continuous flow of buffer.[4-8] One of the interactants (referred to as the ligand) is immobilized on the sensor surface (sensor chip), while the second one (referred to as the analyte) is injected over that surface at a constant flow-rate by an automated fluid-handling system (Figure 1).[9,10] The immobilization of a given molecule leads to the formation of a biospecific sensor surface. The interaction between the analyte and the immobilized ligand is monitored by the optical phenomenon of surface plasmon resonance that occurs when light illuminates certain metals under conditions of total internal reflection.[11] Gold has been chosen in BIAcore™ for its chemical inertness and its good surface plasmon resonance response. The response depends on a change in refractive index in the close vicinity of the sensor chip surface and is proportional to the mass of the analyte bound to the surface. The BIAcore™ uses the principle of a biosensor since it combines a specific biological recognition system with a sensing device or transducer.[12] In biospecific interaction analysis (BIA) the sensor chip is the signal transducer, whereas surface plasmon resonance (SPR) converts changes in mass concentration on the biospecific surface into optical signals. Any change in surface concentration resulting from interaction is detected as a surface plasmon resonance signal, expressed in resonance units (RU). This response, reflecting the binding of the analyte to the immobilized interactant, is recorded as a function of time and is referred to as a sensorgram as illustrated in Figure 2, with the following Protocol.

---

* Registered trademark of Pharmacia Biosensor AB, Uppsala, Sweden.

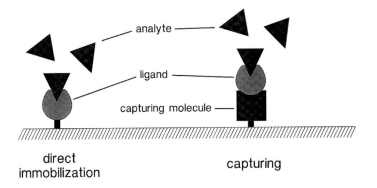

FIGURE 1 Terminology used for interactants in real time biomolecular interaction analysis. The ligand (in direct immobilization) or the capturing molecule (in capturing) was covalently immobilized on the sensor chip surface whereas the analyte is injected over that surface. (From *BIAapplications Handbook,* Pharmacia Biosensor AB, Uppsala, Sweden, 1994. With permission.)

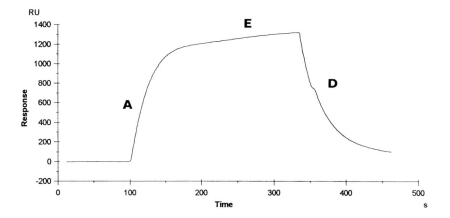

FIGURE 2 IFNγ (5 μg/ml) was injected at a flow rate of 5 μl/min over an "IFNγ-binding surface". The response in resonance units (RU) was recorded as a function of time, and showed the association phase (A), the equilibrium (E), and the dissociation phase (D).

PROTOCOL 1: STANDARD CONDITIONS OF ANALYTE INJECTION

1. Flow rate of 5 μl/min over an "IFNγ-binding surface."
2. Injection of 25 μl of human IFNγ (5 μg/ml) in Hepes Buffer Saline (HBS): 10 mM Hepes pH 7.4; 150 mM NaCl; 3.4 mM EDTA; 0.00 5% v/v surfactant P 20.

The IFNγ injection started at 100 s, and stopped at 320 s. When an analyte is injected over an immobilized ligand, the resulting sensorgram can be divided into three essential phases (Figure 2): (1) the association of analyte with immobilized ligand during injection (from 100 to 180 s), (2) equilibrium or steady-state, where the rate of analyte binding is balanced by dissociation from the complex (from 180 to 320 s), and (3) a dissociation of the analyte from the surface-bound complex at the end of the injection when the analyte is replaced by buffer flow (from 320 to 500 s). Typically, a response of 1000 resonance units corresponds to a change in surface concentration of about 1 ng/mm^2 for proteins and therefore approximately 1.2 ng of IFNγ had been bound per square millimeter in the above experiment.[13] The association and the dissociation phases provide information on the kinetics of the analyte-ligand interaction (the rate of complex formation and dissociation) and the equilibrium phase provides information on the affinity of the analyte-ligand interaction (the strength of binding). After analysis, noncovalently bound interactants may be removed from the sensor chip surface by appropriate regeneration so that the surface can be used for a new analysis. In some cases, BIAcore™ sensor chips fouled with covalently immobilized protein/peptide can be reconditioned by a combination of enzymatic and chemical treatments and reused.[14]

III. DESCRIPTION OF THE SYSTEM

The sensor chip is a gold-coated glass slide consisting of three layers: glass, a thin gold film (50 nm), and a matrix layer made of carboxymethylated dextran (100 nm in its swollen state) for covalent immobilization of one of the interactants.[5] The layer of linear carboxymethylated dextran is covalently attached to the gold via a self-assembled monolayer of an ω-hydroxyalkanethiol and provides a hydrophilic environment for the interaction (Figure 3). Samples, loaded from an autosampler, and buffer are delivered to the sensor chip by a liquid handling system comprising an integrated μ-microfluidic cartridge and two pumps, one for buffer flow and the other one for sample handling (transfer, dilution, and injection).[9] The integrated μ-fluidic cartridge contains flow channels, sample loops, and pneumatic valves to control sample and buffer flows. When it is pressed against the sensor chip, it forms four separate flow cells with the sensor chip dextran surface as one wall. The other side of the sensor chip, the glass side, is pressed into contact with a glass prism. A silicone opto-interface between the sensor chip and the prism ensures good optical coupling between the prism and the sensor chip (Figure 4).[9] Pumps, sample handling, and data collection are controlled by the BIAlogue control software, whereas sensorgram data are analyzed by the BIAevaluation software to test the validity of a model for a given interaction and/or to find the model which best fits the experimental data. There are several methods to analyze complex binding kinetics from data obtained in optical biosensors and to calculate kinetic constants.[15] Another software program, the BIAsimulation package, simulates sensorgrams based on user-defined interaction constants and experimental conditions.

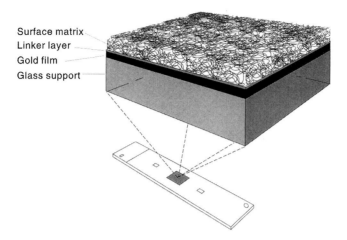

FIGURE 3   The surface of a sensor chip CM5 consists of three layers: glass, a thin gold film and a carboxymethylated dextran layer that is linked to the gold surface through an inert layer. (From *BIAtechnology Handbook,* Pharmacia Biosensor AB, Uppsala, Sweden, 1994. With permission.)

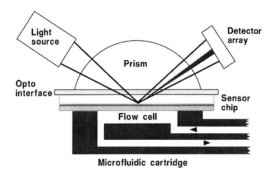

FIGURE 4   Schematic diagram of the optical system and flow-cell in real time BIA instruments. When the integrated μ-fluidic cartridge which delivers samples and buffer is pressed against the sensor chip, it forms four separate flow cells with the sensor chip dextran surface as one wall. The other side of the sensor chip, the glass side, is pressed into contact with a glass prism. (From *BIAtechnology Handbook,* Pharmacia Biosensor AB, Uppsala, Sweden, 1994. With permission.)

IV. FORMATION OF A BIOSPECIFIC SENSOR SURFACE

A. DIRECT IMMOBILIZATION

Immobilization is usually carried inside the BIAcore™, but it can be performed outside the system.[16] Several factors have to be considered in choosing the interactant to immobilize on the sensor chip. First, the molecule to be immobilized must be stable and available in pure form. The immobilization of the smallest interactants leads to the best sensitivity. If the concentration of one of the interactants is unknown, it is better to immobilize it since the concentration of analyte must be known for affinity and kinetics analysis.

Several chemical procedures can be used to immobilize the ligand on the sensor chip surface (Figure 5).[10] The most commonly applied method is amine coupling via primary amino groups of the ligand. The carboxymethylated dextran hydrogel of the sensor chip was activated by derivatization with N-hydroxysuccinimide (NHS) mediated by N-ethyl-N'-(dimethylaminopropyl) carbodiimide hydrochloride (EDC). Other chemistries for ligand immobilization include coupling by thiol disulfide exchange and coupling of aldehyde groups to a hydrazine-activated surface.[17-21] Aldehyde coupling is a method of interest to immobilize polysaccharides and glycoconjugates.[10]

1. Amine Coupling

The carboxymethylated dextran surface is first activated by derivatization with EDC/NHS. Under normal activation conditions, 30 to 40% of the dextran carboxyl groups is converted to reactive N-hydroxysuccinimide esters. These formed NHS-ester groups can react with primary amino groups in proteins or in other molecules (see Protocol 2).

PROTOCOL 2: STANDARD CONDITIONS OF IMMOBILIZATION

1. Inject 35 µl of a mixture of EDC/NHS (200 mM/50 mM in water) over the dextran surface at a flow rate of 5 µl/min.
2. Inject the molecule to be immobilized.
3. Deactivate unreacted NHS-ester with ethanolamine, or other amine-containing molecules.

The extent of dextran activation and of immobilization can be controlled by the concentration of EDC, NHS, or of the molecule to be immobilized such as by their contact time with the sensor chip surface.

In the following example (Figure 6), streptavidin has been covalently immobilized using the amine coupling in standard conditions. The sensorgram shows changes in resonance signal during EDC/NHS, streptavidin, and ethanolamine injections. Streptavidin was diluted to 100 µg/ml in 10 mM acetate buffer pH 4.2, and 35 µl was injected over the activated dextran surface at a flow rate of 5 µl/min. Approximately 3000 resonance units of streptavidin, corresponding to 3 ng/mm^2 streptavidin, have been immobilized. Since streptavidin is a 60-kDa protein, it can be calculated that 3×10^{10} molecules are linked to the sensor chip.

2. Coupling by Thiol-Disulfide Exchange

Thiol coupling can be used to immobilize ligand containing native thiol groups and the immobilization procedure is described below in Protocol 3. Alternatively, reactive disulfide groups may be introduced onto carboxy groups of the ligand or onto the sensor chip surface.[10,22]

PROTOCOL 3: COUPLING BY THIOL-DISULFIDE EXCHANGE

1. Continuously flow buffer at a flow rate of 5 μl/min.
2. Inject 10 μl of a mixture EDC 50 mM/NHS 200 mM in water to activate the carboxyl groups of the sensor chip.
3. Inject 20 μl of a 80-mM PDEA (2-(2-pyridinyldithio)ethaneamine hydrochloride) activation solution in 0.1 M borate buffer pH 8.5 to introduce a reactive disulfide group onto carboxyl groups on the sensor chip matrix.
4. Inject 35 μl of ligand in appropriate buffer (the pH of the buffer must be below the isoelectric point of the ligand).
5. Inject 20 μl of a 50-mM l-cysteine-1 M NaCl deactivation solution in 0.1 M formate buffer pH 4.3 to deactivate unreacted disulfide.

B. LIGAND CAPTURING

An alternative to direct immobilization is ligand capturing: a capturing molecule is immobilized on the sensor chip and binds the ligand which is captured in a noncovalent manner (Figure 1). Sensor chips with immobilized streptavidin (sensor chip SA5, commercially available from Pharmacia Biosensor AB) are used to capture biotinylated molecules such as biotinylated oligonucleotides or biotin-derivatized oligosaccharides.[23-25]

The sensorgram presented in Figure 7 shows the capturing of biotinylated heparin with a streptavidin-activated surface. Using the amine coupling chemistry, 3000 resonance unit of streptavidin have been immobilized on the sensor chip (see Section IV.A.1).

PROTOCOL 4: IMMOBILIZATION OF HEPARIN

1. Inject biotinylated heparin (0.5 mg/ml in HBS/0.3 M NaCl) over the streptavidin surface at a flow rate of 5 μl/min.
2. Perform three pulses of HBS/1.5 M NaCl to remove any unspecific bound material.

The amount of immobilized heparin was approximately 250 resonance units. Since the molecular weight of the heparin we used averaged 15 kDa, approximately 10^{10} molecules per square millimeter were immobilized.

Immobilized antibodies also serve as capturing agents for specific antigens. Polyclonal rabbit anti-mouse $F_c\gamma$ antibodies, immobilized by amine coupling, are used to capture monoclonal mouse antibodies of all subclasses.[26-28]

A sensor chip surface specific for a chosen affinity tail can also be prepared. A reagent kit for capturing proteins fused to glutathione-S-transferase is commercially available (Pharmacia Biosensor AB). An anti-glutathione-S-transferase antibody is covalently immobilized on the sensor chip for capturing recombinant glutathione-S-transferase fusion proteins. Recombinant proteins tagged at their termini with oligohistidine can be immobilized in a reversible manner via a stable chelating

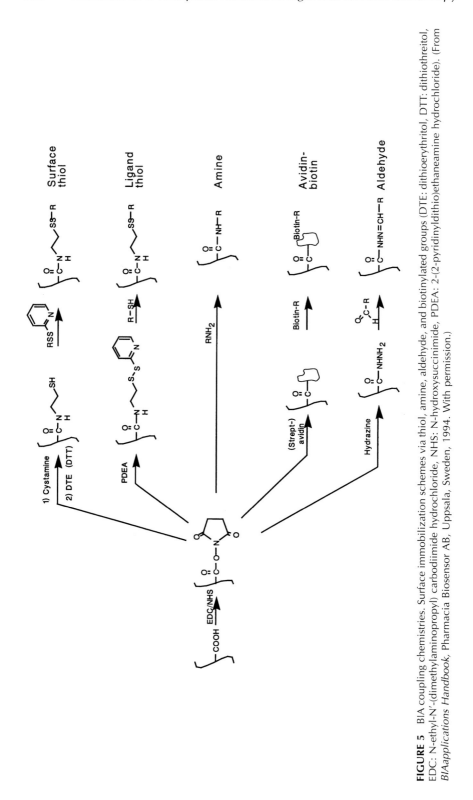

FIGURE 5 BIA coupling chemistries. Surface immobilization schemes via thiol, amine, aldehyde, and biotinylated groups (DTE: dithioerythritol, DTT: dithiothreitol, EDC: N-ethyl-N′-(dimethylaminopropyl) carbodiimide hydrochloride, NHS: N-hydroxysuccinimide, PDEA: 2-(2-pyridinyldithio)ethaneamine hydrochloride). (From *BIAapplications Handbook*, Pharmacia Biosensor AB, Uppsala, Sweden, 1994. With permission.)

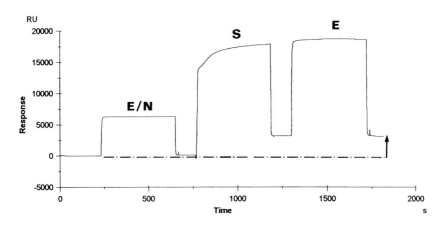

FIGURE 6 Covalent immobilization of streptavidin. The sensor chip was activated with a 7-min pulse of EDC/NHS (E/N, EDC: N-ethyl-N′-(dimethylaminopropyl) carbodiimide hydrochloride, NHS: N-hydroxysuccinimide) followed by the injection of streptavidin (S). Residual activated groups were quenched with ethanolamine (E). The vertical arrow visualize the amount of co-valently immobilized streptavidin (#3000 RU).

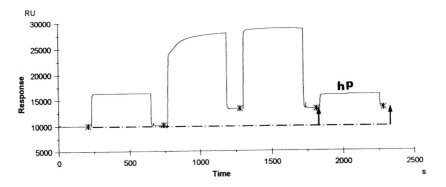

FIGURE 7 Immobilization of biotinylated heparin. Streptavidin was covalently immobilized us-ing the standard amine coupling Protocol (see also Figure 6), after which biotinylated heparin (hp) was injected. This results in the "capturing" of 250 resonance units of heparin, as indicated by the difference between the two vertical arrows.

linkage according to the principle of immobilized metal ion affinity chromatogra-phy.[29] The carboxymethylated dextran is covalently derivatized with the nickel che-lator nitrilotriacetic acid via its single amino group, and the derivatized surface charged with Ni^{2+} solution.[29] The BIAcore™ can thus be used to measure the production of recombinant fusion proteins and to study their interactions.[30]

C. NONCOVALENT IMMOBILIZATION

Another approach to surface derivatization allows the use of membrane-bound species and provides a means to study ligand-cell surface receptor or cell-cell interactions into a physiologically relevant membrane environment. This is achieved by the formation of hybrid bilayer membranes, composed of a monolayer of phospholipid and a monolayer of alkanethiol associated to the thin gold film of the sensor chip.[31] The gold surface is coated with a hydrophobic self-assembled monolayer of octadecanethiol to which phospholipid vesicles are allowed to fuse. A hydrophobic sensor chip (sensor chip HPA) is commercially available and will be very useful to immobilize membrane-bound receptors in a more native environment and to study their interaction with analytes in aqueous buffer.[32]

V. CHARACTERISTIC FEATURES OF THE BIA TECHNOLOGY

The monitoring of interactions in real time allows the collection of kinetic data and hence the determination of the stoichiometry and the kinetic constants (association rate and dissociation rate constants) used to calculate affinity constants K_A in the range 1×10^5 to 5×10^{11} M^{-1}). The BIAcore™ 2000 instrumentation allows the simultaneous monitoring of interaction events on the four sensing surfaces located in the four flow cells on the sensor chip surface with, for example, varying amounts of immobilized ligands. Using this multispot sensing, which appears to be ideal for the characterization of low-affinity interactions, analytes as small as 180 Da were detected and affinities in the 50 μM range could be determined.[33]

By contrast with other techniques used to study macromolecular interactions, the BIAcore™ technology does not require labeling of the interactants. The interactant chosen for immobilization has to be purified to assess the specificity of the surface, but crude preparations such as cell culture supernatants or unfractionated hybridoma culture medium, expected to contain binding activity towards the immobilized ligand, can be injected over the sensor surface. Specific interactions can thus be detected within complex mixtures. Binding site analysis leads to the identification of binding partners to target molecules of interest.

The sample consumption is low and sample recovery facilities allow the collection of the flow through the material during an injection and/or of the material eluted during the dissociation or regeneration steps. Fully automated sample handling and multichannel analysis lead to a higher throughput than conventional techniques. In addition, sensor chips are reusable and typically 50 to 100 analytical cycles can be performed on the same specific surface depending on regeneration conditions.[5]

VI. DOMAINS OF APPLICATIONS

The analysis of the interaction process leads to detailed mechanistic information by detecting functional groups or conformational changes that are important for interactions. It is useful to probe the identity and the role of residues involved in the interaction by monitoring the effect of a single amino acid change in a protein structure. The contribution of individual amino acid residues to monoclonal antibody binding has been evaluated in BIAcore™ using synthetic peptides presenting single amino acid substitutions.[34] Surface plasmon resonance biosensor technology can be used to study complex interactions involving several binding partners, such as those involved in the assembly, structure, and dynamics of a quaternary signal-transducing complex which controls chemotaxis in *Escherichia coli*.[35] The validity of proposed interaction mechanisms can also be tested in this system. The test of two competing models for the function of the chaperonin GroEL using the BIAcore™ system has been helpful to elucidate the ATP-dependent reaction cycle of GroEL.[8,36]

A. ANTIBODY CHARACTERIZATION

The system has been first applied to kinetic analysis of monoclonal antibody-antigen interactions.[26,37] Epitope mapping can be performed using a panel of monoclonal antibodies to probe the surface topology of antigens. This has been done for a number of monoclonal antibodies such as antibodies against the erythropoietin receptor, granulocyte colony-stimulating factor, and insulin.[27,38,39] Viral conformational epitopes have also been mapped using biosensor technology which appears to be superior to classical immunoassays for studying conformational epitopes in viral proteins.[40] The BIAcore™ biosensor is also useful in antibody engineering since it provides a rapid approach for analyzing recombinant antibodies and phage displayed antibody libraries.[41] The interactions of virus particles (cowpea mosaic virus and tobacco mosaic virus) with monoclonal antibodies have been monitored in this system by immobilizing either the antibody or the virus on the sensor chip.[42] Selection of antibodies for immunoassays and their affinity ranking can be also carried out in the BIAcore™ system.

B. RECEPTOR-LIGAND INTERACTIONS
1. Example of Application

Integrins are a family of cell surface receptors involved in both cell-cell and cell-matrix interactions. The BIAcore™ is one of the experimental approaches used to demonstrate that the integrin $\alpha_3\beta_1$ binds in a homophilic manner and can bind to other $\alpha_3\beta_1$ receptors on adjoining cells.[43] It has also been used to measure the binding affinity between integrin $\alpha_v\beta_3$ and osteopontin, an extracellular matrix protein, and to determine the effects of divalent cations on the association and dissociation rate of ligand binding.[44] The interactions between integrins and laminin isoforms have also been studied by surface plasmon resonance in the BIAcore™.[45]

The complex between human growth hormone and the extracellular domain of its receptor has been investigated using the BIA technology.[17,46] The data generated in the biosensor match affinities obtained by radioimmunoassay in solution.[17] This indicates that the sensor chip matrix is not causing systematic binding artifacts and is convenient to characterize the interactions between growth hormone and its receptor. This information is valuable to design hormone analogues for therapeutic purposes. BIAcore™ has been used to determine the effect of replacing each of the 31 contact residues in the human growth hormone site 1 structural binding domain with alanine on the kinetics and affinity of binding to the receptor. BIA studies have shown that the functional binding domain is considerably smaller than the structural binding domain.[17] Very high-affinity variants of human growth hormone have been assayed for binding to the extracellular domain of the immobilized receptor to better understand the molecular basis for the affinity improvements selected.[46] The interactions of monoclonal antibodies to growth hormone-binding protein and their effect on its binding to growth hormone have been determined by conventional competition binding assays and by surface plasmon resonance.[47]

In the multimeric interleukin-2 receptor system, the contributions of each of the subunits to ligand binding have been investigated by the BIA technology.[48] The binding constant for the interaction between interleukin-6 and its soluble receptor has also been determined by using this technology.[49]

The interactions of mutants of the interferon-γ receptor α chain (IFNγR) in its extracellular domains with neutralizing monoclonal antibodies have been monitored in the BIAcore™ system. The data obtained have provided unambiguous criteria for determining which mutations cause functional or conformational changes deleterious for antibody recognition.[50]

2. Analysis of the IFNγ/Heparan Sulfate Interaction

a. IFNγ

IFNγ is a T-cell-secreted cytokine which displays pleiotropic activities.[51] It exercises its biological functions via cell surface receptors, which have been found on all the cells analyzed so far, and which are widely distributed in almost all tissues.[52,53] IFNγ is generally thought of as a soluble factor, but is believed to act locally, in a paracrine or juxtacrine manner. The highly pleiotropic and overlapping activities of IFNγ, like the broad distribution of its receptor, suggest the existence of mechanisms that regulate and/or localize the activities of this polypeptide.[54] Among other things, it has been shown that this cytokine binds to extracellular matrix and cell surface heparan sulfate proteoglycan, as do a number of other cytokines, growth factors, and chemokines.[55,56] The IFNγ-heparan sulfate interaction involves a basic cluster of amino acids of the C-terminal part of the cytokine (amino acids 125-131), and a peculiar structure of heparan sulfate containing two heparin-like domains. These two domains directly bind to the two C-terminal sequences of an IFNγ dimer.[57,58]

Heparin, which is chemically related to heparan sulfate, also binds to IFNγ in a similar manner. The C-terminal part of IFNγ, the integrity of which is critical for biological activity, is highly susceptible to proteolytic cleavages, and it has been postulated that this domain could be a regulatory element of the cytokine.[59-61] Interestingly, heparan sulfate protects the C-terminal part of the cytokine from degradation and plays a role in the tissular targeting of the cytokine following intravenous injection.[57,62,63] Therefore, heparan sulfate or a heparan sulfate-like molecule could be used to protect and/or to modulate the bioavailability of IFNγ. In this work, we have investigated the use of the BIAcore™ technique to characterize the IFNγ/heparan sulfate interaction.

b.   Reducing the Nonspecific Binding of IFNγ to Sensor Chip

Injecting IFNγ (5 µg/ml) over an inactivated dextran surface produces a high response (Figure 8). Since the binding of IFNγ to heparan sulfate involves carboxylic groups, this nonspecific interaction is likely due to the dextran carboxylic residues.[64] A first way to reduce the nonspecific interactions is to add soluble carboxymethyl dextran to the binding buffer. Figure 8 shows the effect of the increasing concentrations (0 to 1 mg/ml) of dextran on the nonspecific binding of IFNγ. These data

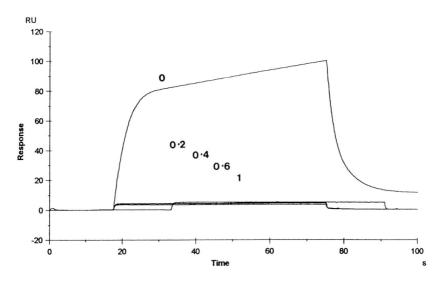

FIGURE 8   Effect of dextran in reducing the nonspecific interaction of IFNγ with the sensor chip. IFNγ (5 μg/ml in Hepes buffer saline containing 0, 0.2, 0.4, 0.6, or 1 mg/ml of carboxymethylated dextran — Fluka) was injected over a sensor chip. The response in resonance units (RU) was expressed as a percentage of maximal binding in the absence of dextran.

indicate that dextran in the binding buffer effectively prevents the interaction of the cytokine with the sensor chip dextran layer itself. We then analyzed the binding of IFNγ to heparin immobilized on a sensor chip as a function of the dextran concentration added in the binding buffer. Dextran significantly reduced the binding of IFNγ (Figure 9), and the remaining response is due to the specific binding of the cytokine to heparin. In order to better characterize the nonspecific interaction of IFNγ with the sensor chip, we also ran C-terminal-deleted IFNγ on an unactivated sensor chip (Figure 10). We found that IFNγ lacking 10 C-terminal amino acids (IFNγ 133), but not IFNγ lacking 4 amino acids (IFNγ 139), displayed a strongly reduced binding to the dextran layer. This indicates that the amino acids involved in the binding of IFNγ to heparin/heparan sulfate are also involved in its interaction with the dextran layer, and therefore the dextran included in the binding buffer may compete with heparin itself to bind to the cytokine.

We next tried to find another way to reduce the nonspecific binding. For that purpose we performed a series of activation/deactivation cycles of the sensor chip with alternating pulses (20 μl each) of EDC-NHS/ethanolamine. Between each cycle the interaction of IFNγ with the chip was determined. This experiment shows that after three to four EDC-NHS/ethanolamine cycles the nonspecific binding was eliminated (Figure 11). Free carboxylic groups were "deactivated" as described above, after which a new EDC-NHS injection was performed followed by streptavidin immobilization. Biotinylated glycosaminoglycan (heparin) was then immobilized onto the streptavidin-activated sensor chip. This results in approximately 50 resonance units of immobilized heparin (Figure 12). At saturating concentration, this heparin-activated sensor chip binds 300 to 400 resonance units of human IFNγ, and is suitable for kinetics analysis of the binding.

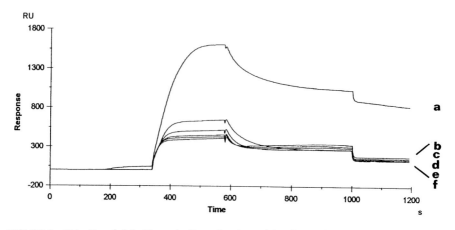

FIGURE 9   IFNγ (5 μg/ml in Hepes buffer saline (containing increasing concentrations of carboxymethylated dextran) was injected over a heparin-activated sensor chip. Carboxymethylated dextran was used at (a): 0, (b): 0.2, (c): 0.4, (d): 0.6, (e): 0.8, and (f): 1 mg/ml.

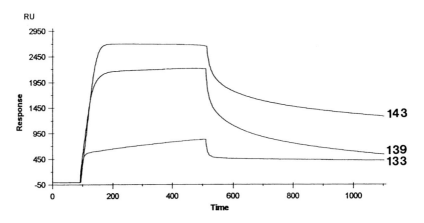

FIGURE 10   The C-terminal sequence of IFNγ is involved in the nonspecific binding of IFNγ. Full-length IFNγ (143) or IFNγ lacking 4 (139) or 10 (133) C-terminal amino acids were injected at 10 μg/ml over a non activated sensor chip.

C. SIGNAL TRANSDUCTION

Numerous intracellular signal-transducing proteins involved in the regulation of cellular events that occur in response to extracellular signals contain domain(s) of conserved sequence, such as the Src homology 2 and 3 (SH2 and SH3) domains. SH2 domains associate specifically with phosphotyrosine while SH3 domains appear to have preference for proline-rich motifs.[65]

Surface plasmon resonance has been used to study the interactions of SH2 domains with their phosphotyrosine-containing ligands. The interactions of the p85 subunit of phosphatidylinositol 3-kinase with various phosphoproteins provide a model for analyzing SH2 protein/phosphoprotein interactions.[66] A panel of monoclonal antibodies against the subunit p85α has been used to map its domain structure

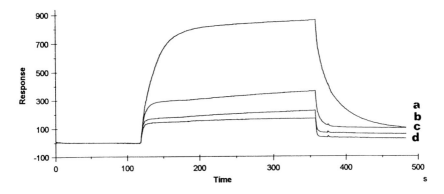

FIGURE 11 A sensor chip was "activated/deactivated" with a series of four pulses of "EDC/NHS-ethanolamine" (EDC: N-ethyl-N′-(dimethylaminopropyl) carbodiimide hydrochloride, NHS: N-hydroxysuccinimide). IFNγ (2.5 μg/ml) was injected over the sensor chip after each EDC/NHS-ethanolamine pulse (a,b,c,d).

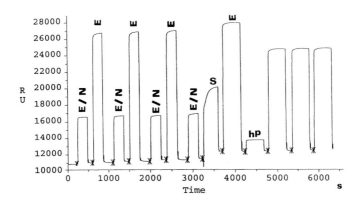

FIGURE 12 The number of anionic groups of the sensor chip was reduced with three pulses of EDC/NHS-ethanolamine (EDC: N-ethyl-N′-(dimethylaminopropyl) carbodiimide hydrochloride, NHS: N-hydroxysuccinimide). A fourth pulse of EDC/NHS was followed by the injection of streptavidin (S), ethanolamine (E), and biotinylated heparin (hp). The heparin injection was followed by three pulses of 1.5 M NaCl.

and to probe its interactions with tyrosine-phosphorylated peptides and growth factor receptors using the BIAtechnology.[67] The analysis of SH2 domain association with phosphotyrosine-containing peptides has demonstrated that SH2 domain specificity and activity can be modified by a single amino acid residue.[68] The binding affinity and specificity of various phosphopeptides by Lck (lymphocyte-specific tyrosine kinase) and Src SH2 domains have been assessed by surface plasmon resonance.[66] The very fast dissociation rates of the interactions between SH2 domains and the hepatocyte growth factor/scatter factor receptor measured in BIAcore™ are consistent with the notion of multiple signaling effectors interacting with the same docking site.[69] Surface plasmon resonance binding assays has also been helpful to define SH3 ligand specificity of different proteins (c-Src, c-Fyn, Lyn, c-Abl, and phosphatidylinositol 3-kinase).[65]

Another field of investigation concerns protein-protein interactions contributing to the specificity of intracellular vesicular trafficking with the demonstration that specific interactions occur between vesicular proteins and plasma membrane proteins.[70]

D. OLIGOSACCHARIDES

The BIAcore™ is also convenient for the characterization of the oligosaccharide structures of glycoproteins. Monosaccharide order and linkages have been identified in fetuin using a combination of lectin probes and *in situ* sequential digestion by specific exoglycosidases.[71]

E. MOLECULAR BIOLOGY

The BIAcore™ technology can be used for the study of DNA-protein interactions, such as the binding of ETS1 oncoproteins and of zinc finger protein to DNA.[23,24] It is thus very useful to elucidate the mechanisms of gene expression and has been used to compare two models for transcriptional activation.[8,72,73] It may also be used to monitor DNA-DNA hybridization.[73] The potential of this technology for applications in molecular biology comprises enzymatic modifications, such as ligation and endonuclease cleavage, and DNA synthesis.[74] In addition, a concept for the determination of single-point mutations in DNA samples has been reported.[74]

VII. CONCLUSION

Recent developments include the use of the biomolecular interaction analysis (BIA) technology for studying interactions between an immobilized ligand and cell suspensions or crude extracts, such as whole blood samples, hybridoma cell cultures to monitor antibody production, or bacterial extracts for monitoring recombinant protein production.[75,76] No purification of crude preparation is required for detecting soluble analytes. Cell culture supernatants have been screened for receptor binding activity by injection over the immobilized extracellular domain of the ECK receptor protein-kinase.[77] The system can also be used to detect specific binding of whole cells (bacteria or mammalian cells such as *Staphylococcus aureus* and erythrocytes) to immobilized ligands on the sensor chip surface.[75] There are some limitations to the study of these interactions. Indeed, the major contribution to the surface plasmon resonance response comes from refractive index changes at distances less than 0.3 µm from the sensor surface.[76] Interactions between immobilized protein and phospholipid-containing vesicles have also been investigated.[67]

REFERENCES

1. Hulme, E. C., *Receptor-Ligand Interactions. A Practical Approach*, The practical approach series, D. Rickwood and B. D. Hames, Eds., IRL Press, Oxford, 1992.
2. Phizicky, E. M. and Fields, S. Protein-protein interactions: methods for detection and analysis, *Microbiol. Rev.*, 59, 94, 1995.

3. Chaiken, I., Rosé, S., and Karlsson, R., Analysis of macromolecular interactions using immobilized ligands, *Anal. Biochem.*, 201, 197, 1992.

4. Jönsson, U., Fägerstam, L., Ivarsson, B., Johnsson, B., Karlsson, R., Lundh, K., Löfas, S., Persson, B., Roos, H., Rönnberg, I., Sjölander, S., Stenberg, E., Stahlberg, R., Urbaniczky, C., Ostlin, H., and Malmqvist, M., Real-time biospecific interaction analysis using surface plasmon resonance and a sensor chip technology, *BioTechniques,* 11, 620, 1991.

5. Malmqvist, M., Biospecific interactions using biosensor technology, *Nature*, 361, 186, 1993.

6. Panayotou, G., Waterfield, M. D., and End, P., Riding the evanescent wave. Recent advances in optical biosensors allow biologists to monitor molecular interactions in real time on a sensor surface by surface plasmon resonance or waveguiding techniques, *Curr. Biol.*, 3, 913, 1993.

7. Markey, F., Biomolecular interaction analysis, *Pharmaceutical Manufacturing International*, 47, 1995.

8. Szabo, A., Stolz, L., and Granzow, R., Surface plasmon resonance and its use in biomolecular interaction analysis (BIA), *Curr. Opin. Struct. Biol.*, 5, 699, 1995.

9. Anon., *BIAtechnology Handbook,* Pharmacia Biosensor AB, Uppsala, Sweden, 1994.

10. Anon., *BIAapplications Handbook,* Pharmacia Biosensor AB, Uppsala, Sweden, 1994.

11. Liedberg, B., Lundström, I., and Stenberg, E., Principles of biosensing with an extended coupling matrix and surface plasmon resonance, *Sensors Actuators B*, 11, 63, 1993.

12. Blum, L. and Coulet, P., *Biosensor Principles and Applications*, Marcel Dekker, New York, 1991.

13. Stenberg, E., Persson, B., Roos, H., and Urbaniczky, C., Quantitative determination of surface concentration of protein with surface plasmon resonance by using radiolabeled proteins, *J. Colloid Interface Sci.*, 143, 513, 1991.

14. Chatelier, R. C., Gengenbach, T. R., Griesser, H. J., Brighamburke, M., and Oshannessy, D. J., A general method to recondition and reuse BIAcore sensor chips fouled with covalently immobilized protein/peptide, *Anal. Biochem.*, 229, 112, 1995.

15. Morton, T. A., Myszka, D. G., and Chaiken, I. M., Interpreting complex binding kinetics from optical biosensors: a comparison of analysis by linearization, the integrated rate equation and numerical integration, *Anal. Biochem.*, 227, 176, 1995.

16. Sternesjö, A., Mellgren, C., and Björck, L., Determination of sulfamethazine residues in milk by a surface plasmon resonance-based biosensor assay, *Anal. Biochem.*, 226, 175, 1995.

17. Cunningham, B. C. and Wells, J. A., Comparison of a structural and functional epitope, *J. Mol. Biol.*, 234, 554, 1993.

18. Khilko, S. N., Corr, M., Boyd, L. F., Lees, A., Inman, J. K., and Margulies, D. H., Direct detection of major histocompatibilty complex class-I binding to antigenic peptides using surface plasmon resonance — peptide immobilization and characterization of binding specificity, *J. Biol. Chem.*, 268, 15425, 1993.

19. Mani, J. C., Marchi, V., and Cucurou, C., Effect of HIV-1 peptide presentation on the affinity constants of two monoclonal antibodies determined by BIAcore™ technology, *Mol. Immunol.*, 31, 439, 1994.

20. Benkirane, N., Guichard, G., Van Regenmortel, M. H. V., Briand, J. P., and Muller, S., Cross-reactivity of antibodies to retro-inverso peptidomimetics with the parent protein histone H3 and chromatin core particle, *J. Biol. Chem.*, 270, 11921, 1995.

21. Abraham, R., Buxbaum, S., Link, J., Smith, R., Venti, C., and Darsley, M., Screening and kinetic analysis of recombinant anti-CEA antibody fragments, *J. Immunol. Methods*, 183, 119, 1995.

22. Anon., *Ligand Immobilization for Real Time BIA Using Thiol-Disulphide Exchange,* Application Note 601, Pharmacia Biosensor AB, Uppsala, Sweden, 1992.

23. Fisher, R. J., Fivash, M., Casas-Finet, J., Erickson, J. W., Kondoh, A., Bladen, S. V., Fisher, C., Watson, D. K., and Papas, T., Real-time DNA binding measurements of the ETS1 recombinant oncoproteins reveal significant kinetic differences between the p42 and p51 isoforms, *Protein Sci.*, 3, 257, 1994.

24. Yang, W. P., Wu, H., and Barbas, IC. F., III, Surface plasmon resonance based kinetic studies of zinc finger-DNA interactions, *J. Immunol. Methods*, 183, 175, 1995.

25. Shinohara, Y., Sota, H., Kim, F., Shimizu, M., Gotoh, M., Tosu, M., and Hasegawa, Y., Use of a biosensor based on surface plasmon resonance and biotinyl glycans for analysis of sugar binding specificities of lectins, *J. Biochem.*, 117, 1076, 1995.

26. Karlsson, R., Michaelsson, A., and Mattsson, L., Kinetic analysis of monoclonal antibody-antigen interactions with a new biosensor based analytical system, *J. Immunol. Methods*, 145, 229, 1991.
27. Nice, E., Layton, J., Fabri, L., Hellman, U., Engstrom, A., Persson, B., and Burgess, A. W., Mapping of the antibody- and receptor-binding domains of granulocyte colony-stimulating factor using an optical biosensor. Comparison with enzyme-linked immunosorbent assay competition studies, *J. Chromatogr.*, 646, 159, 1993.
28. Pellequer, J. L. and Van Regenmortel, M. H. V., Measurement of kinetic binding constants of viral antibodies using a new biosensor technology, *J. Immunol. Methods*, 166, 133, 1995.
29. Gershon, P. D. and Khilko, S., Stable chelating linkage for reversible immobilization of oligo-histidine tagged proteins in the BIAcore surface plasmon resonance detector, *J. Immunol. Methods*, 183, 65, 1995.
30. Anon., *Capturing Fusion Proteins on the Sensor Chip Surface,* Application Note 104, Pharmacia Biosensor AB, Uppsala, Sweden, 1995.
31. Plant, A. L., Brigham-Burke, M., Petrella, E. C., and O'Shannessy, D. J., Phospholipid/alkanethiol bilayers for cell-surface receptor studies by surface plasmon resonance, *Anal. Biochem.*, 226, 342, 1995.
32. Anon., *Supported Polar Lipid Monolayers for BIA,* BIAtechnology Note 106, Pharmacia Biosensor AB, Uppsala, Sweden, 1995.
33. Karlsson, R. and Stahlberg, R., Surface plasmon resonance detection and multispot sensing for direct monitoring of interactions involving low-molecular-weight analytes and for determination of low affinities, *Anal. Biochem.*, 228, 274, 1995.
34. Zeder-Lutz, G., Altschuh, D., Denery-Papini, S., Briand, J. P., Tribbick, G., and Van Regenmortel, M. H. V., Epitope analysis using kinetic measurements of antibody binding to synthetic peptides presenting single amino acid substitutions, *J. Mol. Recogn.*, 6, 171, 1993.
35. Schuster, S. C., Swanson, R. V., Alex, L. A., Bourret, R. B., and Simon, M. I., Assembly and function of a quaternary signal transduction complex monitored by surface plasmon resonance, *Nature*, 365, 343, 1993.
36. Hayer-Hartl, M. K., Martin, J., and Hartl, F. U., Asymmetrical interaction of GroEL and GroES in the ATPase cycle of assisted protein folding, *Science*, 269, 836, 1995.
37. Karlsson, R., Altschuh, D., and Van Regenmortel, M. H. V., Measurement of antibody affinity, in *Structure of Antigens*, Vol. 1, Van Regenmortel, M. H. V., Ed., CRC Press, Boca Raton, FL, 1992, 127.
38. D'Andrea, A. D., Rup, B. J., Fisher, M. J., and Jones, S., Anti-erythropoietin receptor (EPO-R) monoclonal antibodies inhibit erythropoietin binding and neutralize bioactivity, *Blood*, 82, 46, 1993.
39. Allauzen, S., Mani, J. C., Granier, C., Pau, B., and Bouanani, M., Epitope mapping and binding analysis of insulin-specific monoclonal antibodies using a biosensor approach, *J. Immunol. Methods*, 183, 27, 1995.
40. Saunal, H. and Van Regenmortel, M. H. V., Mapping of viral conformational epitopes using biosensor measurements, *J. Immunol. Methods*, 183, 33, 1995.
41. Malmborg, A. C. and Borrebaeck, C. A. K., BIAcore as a tool in antibody engineering, *J. Immunol. Methods*, 183, 7, 1995.
42. Dubs, M. C., Altschuh, D., and Van Regenmortel, M. H. V., Interaction between viruses and monoclonal antibodies studied by surface plasmon resonance, *Immunol. Lett.*, 31, 59, 1991.
43. Sriramarao, P., Steffner, P., and Gehlsen, K. R., Biochemical evidence for a homophilic interaction of the $\alpha_3\beta_1$ integrin, *J. Biol. Chem.*, 268, 22036, 1993.
44. Hu, D. D., Hoyer, J. R., and Smith, J. W., Ca^{2+} suppresses cell adhesion to osteopontin by attenuating binding affinity for integrin $\alpha_v\beta_3$, *J. Biol. Chem.*, 270, 9917, 1995.
45. Pfaff, M., Göhring, W., Brown, J. C., and Timpl, R., Binding of purified collagen receptors ($\alpha1\beta1$, $\alpha2\beta1$) and RGD-dependent integrins to laminins and laminin fragments, *Eur. J. Biochem.*, 225, 975, 1994.
46. Lowman, H. A. and Wells, J. B., Affinity maturation of human growth hormone by monovalent phage display, *J. Mol. Biol.*, 234, 564, 1993.
47. Sadeghi, H., Lumanglas, A. L., Baumbach, W. R., and Wang, B. S., Interaction of monoclonal antibodies with growth hormone-binding protein and its complex with growth hormone, *J. Endocrinol.*, 139, 495, 1993.

48. Wu, Z., Johnson, K. W., Choi, Y., and Ciardelli, T. L., Ligand binding analysis of soluble interleukin-2 receptor complexes by surface plasmon resonance, *J. Biol. Chem.*, 270, 16045, 1995.

49. Ward, L. D., Howlett, G. J., Hammacher, A., Weinstock, J., Yasukawa, K., Simpson, R. J., and Winzor, D. J., Use of a biosensor with surface plasmon resonance detection for the determination of binding constants: measurement of interleukin-6 binding to the soluble interleukin-6 receptor, *Biochemistry*, 34, 2901, 1995.

50. Ruegg, N., Williams, G., Birch, A., Robinson, J. A., Schlatter, D., and Huber, W., Mutagenesis of immunoglobulin-like domains from the extracellular human interferon-γ receptor α chain and their recognition by neutralizing antibodies monitored by surface plasmon resonance technology, *J. Immunol. Methods*, 183, 95, 1995.

51. Farrar, M.A. and Schreiber, R.D., The molecular cell biology of interferon-γ and its receptor, *Annu. Rev. Immunol.*, 11, 571, 1993.

52. Van Loon, A. P. G. M., Ozmen, L., Fountoulakis, M., Kania, M., Haiker, M., and Garotta, G., High-affinity receptor for interferon-gamma (IFN-γ), a ubiquitous protein occurring in different molecular forms on human cells, *J. Leukocyte. Biol.*, 49, 462, 1991.

53. Valente, G., Ozmen, L., Novelli, F., Geuna, M., Palestro, G., Forni, G., and Garotta, G., Distribution of interferon-γ receptor in human tissues, *Eur. J. Immunol.*, 22, 2403, 1993.

54. Lortat-Jacob, H. and Grimaud, J. A., The extracellular matrix: from supporting tissue to cytokines regulations, *Pathol. Biol.*, 42, 612, 1994.

55. Lortat-Jacob, H., Kleinmann, H. K., and Grimaud, J. A., High affinity binding of interferon-gamma to a basement membrane complex (matrigel), *J. Clin. Invest.*, 87, 878, 1991.

56. Jackson, R. L., Busch S. J., and Cardin A. D., Glycosaminoglycans: molecular properties, protein interactions, and role in physiological processes, *Physiol. Rev.*, 71, 481, 1991.

57. Lortat-Jacob, H. and Grimaud, J. A., Interferon-gamma binds to heparan sulfate by a cluster of amino acids located on the C-terminal part of the molecule, *FEBS Lett.*, 280, 152, 1991.

58. Lortat-Jacob, H., Turnbull, J. E., and Grimaud, J. A., Molecular organization of the interferon-γ binding domain in heparan sulphate, *Biochem. J.*, 310, 497, 1995.

59. Döbeli, H., Gentz, R., Jucker, W., Garotta, G., Hartmann, D.W., and Hochuli, E., Role of the C-terminal sequence on the biological activity of human interferon-γ, *J. Biotechnol.* 7, 199, 1988.

60. Lortat-Jacob, H. and Grimaud, J. A., IFN-gamma C-terminus function. New working hypothesis. Heparan sulfate and heparin, new targets for IFN-gamma, protect, relax the cytokine and regulate its activity, *Cell. Mol. Biol.*, 37, 253, 1991.

61. Lundell, D. J. and Narula, S. K., Structural elements required for receptor recognition of human interferon-gamma, *Pharmac. Ther.*, 64, 1, 1994.

62. Lortat-Jacob, H., Baltzer, F., and Grimaud, G., Heparin decreases the blood clearance of interferon-gamma and increases its activity by limiting the processing of its C-terminal sequence, *J. Biol. Chem.*, 271, 16139, 1996.

63. Lortat-Jacob, H., Brisson, C., Guerret, S., and Morel, G., Non-receptor-mediated tissular localisation of human interferon-γ: role of heparan sulfate/heparin-like molecules, *Cytokine*, 8, 557, 1996.

64. Lortat-Jacob, H. and Grimaud, J. A., Binding of interferon-gamma to heparan sulfate is restricted to the heparin-like domains and involves carboxylic- but not N-sulfated groups, *Biochim. Biophys. Acta*, 117, 126, 1992.

65. Rickles, R. J., Botfield, M. C., Weng, Z., Taylor, J. A., Green, O. M., Brugge, J. S., and Zoller, M. J., Identification of Src, Fyn, Lyn, PI3K and Abl SH3 domain ligands using phage display libraries, *EMBO J.*, 13, 5598, 1994.

66. Payne, G., Shoelson, S. E., Gish, G.D., Pawson, T., and Walsh, C. T., Kinetics of p56[lck] and p60[src] Src homology 2 domain binding to tyrosine-phosphorylated peptides determined by a competition assay or surface plasmon resonance, *Proc. Natl. Acad. Sci. U.S.A.*, 90, 4902, 1993.

67. End, P., Gout, I., Fry, M. J., Panayotou, G., Dhand, R., Yonezawa, K., Kasuga, M., and Waterfield, M. D., A biosensor approach to probe the structure and function of the p85α subunit of the phosphatidylinositol 3-kinase complex, *J. Biol. Chem.*, 268, 10066, 1993.

68. Marengere, L. E. M., Songyang, Z., Gish, G. D., Schaller, M. D., Parsons, J. T., Stern, M. J., Cantley, L. C., and Pawson, T., SH2 domain specificity and activity modified by a single residue, *Nature*, 369, 502, 1994.

69. Ponzetto, C., Bardelli, A., Zhen, Z., Maina, F., Dalla Zonca, P., Giordano, S., Graziani, A., Panayotou, G., and Comoglio, P. M., A multifunctional docking site mediates signaling and transformation by the hepatocyte growth factor/scatter factor receptor family, *Cell*, 77, 261, 1994.

70. Calakos, N., Bennett, M. K., Peterson, K. E., and Scheller, R. H., Protein-protein interactions contributing to the specificity of intracellular vesicular trafficking, *Science*, 263, 1146, 1994.

71. Hutchinson, A. M., Characterization of glycoprotein oligosaccharides using surface plasmon resonance, *Anal. Biochem.*, 220, 303, 1994.

72. Barberis, A., Pearlberg, J., Simkovich, N., Farrell, S., Reinagel, P., Bamdad, C., Sigal, G., and Ptashne, M., Contact with a component of the polymerase II holoenzyme suffices for gene activation, *Cell*, 81, 359, 1995.

73. Wood, S. J., DNA-DNA hybridization in real time using BIAcore, *Microchem. J.*, 47, 330, 1993.

74. Nilsson, P., Persson, B., Uhlén, M., and Nygren, P. A., Real time monitoring of DNA manipulations using biosensor technology, *Anal. Biochem.*, 224, 400, 1995.

75. Anon., *Working With Cells and Crude Sample Preparations in BIAcore and BIAlite,* BIAtechnology Note 103, Pharmacia Biosensor AB, Uppsala, Sweden, 1994.

76. Anon., *Biomolecular Interaction Analysis With Cells and Vesicles,* BIAtechnology Note 105, Pharmacia Biosensor AB, Uppsala, Sweden, 1995.

77. Bartley, T. D., Hunt, R. W., Welcher A. A., Boyle, W. J., Parker, V. P., Lindberg, R. A., Lu, H. S., Colombero, A. M., Elliott, R. L., Guthrie, B. A., Holst, P. L., Skrine, J. D., Toso, R. J., Zhang, M., Fernandez, E., Trail, G., Varnum, B., Yarden, Y., Hunter, T., and Fox, G. M., B61 is a ligand for the ECK receptor protein-tyrosine kinase, *Nature*, 368, 558, 1994.

Chapter **9**

LIGHT MICROSCOPIC VISUALIZATION OF POLYPEPTIDE LIGAND RECEPTORS

_____ Peter E. Lobie

CONTENTS

0-8493-2644-3/97/$0.00+$.50
© 1997 by CRC Press LLC

I. INTRODUCTION

The hormonal regulation of target cell function has occupied investigators ever since the discovery of hormones. However, the study of receptor molecules is by no means confined to a study of their structure or signal transduction mechanisms. More detailed understanding of receptor function has required a more detailed analysis of receptor localization in target tissues at both the cellular and subcellular levels. Methods have therefore been developed which allow the visualization and detailed localization for various classes of receptors including receptors for polypeptide ligands. This chapter provides a brief discourse on methods available for the localization of receptors for polypeptide ligands in mammalian cells. It is by no means exhaustive nor complete. It provides a basic summary of several commonly used methods, with examples, which should be within the technical reach of the majority of laboratories.

There are two basic methodologies for the localization of receptors for polypeptide ligands. The first of these is indirect and uses the respective high-affinity ligand labeled in such a manner as to allow visualization. The second method is direct and uses an antibody to specifically bind to the receptor molecule, followed by one of several visualization techniques. Lastly, several newer, more advanced techniques are beginning to bridge localization and structure-function studies.

II. GENERAL CONSIDERATIONS

A. FIXATIVE

A prerequisite for the localization of cellular proteins and other macromolecules in microscopy is a fixation procedure that will immobilize the molecules and make them accessible to probes such as described below. Fixatives most commonly used for this purpose are (1) neutral buffered 4% paraformaldehyde; (2) methanol at –20°C; (3) 95° ethanol, and (4) Bouin's fixative.[1] The fixative employed and the time of fixation is entirely dependent on the individual receptor molecule, and the optimal fixation technique needs to be determined for each protein. Receptor molecules in tissue samples which retain their antigenic determinants after fixation may be paraffin embedded to improve morphology and resolution. Certain receptor molecules are sensitive to either fixation (or certain types thereof) or the temperature required for paraffin embedding, and therefore may need to be cryosectioned and/or air dried (nonfixed). We present below a Protocol for the preparation of 4% paraformaldehyde in phosphate-buffered saline (PBS) (see Protocol 1) which has proven suitable for most polypeptide ligand receptors. This fixative should be made fresh before every experiment.

PROTOCOL 1: PREPARATION OF 4% PARAFORMALDEHYDE

1. Mix 4 g paraformaldehyde with 90 ml distilled water.
2. Add 50 µl of 5 M NaOH.

3. Heat to 60°C with stirring until dissolved.
4. Add 10 ml 10 × PBS.
5. Cool and adjust pH to 7.4 with HCl.

B. PERMEABILIZATION OF CELLS

The cell and nuclear membranes form a barrier for diffusion of labeled ligands and antibodies and provide both problems and opportunities for the visualization of polypeptide ligand receptors in the cell. Visualization of polypeptide ligand receptors merely at the cell surface is not sufficient as polypeptide ligand receptors are predominantly located intracellularly, both before and after ligand stimulation and subsequent internalization.[2] For visualization of receptors at the cell surface only, the cells should be fixed with a suitable fixative (e.g., 4% paraformaldehyde in PBS and not alcohol-based fixatives) which does not cause disruption to the permeability of the membrane. Thus, visualization of the receptor with conjugated ligands or receptor antibodies reveals only cell surface receptors in nonpermeabilized cells. This manipulation can be used to advantage to study regulation of cell surface receptor expression and internalization (or degradation) at the cell surface. The most commonly used method for uniform permeabilization of the cell fixed with 4% paraformaldehyde is with a solution of 0.1% Triton X-100 (in PBS, for example) although other detergents can be used.[2]

The relatively high cholesterol content of the plasma membrane enables selective permeabilization with digitonin.[3] Thus, selective permeabilization with digitonin will leave the nuclear membrane and nuclear pore complexes impermeable to immunoglobulin molecules and also, in some cases, the membranes of the Golgi and endoplasmic reticulum.[3] When labeled ligands are utilized for visualization of receptor molecules, prior cooling and incubation of the cell at 4°C with the labeled ligand will label the cell surface receptor only. Incubation of cells with the labeled ligand at 37°C will result in visualization of cell surface-bound ligand and also internalized ligand. Since many ligand receptor complexes dissociate upon internalization,[4] this method may result in the visualization of free ligand and not the ligand receptor complex.

III. INDIRECT LIGAND BASED DETECTION TECHNIQUES

These methods are dependent on the availability of purified or recombinant ligand. They exploit the high affinity of the ligand receptor interaction to produce a specific labeling of the receptor molecule. Visualization of this labeling is performed by a variety of methods, several of which will be discussed here.

A. AUTORADIOGRAPHY

The earliest attempts to localize receptors at the tissue level were performed using radiolabeled ligands as probes and autoradiography as a detecting system.[5] The pioneering efforts of Fitzgerald,[6] and others,[7] resulted in techniques for the localization of steroids in tissue sections by light microscopic autoradiography (see Chapter 3). At present, autoradiography of radiolabeled ligands after incubation of tissue sections or after *in vivo* injection of the ligands remains an important tool in the study of receptor localization and function. The high resolution that can be obtained and the potential for quantification by grain counting/digitization[8] are important characteristics of this approach. Submacroscopic autoradiography can be used to perform autoradiography of thick sections or of a whole organ or even a whole body.[5] A review of recent trends in receptor autoradiography and instrumentation is available.[9]

1. Iodination of Ligands

We present below Protocol 2, based on the use of 1,3,4,6-tetrachloro-3-alpha, 6-alpha-diphenylglycoluril (IODOGEN) for iodination of IGF-I (for Insulin Growth Factor-I).[10] In our experience this is the easiest and most reliable method for iodination of hGH (for Growth Hormone) and can readily be applied for the iodination of other proteins. [125]I is oxidized to the tyrosine residues of the protein.

PROTOCOL 2: ^{125}I-LABELING OF IGF-I

1. Add to a 1-ml Eppendorf tube coated with the IODOGEN reagent:
2. 42 μl PBS
3. 3 μg IGF-1 (3 μl of a 1mg/ml solution)
4. 0.5 mCi [125]I (5 μl)
5. Allow reaction to proceed for 15 min at room temperature
6. Add 150 μl PBS
7. Load onto PD-10 column (Sepharose G-25) prewashed with PBS-2% BSA
8. Elute with PBS-2% BSA
9. Collect 15 × 1 ml fractions
10. Count 10 μl aliquot to detect free and incorporated radioactivity
11. Pool two highest radioactive incorporated fractions (in first peak)
12. Freeze and store at –20°C.

PROTOCOL 3: AUTORADIOGRAPHY OF IGF-I RECEPTOR IN TISSUE SECTION[11]

1. Tissue perfused with ice-cold saline.
2. Tissue removed and immersed in Freon 22 (–40°C), transferred to crushed dry ice for 10 min and stored at –70°C.
3. Mount 20-μm cryostat sections on gelatin-coated slides.
4. ^{125}I-IGF-1 dissolved to final concentration of 0.1 nM in 10 mM HEPES buffer containing 0.5% BSA, 0.025% bacitracin, 0.0125% N-ethylmaleimide, and 100 kIU/ml aprotinin, pH 7.6.
5. For determination of nonspecific binding unlabeled IGF-1 (100 nM) is mixed with the solution in 4 above.
6. Slides incubated in humidified chamber at 4°C for 24 h with solutions described in 4 and 5 above.

7. Slides rinsed in 3 × 1 min (100 ml) in ice-cold HEPES assay buffer, immersed in 5 s rinse of 0°C distilled water.
8. Slides rapidly dried on a warm plate (70°C) for 30 s.
9. Slides placed in direct contact with emulsion of LKB ultrofilm (in an X-ray film cassette) for 48 h.
10. Film developed for 4 min in D-19 at 20°C.

2. Autoradiography of Ligand Binding in Tissue Section

The autoradiographic technique (see Protocol 3) also has significant limitations.[5] Firstly, the technique is laborious and time consuming, requiring exposure times that can take up to several months. Furthermore, radioactive chemicals require special laboratory facilities and their use is becoming more and more restricted. As a consequence, alternative methods of ligand labeling have been developed. For polypeptide ligands, conjugates have been developed which retain a high ligand specificity and affinity and consequently can be used for receptor histochemistry. Some of these will be discussed below.

B. IMMUNOHISTOCHEMISTRY FOR DETECTION OF LIGAND

This technique allowed the first immunocytochemical demonstration of the presence of the GH and PRL receptor in different tissues. It is based on the binding of ligand to receptor in the tissue section or cell and subsequent immunocytochemical localization of the ligand specifically bound to the tissue section (see Protocol 4). One shortfall of this technique is that endogenous hormone, either produced by the cell or from internalized exogenous hormone, may confound the results and lead to a false receptor localization. It is therefore preferable to use a technique where the exogenous ligand is labeled or can be recognized specifically.

PROTOCOL 4: IMMUNOCYTOCHEMICAL DETECTION OF RECEPTOR USING LIGAND[12]

1. Deparaffinize, bring to TBS (Tris-buffered saline: 10 mM Tris-HCl, 150 mM NaCl, pH 7.4).
2. Incubate section with 1:30 dilution of normal sheep serum in TBS (0.05 M, pH 7.6) for 5 min (to reduce nonspecific protein binding).
3. Incubate section with ovine PRL or bovine GH (0.2 μg/ml in saline) for 60 min (to bind to free binding sites).
4. Incubate section with 1:30 dilution of normal sheep serum in TBS (0.05 M, pH 7.6) for 5 min (to reduce nonspecific protein binding).
5. Incubate section with rabbit anti-canine PRL (for prolactin) or anti-canine GH 1:50 dilution in TBS for 30 min.
6. Incubate section with 1:30 dilution of normal sheep serum in TBS (0.05 M, pH 7.6) for 5 min (to reduce nonspecific protein binding).
7. Incubate section with sheep anti-rabbit gamma globulin at 1:10 dilution in TBS for 30 min.

8. Incubate section with 1:30 dilution of normal sheep serum in TBS (0.05 M, pH 7.6) for 5 min (to reduce nonspecific protein binding).
9. Incubate sections with peroxidase-rabbit-anti-peroxidase (PAP) complex diluted 1:50 in TBS for 30 min at 4°C.
10. Incubate sections with 3-3′-diaminobenzidine tetrahydrochloride (0.2 mg/ml) and hydrogen peroxide (final concentration 0.03%) in TBS for 40 min at 4°C.
11. Wash in distilled water 3 × 5 min.
12. Dehydrate and mount.

All incubations with antigens and the different antisera were followed by 3 × 5 min washes in TBS.

C. FLUORESCENT LABELING OF LIGANDS

The bleaching of fluorochromes constitutes one of the most troublesome problems with this technique. The issue is especially important in confocal microscopy since the optical slices are performed sequentially. It is known that photobleaching is merely due to a reaction of molecular oxygen with the triplet excited state of the dye to produce highly reactive singlet oxygen.[13] As a general rule, rhodamine and its derivatives (lissamine rhodamine, Texas red, TRITC) are more photostable than their fluorescein counterparts.[14] However, there are several arguments which lead to preferring the fluorescein derivatives (see Protocol 5) in single labeling experiments:[15]

1. Firstly, the lower quantum yields of rhodamine conjugates result in a dimmer fluorescence than the corresponding fluorescein derivatives;
2. In immunofluorescence the rhodamine derivatives are especially sensitive to quenching effects with concentration when several dye molecules are linked to the protein;[16]
3. The 514-nm radiation of argon ion lasers is far from optimum to excite the rhodamine derivatives because their absorption maxima lies between 550 and 600 nm. This is not relevant if a mixed-gas krypton-argon ion laser is available since it has a 568 nm line;
4. The photomultiplier tubes which are used as detectors in the current laser-scanning confocal microscopes have a sensitivity to green fluorescence which is two to three times larger than sensitivity to green wavelengths.[17]

Attachment of fluorochromes to polypeptide ligands is technically simple and some examples are provided below (see Protocols 5 to 7).

PROTOCOL 5: FITC LABELING OF GROWTH HORMONE[18]

1. Add 50 μl hGH (20 μg GH in 0.05 M borate buffer, pH 9.3) to:
2. Fluorescein isothiocyanate 50 μl (2 μg in 0.05 M borate buffer, pH 9.3).
3. Leave 4 h at room temperature in dark.
4. Terminate reaction with 100 μl (375 μg) glycine.

5. Dialyze against 4 l PBS, pH 7.0 for 24 h at 4°C.
6. Spin at 13,000 × g in microfuge for 1 min to remove any precipitate.

PROTOCOL 6: CY-3 LABELING OF TRANSFERRIN[19]

1. Dissolve aliquot of Cy3.18 (Biological Detection Systems, Pittsburgh, PA) containing 80 nmol reactive dye with:
2. Solution of transferrin 1.9 mg/ml in 50 nM borate, pH 9.2 (750 µl).
3. Allow reaction to proceed for 30 min at room temperature.
4. Remove excess Cy3.18 by gel filtration on a PD-10 column equilibrated with PBS.

PROTOCOL 7: VISUALIZATION OF TRANSFERRIN RECEPTOR WITH FLUORESCENTLY LABELED LIGAND[19]

1. Cells grown on glass coverslips.
2. Cells washed 3 times with Ham's F-12 balanced salt solution with bicarbonate, supplemented with 5 mM HEPES, 100 µM deferoxamine, and 2 mg/ml ovalbumin, pH 7.4 (Ham's binding buffer).
3. Perform subsequent steps at 4°C if cell surface receptor visualization is required, otherwise at 37°C for visualization of internalized ligand receptor complex.
4. Cells incubated in 120 µl (20 µg/ml Cy3-labeled transferrin in Ham's binding buffer).
5. Cells placed on ice and washed 4 times with cold 150 mM NaCl; 20 mM HEPES, pH 7.5; 5 mM KCl; 1 mM $CaCl_2$; 1 mM $MgCl_2$ (medium 1). At this stage cell surface-bound ligand can be removed with mild acid wash if only internalized ligand receptor complex is to be observed.
6. Cells fixed with 2% paraformaldehyde for 3 min and rinsed 4 times with medium 1.
7. Cells mounted and observed under fluorescence microscope.

As oxygen is involved in photobleaching processes, anti-oxidizing substances added to the mounting medium constitute a first class of anti-fading agents; they are useful to lower the molecular oxygen partial pressure. The most widely used of these agents are DABCO (1,4-diazobicyclo-[2,2,2]-octane), n-propyl gallate, hydroquinone, and *p*-phenylenediamine. A second class of antifading agents are direct quenchers of the oxygen singlet state. Carotenoids appear to be the most efficient. Other reagents of this type are ascorbate, histidine, and reduced glutathione. Some commercially available mounting media contain an antifading agent.

D. BIOTINYLATED OF LIGANDS

Many different kinds of biotinyl derivatives are available (see Protocol 8).[20] For biotinylation of proteins, the N-hydroxysuccinimide ester of biotin or its water-soluble analogue N-hydroxysulfo-succinimide biotin are possibly the most often used.[21,22] Both of these biotinyl derivatives bind primarily to lysine residues under alkaline conditions. Variants with an extended spacer arm can be used to reduce the effect of steric hindrance.[23] Other classes of reactive biotin derivatives can be used to biotinylate other functional groups. Tyrosine or histidine residues can be labeled with *p*-diazobenzoyl biocytin.[24] Sulfhydryls can be biotinylated with N-iodoacetyl-N'-biotinyl-hexanedi-amine (iodoacetyl-biotin) or with N-[6-(biotinamido)hexyl]-3-(2-pyridyldithio) propi-onamide (biotin-HPDP).[20] Sugar residues on glycoproteins are biotinylated using biotin hydrazide.[25] Several Protocols are available in order to visualize the receptors with biotinylated ligands (see Protocols 9 to 10).[26,27]

PROTOCOL 8: BIOTINYLATION OF HUMAN GROWTH HORMONE[20]

1. Dilute NHS-LC-biotin (sulfosuccinimidyl-6-(biotinamido) hexanoate in 0.05 M carbonate-bicarbonate buffer pH 9.
2. hGH diluted in 0.05 M carbonate-bicarbonate buffer.
3. Incubate hGH with NHS-LC-biotin at 1:5, 1:20, or 1:100 molar ratio in 0.05 M carbonate-bicarbonate buffer for 2 h at room temperature
4. Free biotin molecules in sample removed by size exclusion chromatography on PD-10 column or by extensive dialysis in PBS.

PROTOCOL 9: DETECTION OF BIOTINYLATED LIGAND (hGH) ON CELLS[26]

1. Grow cells to monolayer on glass coverslips in six-well plates.
2. Cells fixed in 1:1 (v/v) acetone:methanol for 1 min.
3. Incubation with 1% (v/v) hydrogen peroxide in methanol for 20 min at room temperature to block endogenous peroxidase activity.
4. Wash with PBS pH 7.4.
5. Incubation of cells in PBS-1% (w/v) BSA with biotinylated ligand (1 to 10 μM) for 90 min at room temperature.
6. Control slides incubated with:
    (4) above and unlabeled ligand at 50 M excess
    incubation without biotinylated ligand
7. Wash four times with PBS pH 7.4.
8. Incubation with labeled avidin (horse radish peroxidase (HRP) or fluorescent label), e.g., avidin-biotin-streptavidin HRP complex at 1:150 dilution in PBS-1% BSA for 1 h at room temperature.
9. Wash four times with PBS ph 7.4
10. Incubation with 3-3'-diaminobenzidine (DAB) (0.05% w/v) and 0.1% hydro-gen peroxide in PBS for 10 min.
11. Reaction terminated in water.
12. Mount and view.

PROTOCOL 10: VISUALIZATION OF RECEPTORS WITH BIOTINYLATED LIGANDS[27]

1. Cells grown on coverslips in six-well plates.
2. Cells incubated with 2 nM pHAcAL (Btn)VP (biotinylated [1-phenylacetyl, 2-*o*-methyl-d-tyrosine, 6-arginine, 8-arginine, 9-lysinamide] vasopressin) and 0.2 µM DPH (1,6-diphenyl-1,3,5-hexatriene; used as fluorescent membrane marker).
3. Control incubations:
   Absence of biotinylated ligand
   With biotinylated ligand and 20 nM unlabeled ligand
4. Cells fixed by addition of formaldehyde to a final concentration of 1% (v/v).
5. Cultures allowed to cool to room temperature for 20 min.
6. Coverslips removed and washed in PBS pH 7.4
7. Incubation with avidin-Texas Red (20 µg/ml) in PBS for 30 min at room temperature in the dark.
8. Wash four times in PBS.
9. Mount in antifade mounting media (commercially available or glycerol/100 mM sodium phosphate, pH 7.5 (9:1) plus 25 mg/ml 1,4-diazobicyclo-(2,2,2)-octane).

E. DISADVANTAGES OF LIGAND BINDING APPROACH

One limitation of the ligand binding approach remains the inherent impossibility of detecting ligand-bound receptor molecules. Hence a complete image of receptor localization cannot be provided by ligand-binding studies only.[5] This is especially true of some polypeptide ligand receptors where the ligand essentially binds to the receptor molecule in an irreversible manner.[28] This is the pivotal reason why fundamental receptor studies should always combine the ligand binding approach with more direct receptor localization techniques such as immunocytochemistry. Also, the conjugation of ligands with molecules allowing visualization may interfere with ligand conformation and/or alter or abolish its receptor binding characteristics. Such considerations need to be addressed for each individual ligand.

IV. DIRECT RECEPTOR DETECTION TECHNIQUES

The basis of these techniques is the use of a specific receptor probe, namely antibody, to localize the receptor. One variation of this is to introduce an artificial epitope (epitope tag) by cDNA cloning onto the mature receptor molecule and hence localize this epitope with a specific monoclonal antibody. Direct receptor detection has provided a wealth of data on the tissue, cellular and subcellular localization of receptor molecules, and receptor trafficking within the cell. The method of visualization is divided into two broad categories, enzyme-linked immunocytochemistry and immunofluorescence.

A. ENZYME-LINKED IMMUNOCYTOCHEMISTRY

Enzyme-linked immunocytochemistry utilizes a variety of enzymes such as horseradish peroxidase (see Protocol 11), glucose oxidase, and alkaline phosphatase, again with a variety of substrates for different color development. Immunofluorescence also utilizes the same variety of fluorochromes as discussed previously.

PROTOCOL 11: IMMUNOHISTOCHEMISTRY WITH HORSERADISH PEROXIDASE (HRP)

1. Eliminate endogenous peroxidase activity with 0.5% (v/v) H_2O_2 in PBS for 15 min at 20°C.
2. Wash in PBS.
3. Eliminate nonspecific protein binding by incubation with 10% (v/v) normal serum for 1 h at 20°C.
4. Wash in PBS.
5. Incubate overnight with primary antibody at predetermined dilution.
6. Wash four times in PBS.
7. Incubate for 2 h at 4°C with biotinylated secondary antibody diluted according to manufacturer's instructions in PBS-1% BSA.
8. Wash four times in PBS.
9. Incubate with avidin (streptavidin)-biotin HRP complex diluted according to manufacturer's instructions in PBS-1% BSA.
10. Wash four times in PBS.
11. Visualization with 0.05 mg 3-3′-diaminobenzidine (DAB)/ml in PBS containing 0.1% H_2O_2 for 1 to 5 min.

Double-staining with diaminobenzidine can be achieved by performing the first DAB reaction in the presence of 2.5% (w/v) nickel ammonium sulfate to give a black precipitate. The second reaction can be performed with diaminobenzidine in the absence of nickel ammonium sulfate to give a brown precipitate.[29]

B. IMMUNOFLUORESCENCE AND CONFOCAL MICROSCOPY

The use of fluorescent probes allows the analysis and quantification of receptor localization by confocal laser scanning microscopy (CLSM). Compared to conventional transmission light microscopy, CLSM offers several advantages:

1. CLSM provides thin optical (single or serial) sections of the specimen, avoiding the risk of projection artifacts, and also presents all components in the particular focal plane in focus, regardless of size of the component;
2. CLSM provides better resolution than conventional microscopy, especially along the optical axis but also laterally;
3. CLSM presents data in a digitalized form which directly allows various kinds of image analysis such as subtraction of one image from another and quantification (measurement of relative fluorescent intensities);

4. CLSM allows for two- or three-dimensional reconstruction of cells from serial sectioning data;
5. CLSM provides laser beam excitation of the specimen with a separate monochromatic wavelength for each fluorochrome, compared to a wavelength excitation interval in conventional microscopy. This may constitute an advantage in double-staining experiments.

Thus CLSM offers an intermediate step between conventional light and electron microscopy. It can be used to study receptor trafficking within a cell, coupled with double-labeling experiments to identify specific markers for certain subcellular compartments (see Protocol 12).

PROTOCOL 12: IMMUNOFLUORESCENCE FOR CONFOCAL LASER SCANNING MICROSCOPY

1. Cells grown on glass coverslips in six-well plates.
2. Cells washed in PBS and fixed in 2% paraformaldehyde-0.05% glutaraldehyde in 0.1 M phosphate buffer, pH 7.4.
3. Cells permeabilized in 0.1% Triton X-100 for 4 min at room temperature.
4. Cells washed extensively in PBS.
5. Incubation for 10 min in PBS-1% BSA to block nonspecific protein binding.
6. Incubation with primary antibody at predetermined dilution (in PBS-1% BSA) for 1 h at room temperature.
7. Cells washed extensively (4 × 5 min) in PBS-1% BSA.
8. Incubation with fluorescently labeled second antibody directed against species origin of primary antibody (e.g., Cy3 conjugated affinipure goat anti-rabbit IgG) diluted at predetermined dilution in PBS-1% BSA for 1 h at room temperature.
9. Cells washed extensively (4 × 5 min) in PBS-1% BSA.
10. Mount in commercially available antifading mounting medium.
11. View with confocal laser scanning microscope.

C. LOCALIZATION OF RECEPTOR WITH EPITOPE TAG

The technique of epitope tagging a receptor molecule to study receptor trafficking within a cell has become widespread. The technique is based on the fact that a small, unique antigenic peptide sequence can be inserted by cDNA cloning at either the carboxyl or amino terminus of the receptor molecule. Use of this technique therefore requires either transient or stable cDNA transfection into cells or generation of transgenic animals. The peptide sequence can subsequently be detected by immunocytochemistry or immunofluorescence with the use of a specific monoclonal antibody. There are several different epitope tags described to date which have been used successfully in receptor studies.[30,31] Receptor epitope tagging studies are extremely useful when different receptor mutations need to be studied, as the problem of a low-level endogenous background receptor is avoided.

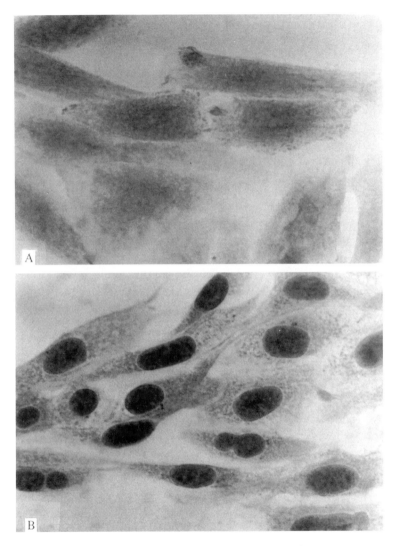

FIGURE 1   Immunocytochemical demonstration of nuclear-associated GH receptor (GHR) and ligand-induced nuclear translocation of GHR in cultured cells transfected with GHR cDNA. The GHR was detected with polyclonal antisera raised against recombinant GHR intracellular domain (ICAb) and use of the biotin-streptavidin-horse-radish peroxidase technique. A: ICAb immunore-activity in CHO-GHR$_{1-638}$ cells cultured in serum-free medium. Note the predominant cytoplasmic immunoreactivity and relative paucity of nuclear immunoreactivity. B: ICAb immunoreactivity in CHO-GHR$_{1-638}$ cells cultured in serum-free medium but treated for 5 min with 100 nM hGH. Note the shift of cytoplasmic immunoreactivity to the nuclear compartment. C: ICAb immunoreactivity in CHO-GHR$_{1-638}$ cells cultured in serum-free medium and treated for 60 min with 100 nM hGH. Note the shift of cytoplasmic immunoreactivity to the nuclear compartment.

The one disadvantage of this technique is that artificial or forced expression of receptor in a cell may lead to overexpression and a localization different to that of the endogenously expressed receptor.

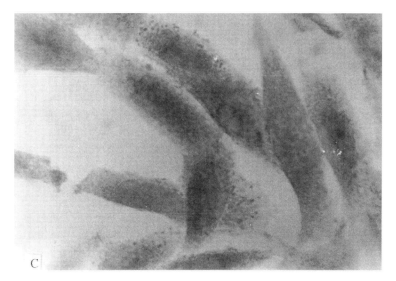

FIGURE 1   (continued)

V. ADVANCED DETECTION TECHNIQUES

It should be pointed out, although it is not in the scope of this article, that several advanced detection techniques exist for the study of receptor molecules by light microscopy. Two of these such techniques are outlined below.

The phenomenon of receptor oligomerization can be studied based on the quantitative determination of fluorescence resonance energy transfer (FRET) in the microscope (i.e., with retention of spatial resolution).[32] The basis of FRET is the transfer of excited-state energy from a donor fluorescent group to an appropriate light-absorbing acceptor molecule. This process takes place only over very short distances (generally less than 10 nm) with an efficiency dependent upon the inverse 6th power of the donor-acceptor separation. Thus, FRET is a sensitive and direct measure of very close molecular interactions. As an example, luminescent probes have been covalently linked to the single terminal amino group of EGF (for Epidermal Growth Factor) as donor and acceptor molecules. This probe has been used in studies on the distribution, internalization, and mechanism of activation of the EGF receptor on living cells or in isolated plasma membranes,[33] and also, using FRET, in studies on EGF receptor aggregation and oligomerization.[34,35]

Development of single particle tracking (SPT) by nanovid microscopy has made it possible to observe the movements of individual (or a small number of) membrane receptor molecules [36] by labeling proteins with colloidal gold particles (20 to 40 nm in diameter) on the living cell surface using video-enhanced contrast microscopy.[37,38] The SPT method is unique in that it can reveal the mechanisms by which the motion of a single protein molecule is regulated in the plasma membrane. The nanometer-level precision of SPT is particularly useful for studying regulation mechanisms that act at the submicron scale, such as the membrane skeleton.[39] Fluorescence SPT has also been developed.[40]

VI. CONCLUSION

Understanding the biology of any polypeptide ligand receptor requires an intimate knowledge of its cellular localization and also the dynamics of its cellular location upon ligand stimulation. The multiplicity of techniques now available for polypeptide ligand receptor localization has greatly facilitated the identification of target cells and the understanding of receptor trafficking within the cell for many polypeptide ligands. However, more complex phenomena remain to be understood. The new technologies promise to bridge both localization and function, to obtain a more complete "picture" of receptor biology.

REFERENCES

1. Gabriel L.K., Franken D.R., van der Horst G. and Kruger T.F., Localization of wheat germ agglutinin receptors on human sperm by fluorescence microscopy: utilization of different fixatives. *Arch. Androl.,* 33, 77, 1994.
2. Lobie P.E., Morel G., Mertani H., Morales-Bustos O., Wood T.T.J., Waters M.J. and Norstedt G., The growth hormone receptor, growth hormone and the nucleus. *Endocrinol. Metab.,* 2, 61, 1994.
3. Söderqvist H. and Hallberg E., The large C-terminal region of the integral pore membrane protein, POM121, is facing the nuclear pore complex. *Eur. J. Cell Biol.,* 64, 186, 1994.
4. Shah N., Zhang S., Harada S., Smith R.M. and Jarrett L., Electron microscopic visualization of insulin translocation into the cytoplasm and nuclei of intact H35 hepatoma cells using covalently linked nanogold insulin. *Endocrinology,* 136, 2825, 1995.
5. Bosman F.T., Histochemical techniques for receptor detection. In. Progress in *Histo- and Cytochemistry,* Vol. 26, Histochemistry of Receptors. Graumann, W. and Drukker, J., Eds., Fischer Verlag, Stuttgart, 1992, pp 30.
6. Fitzgerald P.J., Dry mounting autoradiographic technique for intracellular localization of water soluble compounds in tissue sections. *Lab. Invest.,* 10, 846, 1961.
7. Stumpf W.E. and Roth L.J., Vacuum freeze drying of frozen sections for dry-mounting, high resolution autogradiography. *Stain. Technol.,* 39, 219, 1964.
8. Frederik P.M., Klepper D., Van der Vusse G.J. and Van der Molen J.J., Dynamics of steroid uptake in rat testis studied by quantitive autoradiography. *Mol. Cell. Endocrinol.,* 5, 123, 1976.
9. Palacios J.M., Mengod G., Vilaro M.T. and Ramm P., Recent trends in receptor analysis techniques and instrumentation. *J. Chem. Neuroanat.,* 4, 343, 1991.
10. Salacinski P.R., McLean C., Sykes J.E., Clement-Jones V.V. and Lowry P.J., Iodination of proteins, glycoproteins and peptides using a solid phase oxidizing agent, 1,3,4,6-tetrachloro-3 alpha, 6 alpha-diphenylglycoluril (IODOGEN). *Anal. Biochem.,* 117, 136, 1981.
11. Bohannon N.J., Figlewicz D.P., Corp E.S., Wilcox B.J., Porte D. and Baskin D.G., Identification of binding sites for an insulin like growth factor (IGF1) in the median eminence of the rat brain by quantitive autoradiography. *Endocrinology,* 119, 943, 1986.
12. El Etreby M.F. and Mahrous A.T., Immunocytochemical technique for detection of prolactin (PRL) and growth hormone (GH) in hyperplastic and neoplastic lesions of dog prostate and mammary gland. *Histochemistry,* 64, 279, 1979.
13. Tsien R.Y. and Waggoner A., Flourophores for confocal microscopy: photophysics and photochemistry. In *Handbook of Biological Confocal Microscopy,* Pawley J.B., Ed., IMR Press, Madison, WI, 1990, pp.169.
14. McKay I.C., Forman D. and White R.G., A comparison of fluorescein isothiocyanate and lissamine rhodamine as labels for antibody in the fluorescent antibody technique. *Immunology,* 43, 591, 1981.
15. Laurent M., Johannin G., Gilbert N., Lucas L., Cassio D., Petit P.X. and Fleury A., Power and limits of laser scanning confocal microscopy. *Biol. Cell,* 80, 229, 1994.

16. Clark-Brelje T., Wessendorf M.W. and Sorenson L., Multi-color laser scanning confocal immunofluorescence microscopy: practical application and limitations. *Methods Cell Biol.*, 38, 97, 1993.

17. Majlof L. and Forsgren P.O., Confocal microscopy: important considerations for accurate imaging. *Methods Cell Biol.*, 38, 79, 1993.

18. Eshet R., Peleg S. and Laron Z., Direct visualization of binding, aggregation and internalization of human growth hormone in cultured human lymphocytes. *Acta Endocrinol.*, 107, 9, 1984.

19. Marsh E.W., Leopold P.L., Jones N.L. and Maxfield F.R., Oligomerized transferrin receptors are selectivley retained by a lumenal sorting signal in a long lived endocytic recycling compartment. *J. Cell Biol.*, 129, 1509, 1995.

20. De Jong M.O., Rozemuller H., Bauman J.G.J. and Visser J.W.M., Biotinylation of interleukin 2 (IL-2) for flow cytometric analysis of IL-2 receptor expression. *J. Immunol. Methods,* 184, 101, 1995.

21. Newman W., Beall L.D., Bertolini D.R. and Cone J.L., Modulation of TGF-beta type 1 receptor: flow cytometric detection with biotinylated TGF-beta. *J. Cell. Physiol.*, 141, 170, 1989.

22. Pieri I. and Barritault D., Biotinylated basic fibroblast growth factor is biologically active. *Anal. Biochem.*, 195, 214, 1991.

23. Hnatowich D.J., Virzi F. and Rusckowski M., Investigations of avidin and biotin for imaging applications. *J. Nucl. Med.*, 28, 1294, 1987.

24. Wilcheck M., Ben H.H. and Bayer E.A., *p*-Diazobenzoyl biocytin — a new biotinylating reagent for the labeling of tyrosines and histidines in proteins. *Biochem. Biophys. Res. Commun.*, 138, 872, 1986.

25. Wognum A.W., Lansdorp P.M., Humphries R.K. and Krystal G., Detection and isolation of the erythropoietin receptor using biotinylated erythropoietin. *Blood*, 76, 697, 1990.

26. Bentham J., Ohlsson C., Lindahl A., Isaksson O. and Nilsson A., A double staining technique for detection of growth hormone and insulin like growth factor 1 binding to rat tibial epiphyseal chondrocytes. *J. Endocrinol.*, 137, 361,1993.

27. Howl J., Wang X., Kirk C.J. and Wheatley M., Fluorescent and biotinylated linear peptides as selective bifunctional ligands for the V1a vassopressin receptor. *Eur. J. Biochem.*, 213, 711, 1993.

28. Waters M.J., Lusins S. and Friesen H.G., Immunological and physicochemical evidence for tissue specific prolactin receptors in the rabbit. *Endocrinology,* 115, 1, 1984.

29. Hancock M.B., Visualization of peptide immunoreactive processes on serotonin immunoreactive cells using two-colour immunoperoxidase staining. *J. Histochem. Cytochem.*, 32, 311,1984.

30. Arden J.R., Segredo V., Wang Z., Lameh J. and Sadee W., Phosphorylation and agonist-specific intracellular trafficking of an epitope tagged µ-opioid receptor expressed in HEK 293 cells. *J. Neurochem.*, 65, 1636, 1995.

31. Brown P.M., Tagari P., Rowan K.R., Yu V.L., O'Neill G.P., Russel-Middaugh C., Sanyal, G., Ford-Hutchinson A.W. and Nicholson D.W., Epitope labeled soluble human interleukin-5 (IL-5) receptors. *J. Biol. Chem.*, 270, 29236, 1995.

32. Jovin T.M. and Arndt-Jovin D.J., Luminescence digital imaging microscopy. *Annu. Rev. Biophys. Biophys. Chem.*, 18, 271, 1989.

33. Greenfield C., Hiles I., Waterfield M.D., Federwisch M., Wollmer A., Blundell T.L. and McDonald N., Epidermal growth factor binding induces a conformational change in the external domain of its receptor. *EMBO J.,* 8, 4115, 1989.

34. Zidovetski R., Johnson D.A., Arndt-Jovin D.J. and Jovin T.M., Rotational mobility of high affinity epidermal growth factor receptors on the surface of living A431 cells. *Biochemistry,* 30, 6162, 1991.

35. Gadella T.W.J. and Jovin T.M., Oligomerization of epidermal growth factors on A431 cells studied by time resolved fluorescence imaging microscopy. *J. Cell Biol.*, 129, 1543, 1995.

36. Sako Y. and Kusumi A., Barriers for lateral diffusion of transferrin receptor in the plasma membrane as characterized by receptor dragging by laser tweezers: fence versus tether. *J. Cell Biol.*, 129, 1559, 1995.

37. DeBrabander M., Nuydens R., Geuens G., Moeremans M. and De Mey J., The use of submicroscopic particles combined with video contrast enhancement as a simple molecular probe for the living cell. *Cell Motil. Cytoskeleton*, 6, 105, 1986.

38. Geerts H., DeBrabander M. and Nuydens R., Nanovid microscopy. *Nature*, 351, 765, 1991.

39. Sako Y. and Kusumi A., Compartmentalized structure of the plasma membrane for lateral diffusion of receptors as revealed by nanometer level motion analysis. *J. Cell Biol.*, 125, 1251, 1994.
40. Ghosh R.N. and Webb W.W., Automated detection and trafficking of individual and clustered cell surface low density lipoprotein receptor molecules. *Biophys. J.* 66, 1301, 1994.

Chapter 10

VISUALIZATION OF PEPTIDE RECEPTORS IN TUMORS BY IMMUNOHISTOCHEMISTRY

Tomás García-Caballero
Rosalía Gallego
Máximo Fraga
Andrés Beiras

CONTENTS

I. INTRODUCTION

The presence of receptors for different peptides has been demonstrated in endocrine and nonendocrine tumors by classical biochemical,[1-3] immunologic,[4,5] and molecular biologic techniques.[6-15] In the last few years, even a technique that allows the *in vivo* visualization of peptide receptors by radiodinated peptide analogs has been developed.[16-18]

The morphologic techniques, unlike methods that deal with tissue homogenates, allow the study of the distribution of peptide receptors in tumors. The morphological techniques commonly employed for this purpose were radioautography[18,19] and *in situ* hybridization.[13,20,21] In recent years, the introduction of polyclonal and mono-clonal antibodies to peptide receptors allowed the immunohistochemical detection of these receptors at cellular and subcellular levels with a high morphologic resolution.[21-27] An important advantage of this technique is the use of the tissue samples routinely processed for microscopic diagnosis.

The interest of the study of peptide hormone receptors in human tumors lies in the fact that these hormones, as well as growth factors and cytokines, can interact with specific membrane receptors of tumor cells. This interaction triggers a cascade of intracellular signals resulting in activation or repression of genes which control cell proliferation.[28,29] In this way, it was reported that various peptide hormones may stimulate (prolactin, growth hormone, growth hormone-releasing hormone, gastrin-releasing peptide, gastrin/cholecystokinin, arginine vasopressin, etc.)[13,30-38] or inhibit (somatostatin)[8,39,40] the growth of different types of tumors. This fact raises the possibility for using antagonists (if the peptide enhances the growth) or analogs (if it is inhibitory) of these peptides in the treatment of tumors.[8,17,37,41-43] Moreover, it was demonstrated that peptide analogs can be internalized by tumor cells *in vitro*.[17,44] This important finding provides a potential method to target therapeutic agents (radionuclides or bacterial toxins, for example) to tumor cells that overexpress peptide receptors.[29,45] In fact, radiotherapy using peptide analogs coupled to alpha- or beta-emitting radionuclides has been carried out.[46] In these novel therapies, the responsiveness of the tumors depends on the presence of the peptide receptors. Hence, the demonstration of the receptors will be necessary before the establishment of therapeutic protocols, as is routinely performed for estrogen and progesterone receptors in breast carcinomas.

In this chapter an immunohistochemical procedure suitable for both cells and tissue samples will be given to demonstrate peptide receptors in tumors.

II. PREPARATION OF SAMPLES

A. CELLS

Cytospin preparations of tissue culture cells (MCF-7 human breast cancer cells) or fine-needle aspirations suspended in 1.5 ml of 0.1 M phosphate-buffered saline (PBS), pH 7.6, are made in a Shandon centrifuge at 1500 rpm for 10 min.

Slides coated with 3-aminopropyl-triethoxysilane (APES, Fluka No. 09324) are used to prevent cells from becoming detached (see Protocol 1).

PROTOCOL 1: SLIDE PREPARATION

1. Dip slides in 2% APES in acetone.
2. Dip in acetone (2 × 2 min).
3. Dip in distilled water (2 × 2 min).
4. Dry overnight at 50°C.

The cytospin preparations are air dried and fixed in:

1. Buffered neutral formalin solution 10% (15 min):
   Sodium dihydrogen phosphate monohydrate (Merck No. 106346) 4 g
   Disodium hydrogen phosphate anhydrous (Merck No. 106586)   6.5 g
   37 to 40% formalin (Merck No. 104003)   100 ml
   Distilled water   900 ml
2. PBS (5 min).
3. Absolute methanol (4 min at –20°C).
4. Acetone (2 min at -20 C).
5. PBS (2 × 5 min).

After fixation, the cytospin preparations are ready for the immunohistochemical procedure. For convenience, they may be stored for up to 4 months at –20°C (see Protocol 2).

PROTOCOL 2: SOLUTION FOR CYTOSPIN PREPARATION STORAGE

Sucrose (Merck No. 107687)   42.8 g
Magnesium chloride hexahydrate (Merck No. 105833) 0.7 g
PBS   Adjust to 250 ml
Glycerol (Sigma No. G-5516)   250 ml

B. TISSUE SAMPLES
The specimens are immersion fixed (see Protocol 3).

PROTOCOL 3: IMMERSION FIXATION

1. Buffered neutral formalin solution 10% for 24 h, or
2. Bouin's solution for 4 to 24 h (depending on the size of the sample), prepared as follows:
   Picric acid (Merck No.100623), saturated aqueous solution   75 ml
   Formalin 37 to 40% (Merck No. 104003)   25 ml
   Glacial acetic acid (Merck No. 100063)   5 ml

After fixation, the samples are dehydrated in a graded ethanol series, cleared in xylene, and embedded in paraffin (see Protocol 4).

PROTOCOL 4: PARAFFIN EMBEDDING

1. 50° Ethanol (4 h) (only for samples fixed in Bouin's solution in order to remove the picric acid)
2. 70° Ethanol (30 min)
3. 96° Ethanol (30 min)
4. 100° Ethanol (30 min)
5. 100° Ethanol (2 × 1 h)
6. Xylene (2 × 30 min)
7. Paraffin (2 × 2 h at 60°C)

Sections 5 μm thick are mounted on silanized slides (see Protocol 1) and dried overnight at 50°C.

III. IMMUNOHISTOCHEMICAL PROCEDURE

A. ANTIGEN RETRIEVAL

Antigen retrieval or heat-induced epitope retrieval is a well-established method of improving the detectability of antigenic determinants masked by formalin fixation (see Protocol 5).[47-49]

PROTOCOL 5: ANTIGEN RETRIEVAL USING MICROWAVES

1. Cytospin preparations are washed in PBS (2 x 5 min) and tissue sections are dewaxed in xylol and hydrated through graded ethanols.
2. Slides are then placed in plastic Coplin jars containing 0.01 M sodium citrate buffer (pH 6.0) prepared as follows:
   Trisodium citrate (Merck No. 111037) 2.94 g
   Distilled water 1 l
   Adjust pH to 6.0 using a 1 N hydrochloric acid solution (to prepare this solution add 835 μl of 37% hydrochloric acid to 9.165 ml of distilled water).
3. Jars are heated in a conventional microwave oven for two 5-min cycles at 800 W. Between cycles, it is necessary to check on the fluid level in the jars and to replace the volume evaporated with distilled water.
4. After heating, the slides are allowed to cool down to room temperature over a period of 20 min.

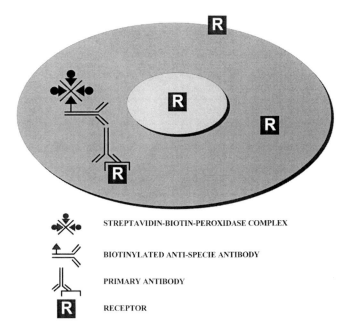

STREPTAVIDIN-BIOTIN-PEROXIDASE COMPLEX

BIOTINYLATED ANTI-SPECIE ANTIBODY

PRIMARY ANTIBODY

R RECEPTOR

FIGURE 1 Schematic representation of the principle of the SABC method.

B. STREPTAVIDIN-BIOTIN-PEROXIDASE COMPLEX METHOD
The method commonly used is described in Protocol 6 (Figure 1).

PROTOCOL 6: STREPTAVIDIN-BIOTIN-PEROXIDASE COMPLEX METHOD

1. Encircle the sections with a water-repellent material (Dako Pen, Dako No. S2002).
2. Rinse in PBS (2 × 5 min).
3. Incubate in monoclonal or polyclonal antibodies to peptide receptors optimally diluted in PBS with 0.1% bovine serum albumin (Sigma No. A-2153) and 0.1% Triton X-100 (Merck No. 108603) (1 h at room temperature). In our case we use: (a) B6.2 mouse monoclonal antibody (IgG1) to prolactin receptor (PRLR), prepared against a membrane enriched fraction of human metastatic breast cancer,[23] and (b) murine monoclonal antibody (MAb) 263 (IgG1) raised against the extracellular domain of the growth hormone receptor (GHR).[50] The dilution used for both antibodies was 1:500.
4. Rinse in PBS (2 × 5 min).
5. Block endogenous peroxidase by 3% hydrogen peroxide (Merck No. 107209) in PBS (10 min).
6. Rinse in PBS (2 × 5 min).

7. Incubate in biotinylated goat antibody to mouse/rabbit immunoglobulins (Duet, Dako No. K492) at 1:100 in PBS with 2% normal goat serum (Dako No. X0907) (30 min).
8. Rinse in PBS (2 × 5 min).
9. Incubate in streptavidin-biotin-peroxidase complex (Duet) prepared 30 min before use adding 10 μl of streptavidin and 10 μl of biotin-peroxidase to 1 ml PBS (30 min).
10. Rinse in PBS (2 × 5 min).
11. Develop in 0.06% (w/v) 3,3' diaminobenzidine tetrahydrochloride (DAB) containing 0.003% (v/v) hydrogen peroxide (10 min). DAB solution is prepared by dissolving 1 DAB-buffer tablet (Merck No. 102924) in 10 ml of distilled water.
12. Rinse in running water.
13. Dehydrate (in ethanol), clear (with xylene), and mount (with a permanent medium).

C. CONTROLS

Adequate specificity controls are essential for validation of the immunostaining results. The most relevant controls that must be performed in parallel with the test sections are the following:[51]

1. Substitution of the monoclonal antibody with ascites or supernatant obtained from nonsecreting hybridomas, and substitution of the polyclonal antibody with the preimmune serum from the animal employed to obtain the antiserum.
2. Substitution of the monoclonal antibody with unrelated monoclonal antibodies of the same immunoglobulin class, and substitution of the polyclonal antibody with unrelated antisera.
3. Preadsorption of the diluted antibody (overnight at 4°C) with an excess of its specific antigen (immunostaining should be abolished) and similar quantities of related substances (a positive immunostaining will be expected). If pure antigen is unavailable, preadsorption can be achieved by incubation with a tissue which contains a great concentration of the antigen.
4. Incubate only with DAB-hydrogen peroxide to test for endogenous peroxidase.
5. Incubate only with streptavidin-biotin-peroxidase complex revealed to test for nonspecific binding of streptavidin.

IV. TECHNICAL COMMENTS

For the immunocytological study, culture cells grown on glass coverslips, smears, and cytospin preparations can be used. In our experience, cytospin preparations are more convenient for routine use than glass coverslips and they are found to have no background staining, unlike smears of aspirated cells.

The fixation Protocol proposed for cytospin preparations is the same one that we routinely use for detection of estrogen and progesterone receptors in frozen sections of breast carcinomas. Fixation of cytological specimens with only 95° ethanol (for 30 min) or acetone (for 10 min) resulted in worse immunostaining. For tissue samples, good results were obtained with both fixatives employed, but a more intense immunostaining was observed with Bouin's liquid than with 10% formalin. On the other hand, no good results were obtained in lymphoid tissues fixed in B5 (buffered formaldehyde sublimate).

The high-temperature antigen retrieval procedure is not essential, but more intense immunostaining was obtained when it was performed, even in cytospin preparations. For this purpose, a pressure cooker or a steam sterilizer instead of the microwave were also used. The results obtained were of similar intensity as immunostaining, but in our hands the tissue preservation was in general poorer.

The incorporation of 0.1% Triton X-100 into the primary antibody diluent enhances the penetration of the immunoreagents and improves the immunostaining, not only in cytospin preparations, but also in ordinary paraffin sections.

To avoid repeatedly thawing and refreezing of primary antibodies they can be mixed with glycerol in equal shares and stored at −20°C (not colder because they will become frozen). The advantages of this kind of antibody storage over division into small aliquots are

1. To save time,
2. To save space in the freezer, and
3. To avoid the risk of thawing and refreezing if the electricity supply fails.

For the primary antibody we prefer to incubate sections for 1 h at room temperature instead of overnight incubation at 4°C in order to save time. The results in our experience are similar, with a very good ratio between signal and background noise. The advantage of overnight incubation is the possibility of increasing the dilution of the primary antibody. On the other hand, incubation at 37°C (for 30 min or 1 h) in this step amplifies the immunostaining but it also increases background noise.

The use of a biotinylated secondary antibody that reacts equally well with mouse and rabbit immunoglobulins means that only one secondary reagent is required to detect mouse or rabbit primary antibodies (we even used it with rat primary antibodies with good performance). The results were identical to those obtained with specific biotinylated antibodies. Apart from the economy, the advantage of this universal antibody lies in the fact that it is unnecessary to separate sections to be incubated with monoclonal and polyclonal antibodies, avoiding possible mistakes. The addition of normal serum from the same species used to obtain the secondary antibody (in our case normal goat serum) reduces background without affecting immunostaining.

We usually quench endogenous peroxidase after the primary antibody incubation to avoid the risk that the treatment with hydrogen peroxide could damage the antigen epitopes.

The use of DAB tablets is a convenient and safe method of dealing with DAB, avoiding the necessity of continually weighing this carcinogenic agent (and the risk of inhalation of this powder).

Controls performed by replacement of the primary antibody with PBS in our experience always result in a negative immunostaining that not necessarily validates the specificity of the antibody.

Endogenous biotin is naturally present in human tissues, particularly in liver, lung, spleen, kidney, breast, adipose tissue, and brain. Binding of avidin or strepta-vidin to endogenous biotin could result in false-positive staining. However, it was reported that this problem has not occurred when using the avidin-biotin complex method[52] and our experience with SABC confirms this observation.

Unspecific immunostaining of eosinophil granulocytes is generally observed in spite of the quench of endogenous peroxidase with highly concentrated hydrogen peroxide (even mixed with methanol and periodic acid). However, the myeloperox-idase activity of granulocytes is so obvious that blocking may be unnecessary. In case it is very troublesome, we recommend the substitution of peroxidase for alkaline phosphatase as the reporter molecule.

V. RESULTS AND DISCUSSION

GHR and PRLR immunoreactivity was found in cytospin preparations and paraffin sections of a variety of tumors tested. Coexpression of both receptors was the rule in the samples studied.

In cytospin preparations of MCF-7 culture cells of human breast carcinoma, immunoreactivity was intense in the cytoplasms and weak in the nuclei (Figure 2). No immunostaining was found in controls performed by substitution of the primary antibody with unrelated antibodies or normal mouse serum (Figure 3).

In paraffin sections of hepatocarcinomas, positivity for GHR and PRLR was also observed in both cytoplasms and nuclei (Figures 4 and 5), although in some cases only cytoplasmic immunoreactivity was found (not shown).

Pituitary adenomas also presented differences in the cellular distribution pattern of immunostaining. In some cases the immunoreactivity was primarily found in the nuclei (Figure 6), whereas in other cases it was mainly found in the cytoplasms (Figure 7).

Meningiomas showed GH- and PRL-receptor immunoreactivity primarily in the cytoplasms (Figure 8) and leiomyomas of the uterus showed positivity both in cytoplasms and nuclei (Figure 9).

The expression of GHR and/or PRLR has been previously demonstrated in breast carcinomas,[7,10,21,23,24,31,53] hepatocarcinomas,[6,53] pituitary adenomas,[3] meningi-omas,[15] and leiomyomas.[13] The greater part of these studies were performed by molecular biologic or *in situ* hybridization techniques, but several authors have also employed immunohistochemical techniques in the study of breast carcinomas.[21,23,24] Immunohistochemistry allows not only the demonstration of the expression of the peptide receptors in the tumors (like the techniques mentioned above), but also the study of their cellular distribution.

Peptide hormones were classically believed to act exclusively by binding with plasma membrane receptors, unlike steroid hormones which present intracellular receptors. However, the cytoplasmic localization of GHR and PRLR in tumors is at present not surprising, since it is well known that normally a large pool of these

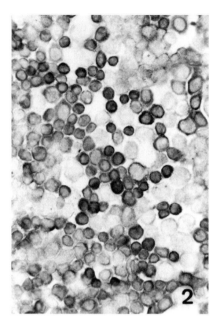

FIGURE 2   Cytospin preparation of MCF-7 breast cancer cells. GHR immunoreactivity was found in approximately 75% of the cells, being the smaller ones more intensely immunostained. Immunoreactivity was higher in cytoplasm, but nuclei were also positive (compare the staining with negative cells). Original magnification ×250.

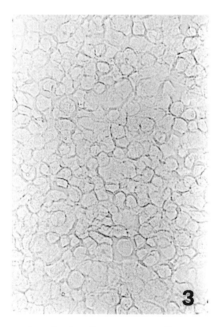

FIGURE 3   Cytospin preparation of MCF-7 breast cancer cells. No immunoreactivity was found in the negative control performed by substitution of the primary antibody with B1.1 (an unrelated monoclonal antibody of the same class as Mab 263, that recognizes normal cross-reacting antigen, a carcinoembryonic-related protein). Original magnification ×250.

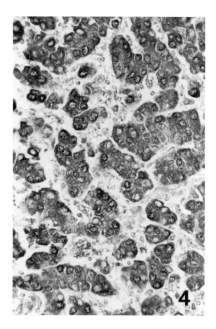

FIGURE 4   Hepatocarcinoma cells were immunoreactive for PRLR. Immunostaining was intense both in the nucleus and the cytoplasm. Note that not all the nuclei were positive. Original magnification ×250.

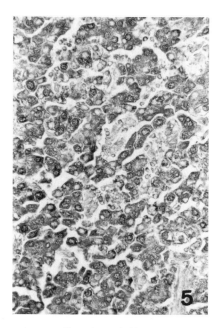

FIGURE 5   Same hepatocarcinoma cells as shown in Figure 4. Immunostaining for GHR presented the same immunoreactivity pattern to that obtained for PRLR. Original magnification ×250.

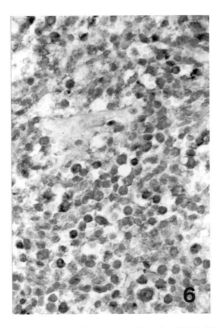

FIGURE 6 FSH/LH producing pituitary adenoma positive for PRLR. The DAB product was principally localized over the nuclei of tumor cells. Note that erythrocytes in the blood vessels around the cells were not immunostained. Original magnification ×300.

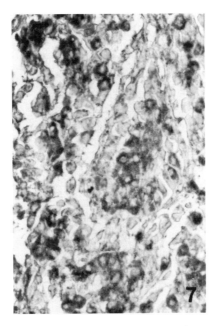

FIGURE 7 Nonsecreting pituitary adenoma immunoreactive for GHR. Immunostaining was found in the cytoplasms. Most nuclei were negative, nevertheless, some of them seem weakly stained. Original magnification ×300.

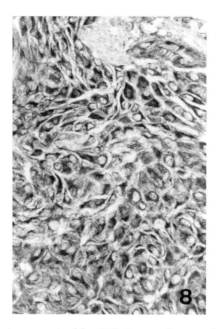

FIGURE 8   Meningioma immunostained for GHR. Tumor cells were clearly delineated because cytoplasmic immunoreactivity was intense. Original magnification ×300.

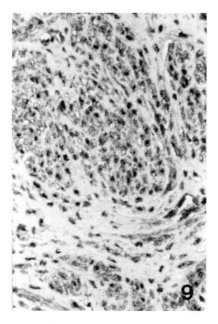

FIGURE 9   PRLR immunoreactive leiomyoma of the uterus. Positivity was found in tumor cells (note that the connective tissue was negative). Both nuclei and cytoplasms were immunostained, the nuclei being more intensely labeled. Original magnification ×300.

receptors exist at the intracellular level.[54] This distribution may be explained by the rapid internalization of the peptide-receptor complexes by endocytosis.[55] This rapid turnover could also explain the absence of plasma membrane positivity in the current study.

On the other hand, the immunoreactivity found in the nuclei is not surprising since GHR[56,57] and PRLR[5,58-60] also are normally localized in the nuclei. The translocation of these peptides to the nucleus and their interaction with nuclear receptors may provide a novel mechanism whereby these hormones regulate the transcription of growth-related genes.[5]

VI. CONCLUSION

In this chapter we have presented a simple method to localize peptide receptors in routinely processed tumors. The presence of peptide receptors in human tumors appears to play an important role in the cell proliferation control. Some novel therapies based on the use of agonists or antagonists of the peptides intend to limit the growth of the peptide receptor-positive tumors. Moreover, the internalization of the peptide receptors could open a way to introduce therapeutic agents into tumor cells.

Immunohistochemical techniques appear to be a reliable and economic method to study the cellular localization of peptide receptors in cytologic and histologic samples of tumors processed for pathologic diagnosis. By these techniques we were able to demonstrate the presence of GHR and PRLR in the cytoplasms and nuclei of a variety of tumors including breast carcinomas, hepatocarcinomas, pituitary adenomas, meningiomas, and leiomyomas.

ACKNOWLEDGMENTS

This work was supported by the Xunta de Galicia (Grant no. 20811B94). The authors thank Dr. B. K. Vonderhaar (National Cancer Institute, Bethesda, MD) and Dr. M. J. Waters (University of Queensland, Australia) for the generous gift of the B6.2 and MAb 263 antibodies, respectively. We are indebted to Prof. J. Forteza, Dr. E. Pintos, and Dr. L. Loidi (Hospital General de Galicia and School of Medicine, Santiago) for providing us with the human samples and culture cells. The expert technical help of Ms. D. Fernández-Roel and Ms. S. Pérez-Peña is gratefully acknowledged.

REFERENCES

1. **Schönbrunn, A. H. and Tashijan, A. J. R.,** Characterization of functional receptors for somatostatin in rat pituitary cells in culture, *J. Biol. Chem.,* 253, 6473, 1978.
2. **Reubi, J. C., Mauer, R., Von Weder, K., Torhorst, J., Klijn, G. M., and Lamberts, S. W. J.,** Somatostatin receptors in human endocrine tumors, *Cancer Res.,* 47, 551, 1987.

3. **Ciccarelli, E., Faccani, G., Longo, A., Ore, G. D., Papotti, M., Grottoli, S., Razzore, P., Ghè, C., and Muccioli, G.,** Prolactin receptors in human pituitary adenomas, *Clin. Endocrinol.,* 42, 487, 1995.

4. **Theveniau, M. A., Yasuda, K., Bell, G. I., and Reisine, T.,** Immunological detection of isoforms of the somatostatin receptor subtype, SSTR2, *J. Neurochem.,* 63, 447, 1994.

5. **Rao, Y. P., Buckley, D. J., and Buckley, A. R.,** The nuclear prolactin receptor: a 62-kDa chromatin-associated protein in rat Nb2 lymphoma cells, *Arch. Biochem. Biophys.,* 322, 506, 1995.

6. **Esposito, N., Paterlini, P., Kelly, P. A., Postel-Vinay, M.-C., and Finidori, J.,** Expression of two isoforms of the human growth hormone receptor in normal liver and hepatocarcinoma, *Mol. Cell. Endocrinol.,* 103, 13, 1994.

7. **Jammes, H., Decouvelaere, C., Bonneterre, J., Fournier, J., Djiane, J., and Peyrat, J. P.,** Expression des récepteurs de l'hormone de croissance humaine dans les cancers du sein, *Bull. Cancer (Paris),* 81, 938, 1994.

8. **Kubota, A., Yamada, Y., Kagimoto, S., Shimatsu, A., Imamura, M., Tsuda, K., Imura, H., Seino, S., and Seino, Y.,** Identification of somatostatin receptor subtypes and an implication for the efficacy of somatostatin analogue SMS 201-995 in treatment of human endocrine tumors, *J. Clin. Invest.,* 93, 1321, 1994.

9. **Greenman, Y. and Melmed, S.,** Expression of three somatostatin receptor subtypes in pituitary adenomas: evidence for preferential SSTR5 expression in the mammosomatotroph lineage, *J. Clin. Endocrinol. Metab.,* 79, 724, 1994.

10. **Decouvelaere, C., Peyrat, J.-P., Bonneterre, J., Djiane, J., and Jammes, H.,** Presence of the growth hormone receptor messenger RNA isoforms in human breast cancer, *Cell Growth Differ.,* 6, 477, 1995.

11. **Hashimoto, K., Koga, M., Motomura, T., Kasayama, S., Kouhara, H., Ohnishi, T., Arita, N., Hayakawa, T., Sato, B., and Kishimoto, T.,** Identification of alternatively spliced messenger ribonucleic acid encoding truncated growth hormone-releasing hormone receptor in human pituitary adenomas, *J. Clin. Endocrinol. Metab.,* 80, 2933, 1995.

12. **Miller, G. M., Alexander, J. M., Bikkal, H. A., Katznelson, L., Zervas, N. T., and Klibanski, A.,** Somatostatin receptor subtype gene expression in pituitary adenomas, *J. Clin. Endocrinol. Metab.,* 80, 1386, 1995.

13. **Sharara, F. I. and Nieman, L. K.,** Growth hormone receptor messenger ribonucleic acid expression in leiomyoma and surrounding myometrium, *Am. J. Obstet. Gynecol.,* 173, 814, 1995.

14. **Vikic-Topic, S., Raisch, K. P., Kvols, L. K., and Vuk-Pavlovic, S.,** Expression of somatostatin receptor subtypes in breast carcinoma, carcinoid tumor, and renal cell carcinoma, *J. Clin. Endocrinol. Metab.,* 80, 2974, 1995.

15. **Carroll, R. S., Schrell, U. M. H., Zhang, J., Dashner, K., Nomikos, P., Fahlbusch, R., and Black, P. McL.,** Dopamine D1, dopamine D2, and prolactin receptor messenger ribonucleic acid expression by the polymerase chain reaction in human meningiomas, *Neurosurgery,* 38, 367, 1996.

16. **Krenning, E. P., Kwekkeboom, D. J., Bakker, W. H., Breeman, W. A. P., Kooiji, P. P. M., Oei, H. Y., van Hagen, M., Postema, P. T. E., de Jong, M., Reubi, J. C., Visser, T. J., Reijs, A. E. M., Hofland, L. J., Koper, J. W., and Lamberts, S. W. J.,** Somatostatin receptor scintigraphy with [¹¹¹In-DTPA-D-Phe¹]- and [¹²⁵I-Tyr³]-octreotide: the Rotterdam experience with more than 1000 patients, *Eur. J. Nucl. Med.,* 20, 716, 1993.

17. **Hofland, L. J., van Koetsveld, P. M., Waaijers, M., Zuyderwijk, J., Breeman, W. A. P., and Lamberts, S. W. J.,** Internalization of the radioiodinated somatostatin analog [¹²⁵I-Tyr³]octreotide by mouse and human pituitary tumor cells: increase by unlabeled octreotide, *Endocrinology,* 136, 3698, 1995.

18. **Reubi, J. C.,** Neuropeptide receptors in health and disease: the molecular basis for *in vivo* imaging, *J. Nucl. Med.* 36, 1825, 1995.

19. **Reubi, J. C., Krenning, E., Lamberts, S. W. J., and Kvols, L.,** *In vitro* detection of somatostatin receptors in human tumors, *Metabolism,* 41, 104, 1992.

20. **Reubi, J. C., Schaer, J. C., Waser, B., and Mengod, G.,** Expression and localization of somatostatin receptor SSTR1, SSTR2 and SSTR3 mRNAs in primary human tumors using *in situ* hybridization, *Cancer Res.,* 54, 3455, 1994.

21. **Clevenger, C. V., Chang, W. P., Ngo, W., Pasha, L. M., Montone, K. T., and Tomaszewski, J. E.,** Expression of prolactin and prolactin receptor in human breast carcinoma. Evidence for an autocrine/paracrine loop, *Am. J. Pathol.,* 146, 695, 1995.

22. **Michel, E. and Parsons, J. A.,** Histochemical and immunocytochemical localization of prolactin receptors on Nb2 lymphoma cells: application of confocal microscopy, *J. Histochem. Cytochem.,* 38, 965, 1990.

23. **Banerjee, R., Ginsburg, E., and Vonderhaar, B. K.,** Characterization of a monoclonal antibody against human prolactin receptors, *Int. J. Cancer,* 55, 712, 1993.

24. **Leroy-Martin, B., Peyrat, J. Ph., Amrani, S., Lorthioir, M., and Leonardelli, J.,** Analyse immunocytochimique des récepteurs prolactiniques (R-PRL) humains à l'aide d'anticorps antiidiotypes dans les cancers du sein humain, *Ann. Pathol.,* 15, 192, 1995.

25. **Lincoln, D. T., Temmin, L., Aljarallah, M. A., Mathew, T. C., and Dashti, H.,** Primary Ki-1 lymphoma of the skin: expression of growth hormone receptors, *Nutrition,* 11, 627, 1995.

26. **Rao, Y. P., Buckley, D., Olson, M. D., and Buckley, A. R.,** Nuclear translocation of prolactin: collaboration of tyrosine kinase and protein kinase C activation in rat Nb2 node lymphoma cells, *J. Cell. Physiol.,* 163, 266, 1995.

27. **Xia, Y., Skoog, V., Muceniece, R., Chhajlani, V., and Wikberg, J. E. S.,** Polyclonal antibodies against human melanocortin MC1 receptor: preliminary immunohistochemical localisation of melanocortin MC1 receptor to malignant melanoma cells, *Eur. J. Pharmacol.,* 288, 277, 1995.

28. **Aaronson, S. A.,** Growth factors and cancer, *Science,* 254, 1146, 1991.

29. **Dickson, R. D. and Lippman, M. E.,** Growth factors in breast cancer, *Endocrinol. Rev.,* 16, 559, 1995.

30. **Gout, P. W., Beer, C. T., and Noble, R. L.,** Prolactin-stimulated growth of cell cultures established from malignant Nb rat lymphomas, *Cancer Res.,* 40, 2433, 1980.

31. **Bonneterre, J., Peyrat, J. P., Beuscart, R., and Demaille, A.,** Biological and clinical aspects of prolactin receptors (PRL-R) in human breast cancer, *J. Steroid Biochem. Mol. Biol.,* 37, 977, 1990.

32. **Jimenez-Hakin, E., El-Azouzi, M., and Black, P. McL.,** The effect of prolactin and bombesin on the growth of meningioma-derived cells in monolayer culture, *J. Neurooncol.,* 16, 185, 1993.

33. **Bhatavdekar, J. M., Patel, D. D., Karelia, N. H., Vora, H. H., Ghosh, N., Shah, N. G., Balar, D. B., and Trivedi, S. N.,** Tumor markers in patients with advanced breast cancer as prognosticators. A preliminary study, *Breast Cancer Res. Treat.,* 30, 193, 1994.

34. **Lebrun, J. J., Ali, S., Sofer, L., Ullrich, A., and Kelly, P. A.,** Prolactin-induced proliferation of Nb2 cells involves tyrosine phophorylation of the prolactin receptor and its associated tyrosine kinase JAK2, *J. Biol. Chem.,* 269, 14021, 1994.

35. **Mershon, J., Sall, W., Mitchner, N., and Ben-Jonathan, N.,** Prolactin is a local growth factor in rat mammary tumors, *Endocrinology,* 136, 3619, 1995.

36. **Dhanasekaran, N., Heasley, L. E., and Johnson, G. L.,** G protein-coupled receptor systems involved in cell growth and oncogenesis, *Endocrinol. Rev.,* 16, 259, 1995.

37. **Fuh, G. and Wells, J. A.,** Prolactin receptor antagonists that inhibit the growth of breast cancer cell lines, *J. Biol. Chem.,* 270, 13133, 1995.

38. **Horseman, N. D.,** Prolactin, proliferation, and protooncogenes, *Endocrinology,* 136, 5249, 1995.

39. **Taylor, J. E., Bogden, A. E., Moreau, J. P., and Coy, D. H.,** *In vitro* and *in vivo* inhibition of human small cell lung carcinoma (NCI-H69) growth by a somatostatin analogue, *Biochem. Biophys. Res. Commun.,* 153, 81, 1988.

40. **Liebow, C., Reilly, C., Serrano, M., and Schally, A. V.,** Somatostatin analogues inhibit growth of pancreatic cancer by stimulating tyrosine phosphatase, *Proc. Natl. Acad. Sci. U.S.A.,* 86, 2003, 1989.

41. **Kvols, L. K., Moertel, C. G., O'Connell, M. J., Schutt, A. J., Rubin, J., and Hahn, R. G.,** Treatment of the malignant carcinoid syndrome: evaluation of a long-lasting somatostatin analogue, *N. Engl. J. Med.,* 315, 663, 1986.

42. **Lamberts, S. W. J., Krenning, E. P., and Reubi, J. C.,** The role of somatostatin and its analogs in the diagnosis and treatment of tumors, *Endocrinol. Rev.,* 12, 450, 1991.

43. **Pinski, J., Schally, A. V., Groot, K., Halmos, G., Szepeshazi, K., Zarandi, M., and Armatis, P.,** Inhibition of growth of human osteosarcomas by antagonists of growth hormone releasing hormone, *J. Natl. Cancer Inst.,* 87, 1787, 1995.

44. **Lamberts, S. W. J.,** Internalization of the radioiodinated somatostatin analog [^{125}I-Tyr^{3}]octreotide by mouse and human pituitary tumor cells: increase by unlabeled octreotide, *Endocrinology,* 136, 3698, 1995.

45. **Berelowitz, M.,** The somatostatin receptor. A window of therapeutic opportunity, *Endocrinology,* 136, 3695, 1995.

46. **Krenning, E. P., Kooiji, P. P. M., Bakker, W. H. B., Breeman, W. A. P., Postema, P. T. E., Kwekkeboom, D. J., Oei, H. Y., de Jong, M., Visser, T. J., Reijs, A. E. M., and Lamberts, S. W. J.,** Radiotherapy with a radiolabeled somatostatin analogue [^{111}In-DTPA-D-Phe^{1}]octreotide: a case history, *Ann. N.Y. Acad. Sci.,* 733, 496, 1994.

47. **Shi, S.-R., Key, M. E., and Kalra, K. L.,** Antigen retrieval in formalin-fixed, paraffin-embedded tissues: an enhancement method for immunohistochemical staining based on microwave oven heating of tissue sections, *J. Histochem. Cytochem.,* 39, 741, 1991.

48. **Beckstead, J. H.,** Improved antigen retrieval in formalin-fixed, paraffin-embedded tissues, *Appl. Immunochem.,* 2, 274, 1994.

49. **Miller, R. T. and Estran, C.,** Heat-induced epitope retrieval with a pressure cooker, *Appl. Immunochem.,* 3, 190, 1995.

50. **Barnard, R., Bundesen, P. G., Rylatt, D. B., and Waters, M. J.,** Evidence from the use of monoclonal antibody probes for structural heterogeneity of the growth hormone receptor, *Biochem. J.,* 231, 459, 1985.

51. **Elias, J. M.,** Immunohistochemical methods, in *Immunohistopathology. A Practical Approach to Diagnosis,* Elias, J. M., Ed., ASCP Press, Chicago, 1990, 53.

52. **Coggi, G., Dell'Orto, P., and Viale, G.,** Avidin-biotin methods, in *Immunocytochemistry. Modern Methods and Applications,* Polak, J. M. and van Noorden, S., Ed., Wright, Bristol, 1986, 64.

53. **Boutin, J. M., Edery, M., Shirota, M., Jolicoeur, C., Lesueur, L., Ali, S., Gould, D., Djiane, J., and Kelly, P. A.,** Identification of a cDNA encoding a long form of PRL receptor in human hepatoma and breast cancer cells, *Mol. Endocrinol.,* 3, 1455, 1989.

54. **Boutin, J. M., Jolicoeur, C., Okamura, H., Gagnon, J., Edery, M., Shirota, M., Banville, M., Dusanter-Fourt, I., Djiane, J., and Kelly, P.A.,** Cloning expression of the rat prolactin receptor, a member of the GH/PRL receptor gene family, *Cell,* 53, 69, 1988.

55. **Genty, N., Paly, J., Edery, M., Kelly, P. A., Djiane, J., and Salesse, R.,** Endocytosis and degradation of prolactin and its receptor in Chinese hamster ovary cells stably transfected with prolactin receptor cDNA, *Mol. Cell. Endocrinol.,* 99, 221, 1994.

56. **Lobie, P. E., Wood, T. J. J., Chen, C. M., Waters, M. J., and Norstedt, G.,** Nuclear translocation and anchorage of the growth hormone receptor, *J. Biol. Chem.,* 269, 31735, 1994.

57. **Mertani, H. C., Waters, M. J., Jambou, R., Gossard, F., and Morel, G.,** Growth hormone receptor binding protein in rat anterior pituitary, *Neuroendocrinology,* 59, 483, 1994.

58. **Bucley, A. R., Montgomery, D. W., Hendrix, M. J., Zukoski, C. F., and Putnam, C. W.,** Identification of prolactin receptors in hepatic nuclei, *Arch. Biochem. Biophys.,* 296, 198, 1992.

59. **Morel, G., Ouhtit, A., and Kelly, P. A.,** Prolactin receptor immunoreactivity in rat anterior pituitary, *Neuroendocrinology,* 59, 78, 1994.

60. **García-Caballero, T., Morel, G., Gallego, R., Fraga, M., Pintos, E., Gago, D., Vonderhaar, B. K., and Beiras, A.,** Cellular distribution of prolactin receptors in human digestive tissues, *J. Clin. Endocrinol. Metab.,* 81, 1861, 1996.

Chapter **11**

VISUALIZATION OF STEROID RECEPTORS: A CLINICAL APPROACH

France Wallet
Daniel Seigneurin
Pierre-Marie Martin

CONTENTS

I. INTRODUCTION

A. TECHNICAL CONSIDERATIONS

In the past 30 years, scientific investigations in the area of molecular endocrinology have revealed the presence of steroid hormone receptor proteins in both normal and abnormal human and animal tissues (see Chapter 2). Steroid hormone receptors have been described in breast cancer, prostatic carcinoma and hyperplasia, lung, kidney, and central nervous system, among others.[1] Developments of histologic methods for detection of steroid receptors in intact tissue sections provides a means for exact anatomical localization of these proteins.[2] Three models have been developed for steroid hormone intracellular interaction (Figure 1). In the two-step model proposed by Jensen et al.[3] in 1968, estradiol enters the cell, binds to a specific cytoplasmic receptor, and then is translocated to the nucleus by a temperature-dependent activation process. In this model, cytoplasmic receptors can exist either steroid-filled or unfilled, but nuclear receptors can exist only when steroid-filled. However, the observation of unfilled receptors in the nucleus have led to the hypothesis that in intact cells, unfilled receptors may exist in a state of equilibrium between the cytoplasm and nucleus. In the equilibrium model proposed by Sheridan et al.[4] in 1979 the equilibrium is shifted in favor of the cytoplasm by tissue homogenization. Finally, a third model was proposed by King and Greene[5] in 1984: the estrogen receptors may reside exclusively in target cell nuclei of estrogen-sensitive tissues.

The presence of two types of binding sites was shown (Figure 2): one of high affinity (type I) and two of lower affinity (type II A and B). High-affinity receptors are nuclear and perinuclear and can be detected by techniques like radioautography and intrinsic fluorescent ligands as described below (see Chapter 9). Low-affinity receptors are located in cytoplasm (A) and membrane (B) and can be detected by techniques like bipolar tracers. These three techniques can be called functional steroid detection techniques; unlike structural detection by antisteroid antibody this can be a conformational or sequential detection.

B. CLINICAL APPLICATIONS

The value of steroid receptors (SR) as prognostic indicators remains controversial in breast cancer patients but SR status still constitutes the best marker for predicting response to endocrine therapy and may be useful in selecting patients for adjuvant therapy.[6] This fact explains the prognostic function of SR in treated patients.[7,8]

Three major points have to be considered before using the visual determination of intracellular hormone receptors as a test for the hormone dependence of human tumors:

1. Tumoral tissue and cells are of course needed; this tissue must be first used for pathological examination, which is the basis of diagnosis, and can be used for complementary techniques. Due to screening Protocols, breast tumors are more and more diagnosed as small T1 tumors and enough tissue may be lacking for biochemical steroid receptors assays. Furthermore, in some cases, only micro-biopsies or even cytopunctions[9] are available; in these cases, as in the case of pleural metastasis, SR status can be only determined by visual methods.

2. Tumors are heterogeneous for SR expression; every breast tumor is made of epithelial and conjunctive cells; visual assays can distinguish between epithelial

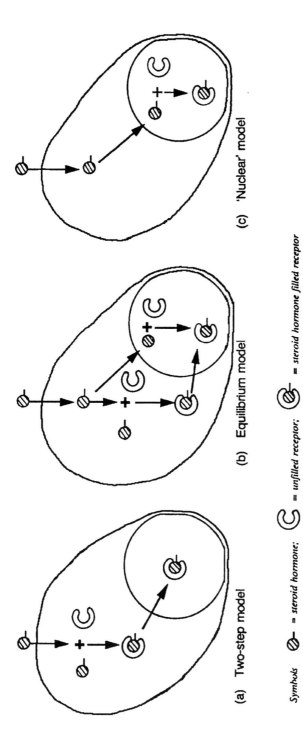

(a) Two-step model **(b)** Equilibrium model **(c)** 'Nuclear' model

Symbols  = *steroid hormone*; = *unfilled receptor*; = *steroid hormone filled receptor*

FIGURE 1 Three models of "high-affinity" receptor localization.

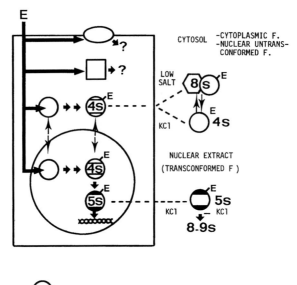

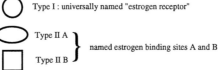

FIGURE 2   Different types of estrogen binding sites.

tumoral or benign cells and stromal cells. It is well-known that tumors are not SR positive or negative but contain a proportion of SR positive cells varying from 0 to 100%; in most of the cases, malignant cell population is heterogeneous for SR expression. Unlike biochemical techniques, visualization methods, by preserving the tissue structure, enable the determination of the percentage of hormone-dependent cells, the target for hormone therapy among cancer cells. On the other hand, SR tumoral heterogeneity may lead to erroneous evaluation of SR status if these methods are done on little pieces of tumoral material as microbiopsies or fine needle aspirates.

3. The third point is the need for adequate standardization and quantitation, which remain limiting factors in the use of histochemical hormonal receptor assays. The interpretations are subjective and the measurements are at best semiquantitative; however, continuous efforts are being made to quantitate hormone receptor proteins.[10,11]

II. VISUALIZATION TECHNIQUES

A. FUNCTIONAL DETECTION

Functional detection of steroid receptors can visualize the two types of binding site: high affinity (type I) by radioautography and intrinsic fluorescence by the use of low concentrations of ligand (10^{-9} to 10^{-10} M) and low affinity (type II) by bipolar tracers by the use of higher concentration of ligand (10^{-6} M). This is explained by the hormone-receptor equation of interaction:

$$[E + R] \xrightleftharpoons[K_D]{K_A} [ER] \quad \text{with} \quad K_D = \frac{1}{K_A}$$

Where E: estrogen, R: receptor, K_D: dissociation constant, and K_A: association constant.

1. Radiolabeled Ligand

For the past 45 years, radioautography has been one of the tools used to explore the interactions of steroid hormones with their numerous target organs within the body. At first, it was barely possible to demonstrate that the uterine endometrium concentrated circulating radioactive estrogen. Later, conflicting results were published from several laboratories concerning the subcellular localization of steroids. For example, radioactive estrogen was found in the lumen of the uterus, in the cytoplasm, only in the cytoplasm of the eosinophils, at the apical pole of epithelial cells, and over nuclei. After extensive comparative studies utilizing multiple radioactive tracers, it was shown that most of the conflicting data were due to translocation artifacts. Such things as liquid formalin fixation, paraffin embedding, and liquid emulsion dipping were found to cause translocation and leaching of the radioactive probe to varying degrees and an impairment of radioautographic resolution. The technique which had theoretically the smallest chance for producing translocation artifacts and which gave the greatest radioautographic resolution was radioautography using dry-mounted freeze-dried sections (see Protocol 1). Later, the thaw-mount technique (see Protocol 2) was also found to be suitable.[12,13]

PROTOCOL 1: DRY-MOUNTED FREEZE-DRIED SECTIONS

1. Section the tissues between 1 to 4 μm in a cryostat at a knife temperature of between –60 to –30°C.
2. Transfer the sections with a fine brush to small vials within the cryostat.
3. After sectioning, transfer the vials to the specimen chamber of a cryosorption pump in the back of the cryostat.
4. Assemble the cryosorption pump within the cryostat by creating an initial vacuum with a mechanical forepump.
5. Disconnect the forepump after 5 min of evacuation and submerge the portion of the assembled cryosorption pump containing the molecular sieve in a Dewar jar containing liquid nitrogen.
6. After 24 h, remove the assembled cryosorption pump together with the Dewar containing the liquid nitrogen from the cryostat and let come to room temperature.
7. Break the vacuum (about 10^{-5} torr) by slowly introducing nitrogen gas.

8. Under conditions of low relative humidity (20 to 40%), place the freeze-dried sections on Teflon supports.
9. Under safelight conditions, bring a slide precoated with emulsion (Kodak NTB3) in contact with the tissue section by pressing the Teflon support against the slide.
10. When the pressure is released, the Teflon support falls away and the slide with the sections adhering are stored in a desiccated box at −15°C for exposure (a variable depending on the experiment).
11. At the end of the exposure the slides are processed photographically and stained.

PROTOCOL 2: THE THAW-MOUNT TECHNIQUE

1. Cut frozen sections (4 μm) in a cryostat in a darkroom with the use of a fiber light mounted on the cryostat.
2. After sectioning, turn off the fiber light.
3. Mount the section directly onto emulsion-coated slides under safelight conditions.
4. Transfer the sections to a desiccated box.
5. After the appropriate exposure time, the slides are photographically processed and stained.

In a typical radioautographic experiment castrated and/or adrenalectomized animals are injected with ³H-steroid and sacrificed at appropriate time points afterwards (usually between 30 min to 2 h). The tissues of interest are removed, cut into small pieces, mounted onto brass tissue holders using minced liver or ornithine carbamyl transferase (OCT) mounting medium as an adhesive, immersed into liquefied propane, and then stored in liquid nitrogen until sectioning. An alternative to the *in vivo* injection of ³H-steroids was an *in vitro* incubation of small pieces of tissue with a known concentration of ³H-steroid at a controlled temperature (see Protocol 3). After an appropriate period of time the tissues are removed from the medium containing the ³H-steroid and then washed extensively before being frozen and processed for radioautography. Detailed techniques for tissue and cell preparation are described in Protocol 3.

PROTOCOL 3: TISSUE OR CELL PREPARATION

Reagents

- Mimimal essential medium (MEM, Gibco)
- Diethylstilbestrol (DES)
- Labeled estradiol: [1,2,6,7-³H]-estradiol-17β, [³H]-estradiol, [³H]-2858 [6,7-³H]-methoxy-17-ethinylestradiol

- Rabbit anti-Estrogen Receptor
- Sucrose
- Methyl-green-pyronin

Method

A. For tissue:

1. Incubate the tissues in MEM in the presence of 5 nM of [1,2,6,7-^{3}H]-estradiol-17β with or without 500 nM unlabeled diethylstilbestrol for either 5 min or 2 h at either 0°C or 37°C. The tissue is then processed for radioautography.

B. For cultured cells:

1. Incubate cells in MEM in the presence of different concentrations (50 pM to 100 nM) of [^{3}H]-estradiol or [^{3}H]-2858 [6,7-^{3}H]-methoxy-17-ethinylestradiol alone or in the presence of 100-fold excess unlabeled diethylstilbestrol from 2 min to 6 h at 0°C and 37°C.
2. Wash three times in MEM.
3. Wash three times with a rabbit antiserum against estradiol or R 2858 (1:500 dilution in MEM). Washing with antisera to estrogens in these experiments greatly reduced radioautographic background.
4. Spin down in 0.3 M sucrose in MEM.

C. For tissues and cells:

5. Mount tissues or cells on a brass stud with OCT compound.
6. Freeze in liquefied propane for radioautography.
7. Cut frozen sections in a cryostat.
8. Mount for radioautography on emulsion-coated slides.
9. Expose at −15°C for up to 26 weeks.
10. Process photographically and stain with methyl-green-pyronin.

Radioautography compared to biochemical techniques,[14] shows:

- Similitudes: the sensitivity is identical and is obtained with 10^{-10} M radiolabeled tracer displaced by 10^{-8} M radioinert competitor. The specificity is also identical. With respect to quantification, the semiquantitative results obtained with low (10^{-10} M) concentrations of tracer are comparable to the results from biochemical studies.
- Dissimilitudes: radioautography does not distinguish between type I or II binding except on the basis of the concentration of radioactive tracer used. The very high background values prevent any conclusions on studies of low-affinity binding requiring high radiolabeled tracer concentrations. The major differences, however, reside in the intracellular localization of the binding sites:

- At 4°C, when the hormone system is "inactive", biochemical techniques reveal a predominantly cytosolic localization of binding, whereas radioautography reveals both cytoplasmic and nuclear binding. This latter result would seem to imply that the receptor, which is inactive or has not undergone transconformation, is stabilized only in the cytoplasmic compartment although present in both. It could indeed be extracted during homogenization and thus recovered in the cytosolic fraction.
- At 37°C, on the other hand, the activation of the receptor by steroid binding and by increasing the temperature makes the extraction of the intracellular nuclear fraction difficult. This fraction is thus activated or stabilized, whether in radioautography or biochemical studies. The classic concept of nuclear translocation would thus appear to correspond to a stabilization by transconformation into the nucleus compartment.

The radioautography technique was validated to obtain reproducible results, but it requires the preparation of 4-μm frozen sections and long exposure times at low temperatures. These procedural drawbacks make this technique inappropriate for widespread clinical use. However, this technique does demonstrate either steroid or its metabolites depending on whether ^{14}C or ^3H are used for labeling: Ag^{2+} precipitation is a function of energy and different features of precipitation are obtained.[15]

2. Intrinsic Fluorescent Ligand

Described here is the use of ligands for the estrogen receptor that are inherently fluorescent and that have high affinity for the estrogen receptor (see Protocol 4): the natural phytoestrogen coumestrol or the estradiol derivative 12-oxoestradiol (prepared by oxidation of estradiol with dichlorodicyanoquinone).[16-18] Coumestrol is a spontaneously fluorescent molecule which binds to the cytosol and nuclear estrogen receptor, and stimulates the growth of hormone-dependent cells as has been shown on a continuous cell-line: MCF-7. However, coumestrol is a weak estrogen with an affinity 1/100th that of estradiol. Furthermore, its weak fluorescence implies the use of of high (10^{-5} M) concentrations. With such concentrations it is unlikely that it binds uniquely to high-affinity receptor sites. The low concentrations of these ligands required to label the high-affinity binding sites specifically (10^{-9} to 10^{-10} M) make conventional fluorescence microscopy unsuitable. However, observation of this fluorescence required the use of a fluorescence microscope equipped with a microchannel image intensifier and a video camera detector, providing a light amplification of ≈10^4. In every case, the fluorescence could be abolished by a 200-fold excess of diethylstilbestrol (DES). Specific fluorescence is predominantly nuclear, in accord with the nuclear localization of estrogen receptors at 37°C. The subcellular distribution of fluorescence is concentration dependent. While coumestrol fluorescence is nuclear at the nanomolar concentrations observable by image-intensified fluorescence, it is predominantly cytoplasmic when the high concentrations (10^{-5} M) required for conventional fluorescence microscopy are used. This finding suggests that conventional fluorescence microscopy is inadequate for visualizing high-affinity steroid receptors and indicates that its application to the clinical situation should be made with caution.

PROTOCOL 4: FLUORESCENCE LABELING

Reagents

Progesterone
Phosphate Buffer Saline (PBS): NaCl 120 mM, KCl 2.7 mM, phosphate buffer
salts 10 mM
Fluorescent ligand
Buffered glycerol: 90% glycerol, 10% Tris (10 mM)

Method

1. Incubate 20 min at 20°C with 10^{-6} M progesterone prior to fluorescent labeling. This preincubation with progesterone was shown to diminish nonspecific
   steroid labeling, especially on the cell membrane, without interfering with the
   emission spectrum of the fluorescent labels.
2. Wash with PBS.
3. Add either 10^{-8}, 10^{-9}, or 10^{-10} M fluorescent ligand.
4. Incubate 50 min at 37°C.
5. Wash three times with PBS and air dry.
6. The slides are soaked with buffered glycerol (90% glycerol, 10% Tris (10
   mM) at pH 8 for coumestrol, pH 9.75 for 12-oxoestradiol).
7. Keep slides at low temperature (4°C) until examination by fluorescence microscopy performed within 1 h.

3. Bipolar Tracers

The steroid end of the complex binding to the receptor site and the protein
end are coupled to a fluorochrome (FITC or rhodamine). Utilization of an estradiol
molecular complex coupled to a fluorescent molecule permits microscopic observation of the membrane binding (see Protocol 5). The use of fluorochrome with
different spectra enables the simultaneous detection of several classes of hormone
dependence. Examples of such molecules are type E2-17beta-hemisuccinate-
bovine serum albumin-FITC (E2-BSA-FITC), E2-6-carboxymethyloxime-BSA-
FITC, and a derivative of diethylstilbestrol.[19] Their use requires a concentration
of 10^{-6} M in the incubate because of their low affinity (100 to 300 times less than
estradiol). The estradiol-macromolecular-fluorescent complex offers the following
advantages:

- High molecular weight, impeding diffusion across the plasma membrane,
- Rapid binding to living cells confirmed by direct microscopic observation
  and, most importantly
- Highly sensitive quantitative analysis of the membrane binding by spectro-
  fluorimetry.

Evaluation of the extent of fluorescence, which is not influenced by the low
radioactive tracer concentration, revealed a predominantly cytoplasmic localization

as was already observed after incubation with polar tracers alone. Cytoplasmic fluorescence is also heat labile and necessitates fixation at room temperature before analysis by fluorescence microscopy. The result obtained, including the nuclear background, are highly comparable to those obtained from the double-antibody technique. This latter technique, however, is more constant, reliable, and reproducible. The low affinity of the complex limits the specificity studies.

In conclusion, although the utilization of fluorescent complexes has been less successful for the detection of cytosolic receptors by histochemistry on frozen slices, the fluorescent labeling techniques described here have permitted the demonstration of two types of specific estrogen membrane binding sites.

PROTOCOL 5: ESTRADIOL MOLECULAR FLUORESCENT COMPLEX PREPARATION

Reagents

Phosphate Buffer Saline (PBS): NaCl 120 mM, KCl 2.7 mM, phosphate buffer
Bovine serum albumin (BSA) 3% in PBS
17β-Estradiol-6-carboxymethyloxime-BSA complex (E2-BSA) (Steraloid, Inc. Wilton, NH) coupled to the fluorescein isothiocyanate by the technique of Nairn.[20]

Method

1. Incubate twice for 10 min at 37°C with 5 ml PBS.
2. Incubate with 0.04 to 1 µM of E2-BSA-FITC at 4°C or 37°C.
3. Wash three times with cold PBS.
4. Observe with a photomicroscope equipped with epifluorescence and excitation-emission filters of 400 to 490 nm for FITC.

B. STRUCTURAL DETECTION: IMMUNOCYTO(HISTO)CHEMISTRY

Antibody markers have become valuable tools in the analysis of steroid receptors. The indirect immunological staining procedure is normally used where a wide variety of detection systems, including fluorescent markers or enzyme-based markers, are available.

As SR expression needs to be correlated to cell morphology, it is easier to use an enzyme-based detection system than a fluorescent one.

Antibodies can be conformational or sequential for recognizing specific epitopes. They are native proteins tested *in vitro* on known cells in term of specificity and avidity by biosensor technology (see Chapter 8). Their use give rise to several questions, e.g., antibodies interact with macromolecular structures: estrogen receptors, heat shock proteins, calmodulin, etc. What happens when they are used in human tissue: is the specific epitope present? Is it accessible? Is it mutated? Are avidity and affinity of the antibody still available in these conditions? What happens when these proteins are used in a denatured environment like paraffin-embedded sections?

1. Methodology
Primary Antibodies
Several primary antibodies[5,21-29] are available (see Table 1); those most often used are H 222 or ER1D5 for Estrogen Receptor (ER) and JZB 39 or Progesterone Receptor (PR) 10 A9 for PR; they recognize specific epitopes.

TABLE 1 Primary Antibodies Available for SR Visualization in 1996

Antibody	Produced From	Commercialized By	Characteristic	Material Used	Ref.
			ER		
H222	Rat	Abbott	Recognize the hormone binding domain. AA 463-528	Frozen	Greene, 80 King, 84
D547	Rat	Abbott	Recognize the epitope localized between DNA binding domain and the hormone binding domain. AA 282-300	For research only	King, 84 Pertschuk, 85
ER1D5	Mouse	Immunotech	AA 118-140	Paraffin Frozen	Al Saati, 93
B10	Mouse	Cisbio	AA 151-165	For research only	Ali, 93
F3 LH1	Mouse		AA 578-595		Ali, 93 Henry, 93
			PR		
KD68			AA 165-566		Pertschuk, 90
1A6 mPRI					Henry, 93 Seymour, 90
JZB39	Rat	Abbott	AA 165-566		Pertschuk, 88
PR10A9	Mouse	Immunotech	C-terminal domain	Paraffin Frozen	
Li417	Mouse	Cisbio	AA 208-296	For research only	Vu Hai, 89

Secondary Antibodies and Detection System
Having bound a primary antibody, the simplest method is to apply a second antibody recognizing the species in which the primary was raised. As the primary is usually a mouse monoclonal antibody, the second antibody can be sheep anti-mouse immunoglobulins; they are coupled with conjugated markers as fluorescent markers (fluorescein isothiocyanate or rhodamine) or enzymes which will produce an insoluble colored reaction when incubated with their suitable substrates: for peroxidase conjugates, a yellow-brown reaction is obtained by incubation with diaminobenzidine; for

alkaline phosphatase conjugates a red reaction is produced with naphthol AS B acetate and Fast Red.

Fluorescence methods are the simplest; they are extremely sensitive but cell identification is not easy. Enzyme conjugates require a further visualization step and counterstaining which allows correlation of antigen location and cellular morphology. Other methods of detection, such as the peroxidase-anti-peroxidase or the biotin-avidin systems, offer higher sensitivity.

Material and procedures required for SR visualization in cell lines (see Protocol 6), in frozen sections (see Protocol 7), and in paraffin-embedded sections (see Protocol 8) are detailed.

PROTOCOL 6: PROCEDURE FOR SR VISUALIZATION IN CELL LINES

Material and equipment

- Formaldehyde fixative diluted to 4%
- Phosphate buffered saline: 4 g sodium dihydrogen phosphate, 27 g hydrogen-phosphate disodium, 80 g NaCl in 1 l of distilled water
- Absolute methanol
- Acetone
- ER-ICA monoclonal kit (Abbott) with blocking reagent, primary antibody, bridging antibody, PAP complex, DAB tablets, and substrate reagent
- Toluidine blue O CI 52040 (Sigma)
- Ethanol 100°, 95°, 70°
- Toluene
- Neutral mounting medium (XAM)

Cell fixation

1. Rinse the cells in PBS for 5 min.
2. Fix the cells in formaldehyde 4% for 10 min at room temperature.
3. Rinse the cells twice in PBS (5 min × 2).
4. Permeabilize the cells in absolute methanol for 10 min at $-20°C$ and then in acetone for 5 min at $-20°C$.
5. Rinse the cells 3 times in PBS (5 min × 3).

Immunoenzymochemical assay for ER

1. Incubate the cells in a humidified chamber at room temperature for 15 min with blocking reagent.
2. Incubate the cells in a humidified chamber at room temperature for 30 min with anti-ER rat at the recommended dilution.
3. Wash twice in PBS (5 min × 2).

4. Incubate the cells in a humidified chamber at room temperature for 30 min with the bridging antibody at the recommended dilution.
5. Rinse the cells twice in PBS (5 min × 2).
6. Incubate the cells in a humidified chamber at room temperature for 30 min with the PAP complex.
7. Rinse the cells twice in PBS (5 min × 2).
8. Ten to 15 min prior to color development, prepare the chromogen substrate solution by dissolving the DAB tablet in substrate reagent (chromogen substrate solution must be used within 30 min).
9. Incubate the cells in a humidified chamber at room temperature for 6 min with the chromogen substrate for color development.
10. Rinse the cells in gently running distilled water for 10 min.

Nuclear counterstaining
Toluidine blue (spectral peak = 650 nm) is recommended as it allows the ER-negative cells to be seen but avoids spectral confusion with DAB (spectral peak = 450 nm) labeling.

1. Add 0.1 g toluidine blue and 0.1 g disodium tetraborate in 100 ml distilled water, dilute to 1:10 and filter.
2. Stain the cells in this solution for 2 min at room temperature.
3. Rinse in distilled water (5 min).
4. Dehydrate in successive baths of ethanol 70°, 95°, and 100° for 15 s in each bath.
5. Dehydrate in toluene (10 s and 5 min).
6. Mount the slides using a neutral mounting medium (XAM).
7. Store in the dark at 4°C.

PROTOCOL 7: SR VISUALIZATION IN FROZEN SECTIONS

Reagents

Phosphate Buffer Saline (PBS): NaCl 120 mM, KCl 2.7 mM, phosphate buffer
Bovine serum albumin (BSA) 3% in PBS
Formaldehyde fixative diluted to 10% in PBS
Absolute methanol
Acetone
SR-ICA monoclonal kit (Abbott) with blocking reagent, primary antibody, bridging antibody, PAP complex, DAB tablets, and substrate reagent
Hematoxylin
Ethanol 100°, 95°, 70°
Toluene

Method

1. Cut frozen sections (10 μm).
2. Fix immediately for 12 min in formol 10% in PBS.

3. Rinse 5 min in PBS, 4 min in methanol stored at –20°C, 2 min in acetone stored at –20°C and then rinse twice for 5 min in PBS.
4. Incubate 20 min with normal goat serum (1:20 dilution in PBS-BSA).
5. Apply the specific antibody for 30 min.
6. Rinse 5 min in PBS.
7. Incubate with with biotin-conjugated goat anti-mouse for 30 min (1:50 dilution in PBS-BSA).
8. Rinse 5 min in PBS.
9. Incubate with peroxidase anti-peroxidase complex for 30 min (1:25 dilution in PBS-BSA).
10. Rinse 5 min in PBS.
11. Reveal with diaminobenzidine tetrachloride (DAB) for 10 min.
12. Rinse 5 min in distilled water.
13. Counterstain with hematoxylin 2 to 5 min.
14. Rinse in distilled water.
15. Dehydrate by immersing slides successively in 75°, 95°, and 100° ethanols, and then twice in toluene.

Figure 3 shows the correlation between immunohistochemistry on frozen sections and the biochemical assay.

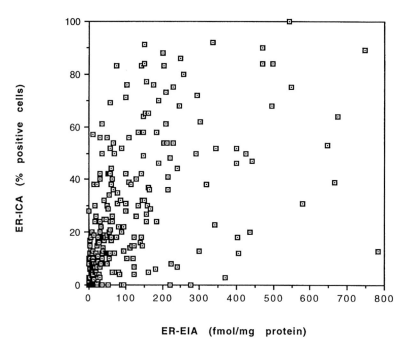

FIGURE 3   Comparison between immunohistochemistry on frozen sections and biochemical assay of estrogen receptors on 317 breast tumors.

PROTOCOL 8: SR VISUALIZATION IN PARAFFIN SECTIONS

Reagents

Bouin's solution
Phosphate Buffer Saline (PBS): NaCl 120 mM, KCl 2.7 mM, phosphate buffer
Bovine serum albumin (BSA) 3% in PBS
Formaldehyde fixative diluted to 10% in PBS
Citrate buffer: 10 mM citric acid monohydrate, pH 6.0
Absolute methanol
Acetone
Mouse specific antibody: anti-RE (Immunotech), anti-RP (Abbott)
DAB tablets
Bridging antibody, PAP complex (Amersham)
Hematoxylin
Ethanol 100°, 95°, 70°
Toluene

Fixation and paraffin embedding

1. Fix in formol 10% for times ranging from a few hours for small samples to 24 h for big samples.
2. Place in a tissue processor for a program which lasts overnight with fixation in Bouin's solution, successive immersions in 70° to 100° ethanol, toluene, and hot paraffin.

Immunohistochemical assay

1. Deparaffinize in toluene twice for 10 min.
2. Dehydrate by immersing slides successively in 100°, 95°, and 75° ethanols.
3. Rinse in water then in PBS.
4. Incubate 3 times, 5 min each time, at 750 W in microwave in 10 mM citrate buffer (10 mM citric acid monohydrate, pH 6.0).*
5. Cool in buffer 20 min at room temperature.
6. Rinse in PBS.
7. Incubate 20 min with normal goat serum diluted 1:20 in PBS-BSA.
8. Apply the mouse anti-SR for 1 h.
9. Rinse 5 min in PBS.
10. Incubate with biotin-conjugated goat anti-mouse for 40 min (1:50 dilution in PBS-BSA).
11. Rinse 5 min in PBS.

---

* Jacquemier et al. tried enzymatic digestion by proteinase K to take the place of renaturation by microwaves.[30]

12. Incubate 40 min with streptavidin-conjugated horseradish peroxidase diluted in 1:25 in PBS-BSA.
13. Rinse 5 min in PBS.
14. Reveal with diaminobenzidine tetrachloride (DAB) for 15 min.
15. Rinse in distilled water.
16. Counterstain with hematoxylin 1 to 5 min.
17. Rinse in distilled water.
18. Dehydrate by immersing slides twice successively in 75°, 95°, and 100° ethanols, and then twice in toluene.

Fixation protocol is a crucial step of this technique. On it depend the results. Figure 4 shows the correlation between immunohistochemistry on paraffin-embedded sections and biochemical assay.

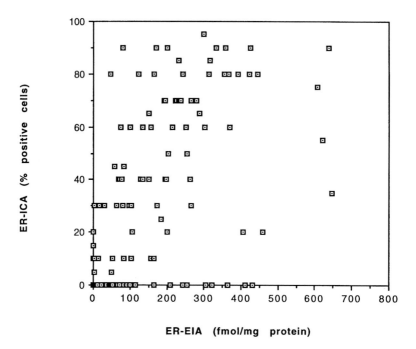

FIGURE 4   Comparison between immunohistochemistry on paraffin-embedded sections and biochemical assay of estrogen receptors on 137 breast tumors.

Quality Control and Standardization

Any labeling experiment should be associated with negative and positive controls. The simplest way to perform a negative control is to omit the primary antibody from the first incubation and replace it by PBS or normal IgG. The positive control needs slides or sections from cell lines (MCF7 for example) or tissues known as positive (positive breast tumor, for example).

Every step of the reaction should be reproducible and thus clearly standardized in terms of incubation time and temperature, dilution of reagents, etc. Antibodies must be used following the manufacturer's instructions: some primary antibodies are not developed for use on paraffin-embedded tissue, others require antigen retrieval by microwave. Confocal microscopy allows us to show what happens during "renaturation" by microwave: Figure 3 shows control (A) and "microwave-renaturated" nuclei (B). However, quality control must take into account not only the immunohistochemical staining itself but the whole technique, especially the fixation of samples. On it depends the accessibility of the antigenic site. Epitope exposure may differ from one procedure to another. Table 2 shows the results obtained with different fixation Protocols.

However, the use of these techniques raises questions concerning the specificity of the observed phenomenon and its relationship to the receptor-specific system as evidenced by well-established biochemical studies.[31]

A good immunohistochemistry procedure is a crucial first step for a good quantification.

Expressing the Results

Results could be expressed as binary information (positive or negative tumor), as the percentage of marked cells, or as a score based on the percentage of positive cells and on the intensity of staining. The threshold of positivity for a biochemical assay takes into account the threshold of detectability and the value detected in normal tissue and benign tumors. Because of tissue heterogeneity, it is difficult to determine the threshold of positivity in immunohistochemistry. However, it varies according to the researchers (for example, it is 20% for Masood).[32]

For scoring, the intensity of staining is assessed semiquantitatively using an index of 0 to 3 corresponding to negative, weak, intermediate, and strong staining intensities, and the percentage of cells stained at each intensity is estimated. Multiplication of these percentages by the scoring index gives an SR score ranging from 0 to 300.

The definition of weak vs. intermediate staining intensity is mainly subjective and poorly reproducible, as well as, in some cases, the diagnosis of negative cells.

2. Results in Breast Disease

Several studies have concluded that very good correlations are obtained between immunocyto(histo)chemical and biochemical assays and that they are comparable in their predictive power of response to endocrine treatment.[33-35] The degree of concordance reached 92 and 95%, respectively, for De Rosa and Masood if only binary information (positive and negative) is taken into account.[32,36] Lower values are obtained when the semiquantitative index is correlated with results of the biochemical assay. Figures 4 and 5 shows the correlation obained between immunohistochemistry on frozen sections and the biochemical assay and immunohistochemistry on paraffin-embedded sections and biochemical assay, respectively.

Discordances can be explained by the paucity of tumor cells due to fibrosis or necrosis, presence of mixed benign epithelial cells, tissue handling and processing, and possible tumor hormone receptor heterogeneity.

Results on various kind of biological material have shown no significant differences between frozen sections, fine needle aspirates, and imprints.[37]

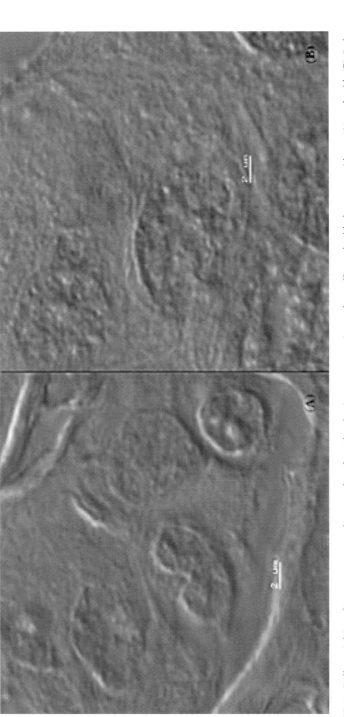

FIGURE 5   Differential interference contrast photographs of confocal microscopy sections of paraffin-embedded tumors without (A) and with (B) "microwave renaturation".

TABLE 2 Percentages of ER Positivity on Paraffin-Embedded Breast Tumor Sections With Same Patient's Characteristics for Three Different Fixation Protocols and Their Repartition on Proportion and Intensity Score

Proportion Score

Fixation Protocol	Positivity	Score 2	Score 3	Score 4
1	67%	0%	38%	62%
2	46%	8%	59%	33%
3	17%	33%	67%	0%

Intensity score

	Score 1	Score 2	Score 3
1	0%	88%	12%
2	33%	42%	25%
3	67%	33%	0%

III. QUANTIFICATION OF SR IMMUNOCYTOCHEMICAL ASSAYS

Computerized quantitation of SR immunocytochemical assays is mainly done through absorption image cytometry; it offers many advantages: objectivity, consistency, numeric scale of results more comparable to the biochemical results, possible automation of the procedure, and clear definition of the tumoral heterogeneity.[38]

A. GENERAL METHODOLOGY

Image cytometry involves four successive steps: image acquisition, image segmentation and object labeling, object featuring, and data analysis.[39]

1. Image Acquisition

The optical image from the microscope is transformed into a digital image by electronic devices (camera, analog/digital converter); this image is composed of numbers which are measurements of the relative intensity of the light transmitted by the image at every pixel. Quantitation of colored labeling takes advantage of color analysis which features hue, saturation, and luminance of the nucleus. The luminance is related to the quantity of light emitted by a colored object, the hue is the attribute of color denoted by blue, green, yellow, or red, and the saturation is the purity factor of the color depending on the amount of white altering the hue.[40] For each nucleus, three images (red, green, and blue) are analyzed in each of the three channels of the color camera.

2. Image Segmentation

This consists of discerning the various objects which may lie within the image. The delineation of whole nuclei is made on the red image; it leads to the computation of a binary image where each pixel is assigned to a class of objects (nucleus or background here) and where cell nuclei are labeled. Various procedures for image cleaning and object isolation can be used.

3. Object Featuring

The nuclear ER or PR content and concentration are estimated, respectively, by the integrated optical density (IOD) and the mean optical density (MOD) measurements performed on the blue image. The labeled vs. nonlabeled cell detection is estimated by the mean parameter computed on the hue image.

4. Data Analysis

Several indices can be computed:[11]

- Labeling index: ratio of the number of labeled nuclei and the total number of nuclei.
- Mean optical density (MOD), which measures the cell SR concentration.
- The Quick score, equal to the product of the labeling index times MOD.
- Standard deviation of MOD represents the heterogeneity of the cell population for SR.
- Skewness, standard deviation, and kurtosis of the nuclear optical densities, which are representative of the heterogeneity of SR-labeled positive sites within the nucleus.[41]
- Percentage of positive surface.

5. Quality Control

An image analyzer must have acceptable levels of linearity, shading correction, stability, and geometric distorsion;[42] the European "PRESS" project is going to define a normalized procedure to test the quality of the instrumentation used to do densitometric measurements. However, to date, the different characteristics of the various analyzers commercially available, the different algorithms used for image segmentation and object featuring, and the specific software modules lead to great variations and difficulties in comparing the studies published.

Comment

Mean optical density can only be measured on entire cells (fine needle aspirates, imprints) and not on frozen or paraffin-embedded sections for obtaining unbiased SR concentration by the cell.

B. RESULTS IN BREAST TUMORS

Preliminary results from Charpin et al.[10] showed that image cytometry provides accurate and reliable data. Using the same image analyzer, Cohen et al.[11] demonstrated, on breast tumor imprints a high correlation (r = 0.74) between the biochemical assay values and the score derived from the mean ER concentration per positive tumor cell and the labeling index (Quick score).

Expressing the results as positive or negative after thresholding of a quantitative immunocytochemical score done on tumor sections, Bacus et al.[43] showed a sensitivity of 98% and a specificity of 100% of image analysis vs. biochemical estrogen receptor analysis. A linear correlation (r = 0.86) was found between this score and the results of biochemical assay.

Simony et al. agreed that "computer assisted image analysis is an effective method of observing and objectively quantitating ER content" in breast carcinoma and emphasized the possible assessment of SR expression heterogeneity by this method.[44]

Other studies demonstrated the same kinds of correlations between biochemical and image analysis methods, both for ER[45] and PR.[38] Our results about PR image cytometry done on breast tumor imprints also showed a high degree of correlation with the biochemical assay. The highest correlation was obtained with the calculated parameter: P = (MOD of marked cells minus MOD of nonmarked cells) multiplied by the percentage of marked cells (r = 0.92). This product take into account the overestimation of the nucleus MOD due to the counterstain. Furthermore, tumors expressing PR heterogeneity, as revealed by high values of SD of MOD or SD and skewness of the optical density histogram, were larger and had higher Scarff-Bloom grades than homogeneous ones.[41]

In the study of McClelland et al., all clinically hormone responding patients were found to have ER-positive cells by image analysis.[45] In a very recent paper, Querzoli et al. demonstrated a longer relapse-free interval for patients with ER and PR positive tumors, the cut-off value being 10% of the positive nuclear area.[46]

IV. CONCLUSION

Visualization of steroid receptors offers many possibilities for functional or structural detection. These techniques are used more and more because of the often smaller size of tumor specimens.

Any method should be standardized and necessitates a good quality control program at each step of its realization: from the fixation to the quantification. Only cultured cells can be quantified for unbiased results.

REFERENCES

1. Jungblut P. W., DeSombre E. R. and Jensen E. V., Estrogen receptors in induced rat mammary tumor. In *Hormone in genese und therapie des mammacarcinoms*, A. d. D. A. d. W. z. Berlin (Eds). Akademie-Verlag-Berlin, Berlin, 1967, 109.
2. Pertschuk L. P., Tobin E. H., Gaetjens E., Brigati D. J., Carter A., Kim D. S., Degenshein G. A. and Bloom N. D., Histochemical assay of steroid hormone receptors. In *Perspectives in Steroid Receptor Research*, F. Bresciani (Ed). Raven Press, New York, 1980, 299.
3. Jensen E. V., Suzuki T., Kawashima T., Stumpf W. E., Jungblut P. W. and De Sombre E. R., A two-step mechanism for the interaction of oestradiol with rat uterus. *Proc. Natl. Acad. Sci. U.S.A.*, 59, 632, 1968.
4. Sheridan P. J., Buchanan M., Anselmo V. C. and Martin P. M., Equilibrium: the intracellular distribution of steroid receptors. *Nature*, 282, 579, 1979.
5. King R. J. and Greene G. L., Monoclonal antibodies localise oestrogen receptor in the nucleus of target cells. *Nature*, 307, 745, 1984.
6. Rayter Z., Steroid receptors in breast cancer. *Br. J. Surg.*, 78, 528, 1991.

7. Heuson J. C., Mattheiem W. H., Longeval E., Deboel M. C. and Leclercq G., Clinical significance of the quantitative assessment of estrogen receptors in breast cancer. In *Hormones and Breast Cancer*, INSERM (Eds). INSERM, Paris, 1975, 57.

8. McGuire W. L., Steroid hormone receptors in breast cancer treatment strategy. In *Recent Progress in Hormone Research*, R. O. Greep (Ed). Academic Press, New York, 1980, 135.

9. Masood S., Estrogen and progesterone receptors in cytology: a comprehensive review. *Diagn. Cytopathol.*, 8, 475, 1992.

10. Charpin C., Martin P. M., Lissitzky J. C., Jacquemier J., Kopp F., Pourreau-Scneider N., Lavaut M. N. and Toga M., Estrogen receptor immunocytochemical assay and laminin detection in 130 breast carcinomas and computerized (Samba 200) multiparametric quantitative analysis on tissue sections. *Bull. Cancer*, 73, 651, 1986.

11. Cohen O., Brugal G., Seigneurin D. and Demongeot J., Image cytometry of estrogen receptors in breast carcinomas. *Cytometry*, 9, 579, 1988.

12. Martin P. M. and Sheridan P. J., Intracellular localization of steroid hormone receptors. *Ann. Pathol.*, 6, 115, 1986.

13. Martin P. M. and Sheridan P. J., Towards a new model for the mechanism of action of steroid. *J. Steroid Biochem.*, 16, 215, 1982.

14. Martin P. M. and Rolland P. H., Human mammary tumours: a comparison of methods for detecting intracellular steroid binding sites. In: *Reviews on Endocrine-Related Cancer*, ICI (Eds). ICI, Nice, 1981, 59.

15. Stumpf W. E. and Sar M., Radioautographic techniques for localizing steroid hormones. In *Methods in Enzymology*, Academic Press, New York, 1975, 135.

16. Martin P. M., Magdelenat H. P., Benyahia B., Rigaud O. and Katzenellenbogen A., New approach for visualizing estrogen receptors in target cells using inherently fluorescent ligands and image intensification. *Cancer Res.*, 43, 4956, 1983.

17. Martin P. M., Horwitz K. B., Ryan D. S. and McGuire W. L., Phytoestrogen interaction with estrogen receptors in human breast cancer cells. *Endocrinology*, 103, 1860, 1978.

18. Katzenellenbogen J. A., Carlson K. E., Bindal R. D., Neeley R. L., Martin P. M. and Magdelenat H. P., Fluorescence-based assay of estrogen receptor using 12-oxo-9(11)-dehydroestradiol-17β. *Anal. Biochem.*, 159, 336, 1986.

19. Berthois Y., Pourreau-Schneider N., Gandilhon P., Mittre H., Tubiana N. and Martin P. M., Estradiol membrane binding sites on human breast cancer cell lines. Use of fluorescent estradiol conjugate to demonstrate plasma membrane binding systems. *J. Steroid Biochem.*, 25, 963, 1986.

20. Nairn R. C., Fluorescent labeling of sera. In *Fluorescent Protein Tracing*, R. C. Nairn (Ed). Williams & Wilkins, Philadelphia, 1969, 303.

21. Greene G. L., Nolan C., Engler J. P. and Jensen E. V., Monoclonal antibodies to human estrogen receptor. *Proc. Natl. Acad. Sci. U.S.A.*, 77, 5115, 1980.

22. Pertschuk L. P., Eisenberg K. B., Carter A. C. and Feldman J. G., Immunohistologic localization of estrogen receptors in breast cancer with monoclonal antibodies. Correlation with biochemistry and clinical endocrine response. *Cancer*, 55, 1513, 1985.

23. Al Saati T., Clamens S., Cohen-Knafo E., Faye J. C., Prats H., Coindre J. M., Wafflart J., Caveriviere P., Bayard F. and Delsol G., Production of monoclonal antibodies to human estrogen-receptor protein (ER) using recombinant ER (RER). *Int. J. Cancer*, 55, 651, 1993.

24. Ali S., Lutz Y., Bellocq J. P., Chenard-Neu M. P., Rouyer N. and Metzger D., Production and characterization of monoclonal antobodies recognising defined regions of the human oestrogen receptor. *Hybridoma*, 12, 391, 1993.

25. Henry L., Angus B., Lennard T. W. J., Hennessy C., Wright C., Henry J. A., Anderson J. and Horne C. H. W., New monoclonal antibodies to estrogen and progesterone receptor: characterization and clinical studies in human breast cancer. *J. Pathol.*, 168, abstract, 1993.

26. Pertschuk L. P., Kim D. S., Nayer K., Feldman J. G., Eisenberg K. B., Carter A. C., Rong Z. T., Thelmo W. L., Fleisher J. and Greene G. L., Immunocytochemical estrogen and progestin receptor assay in breast cancer with monoclonal antibodies. Histologic, demographic, and biochemical correlations and relationship to endocrine response and survival. *Cancer*, 66, 1663, 1990.

27. Pertschuk L. P., Feldman J. G., Eisenberg K. B., Carter A. C., Thelmo W. L., Cruz W. P., Thorpe S. M., Rose C., Christensen I. and Greene G. L., Immunohistological detection of progesterone receptors in breast cancer with monoclonal antibody. Relation to biochemical assay, disease-free interval and clinical endocrine response. *Cancer*, 62, 342, 1988.

28. Seymour L., Meyer K., Eser J., MacPhail A. P., Behr A. and Bezwoda W. R., Estimation of PR and ER by immunocytochemistry in breast cancer : comparison with radioligand binding methods. *Am. J. Clin. Pathol.*, 94, S35, 1990.

29. Vu Hai M. T., Jolivet A., Ravet V., Lorenzo F., Perrot-Applanat M., Citerne M. and Milgrom E., Novel monoclonal antibodies against human uterine progesterone receptor. *Biochem. J.*, 260, 371, 1989.

30. Jacquemier J. D., Hassoun J., Torrente M. and Martin P. M., Distribution of oestrogen and progesterone receptors in healthy tissue adjacent to breast lesions at various stages. Immunohistochemical study of 107 cases. *Breast Cancer Res. Treat.*, 15, 109, 1990.

31. Kopp F., Martin P. M., Rolland P. H. and Bertrand M. F., A preliminary report on the use of immunoperoxidases to study binding of estrogens in rat uteri. *J. Steroid Biochem.*, 11, 1081, 1979.

32. Masood S., Immunocytochemical localization of estrogen and progesterone receptors in imprints preparations of breast carcinomas. *Cancer*, 70, 2109, 1992.

33. McClelland R. A., Berger U. and Miller L. S., Immunocytochemical assay for estrogen receptor: relationship to outcome of therapy in patients with advanced breast cancer. *Cancer Res.*, 46, 4241, 1986.

34. Hawkins R. S., Sangster K. and Tesdale A., The cytochemical detection of oestrogen receptors in fine needle aspirates of breast cancer: correlation with biochemical assay and prediction of response to endocrine therapy. *Br. J. Cancer*, 58, 77, 1988.

35. Jordan V. C., Jacobson H. I. and Keenan E. J., Determination of oestrogen receptor in breast cancer using monoclonal antibody technology: result of a multicentre study in the United States. *Cancer Res.*, 46, 4237, 1986.

36. De Rosa C. M., Ozzello L., Habif D. U., Konrath J. C. and Greene G. L., Immunohistochemical assessment of estrogen and progesterone receptors in stored imprints and cryostat sections of breast carcinoma. *Ann. Surg.*, 212, 224, 1989.

37. Ryde C. M., Smith D., King N., Trott P. A., MacLennan K., McKinna J. A., Minasian H. and Dowsett M., Comparison of four immunochemical methods for the measurement of oestrogen receptor levels in breast cancer. *Cytopathology*, 3, 155, 1992.

38. Auger M., Katz R. L., Johnston D. A., Sneige N., Ordonez N. G. and Fritsche H., Quantitation of immunocytochemical estrogen and progesterone receptor content in fine needle aspirates of breast carcinoma using SAMBA 4000 image analysis system. *Anal. Quant. Cytol. Histol.*, 15, 274, 1993.

39. Brugal G., Image analysis of microscopic preparations. In *Methods and Achievmements in Experimental Pathology*, G. Jasmin and L. Proschek (Eds). Karger, Basel, 1984, 1.

40. Garbay C., Modélisation de la Couleur dans le Cadre de l'Analyse d'Images et de son Application à la Cytologie Automatique. *Ingineer Thesis*, University of Grenoble, France, 1979.

41. Seigneurin D., Cohen O. and Louis J., Immunocytochemistry and image cytometry of progesterone receptors in breast carcinoma imprints. *Anal. Cell. Pathol.*, 1, 97, 1989.

42. Giroud F. and Brugal G., Absorption image cytometry. In *Cell and Tissue Culture: Laboratory Procedures*, A. Doyle, J. B. Griffthic and D. G. Newel (Eds.). John Wiley & sons, Chichester, 1994, 1.

43. Bacus S., Flowers J. L., Press M. F., Bacus J. W. and McCarty K. S., The evaluation of estrogen receptor in primary breast carcinoma by computer assisted image analysis. *Am. J. Clin. Pathol.*, 90, 233, 1988.

44. Simony J., Pujol L. J., Grenier J. and Pujol H., Heterogeneity of estrogen receptor in primary breast carcinoma evaluated by computer image analysis. *Am. J. Clin. Pathol.*, 91, 559, 1989.

45. McClelland R. A., Finlay P., Walker K. J., Nicholson D., Robertson J. F., Blamey R. W. and Nicholson R. I., Automated quantitation immunocytochemically localizes estrogen receptors in human breast cancer. *Cancer Res.*, 50, 3545, 1990.

46. Querzoli P., Ferretti S., Albonico G., Magri E., Scapoli D. and Indelli M., Application of quantitative analysis to biologic profile evaluation in breast cancer. *Cancer*, 76, 2510, 1995.

Chapter **12**

IMMUNOELECTRON MICROSCOPIC LOCALIZATION OF LIGANDS AND RECEPTORS

Christine Brisson
Gérard Morel

CONTENTS

I. INTRODUCTION

The visualization of receptors which have been biochemically characterized in tissues remains an open question. More detailed understanding of this localization is required at both cellular and subcellular level, in order to summarize the function of these receptors.

The methods developed at the ultrastuctural level used ligand detection or direct observation of the receptor molecule. The ligand can be revealed by a marker which is linked to its molecule, or by its antigenicity. The marker used, which cannot modify its biological effect, is a radioactive istope (i.e., iodine-125) revealed by radioautography (see Chapter 7), or by a tag (i.e., biotin, fluorescein, or other antigenic molecule) detected by immunocytology. When it is available, direct detection of the ligand with an anti-ligand serum, by immunocytology remains the best solution. In the same way, direct observation of the receptor molecules by immunocytology gives complementary observations.

The main problem which remains, is the low expression of these receptor molecules in the cells and the low uptake of the ligand in their target cells. Thus, the method of detection at the electron microscopic level must be extremely sensitive. The results obtained are always checked with care and, if possible, correlated with structure-function studies.

II. TISSUE PREPARATION

A. FIXATION
Successful visualization of antigens in tissue or cells depends on the fixation steps. Fixatives induce precipitation and intra- and/or intermolecular bridges, which can affect the antigenicity. In this regard, there are several parameters which must be considered. Cells and tissues are generally composed of aqueous matter, and the fixation step must be able to fix them in a state as close as possible to their native structures. At this point one paradox remains: the fixative solutions involve protein modification, and the majority of detectable antigens are also proteins. Rapid fixation is necessary in order to preserve morphology and avoid any loss of antigenicity. However, there is no ideal fixative which can preserve the totality of the tridimensional relationships between proteins and the specificity of the immune reaction. For each antigen, the best fixative has to be found empirically. Some compromises have to be reached between morphological preservation and antigen detection.

Fixative solutions such as alcohols or acetone, which precipitate proteins without affecting their immunoreactivity, are not compatible with the preservation of ultrastructural morphology. Aldehydes must be used; they induce the formation of crosslinks between proteins with good morphological preservation. Paraformaldehyde (see Protocol 1) is the one most commonly used in association with phosphate or cacodylate buffer (see appendix). Preservation of subcellular structures is compatible with the immunoreactivity of most antigens.

PROTOCOL 1: PARAFORMALDEHYDE FIXATION

- Tissues and cells are fixed by immersion (1 to 2 h) in 2 to 4% freshly prepared paraformaldehyde in 0.1 *M* phosphate buffer (pH 7.4), (see Appendix).
  - *Add 40 g of paraformaldehyde to 100 ml of water.*
  - *Heat to 60°C and alkalinize with drops of 1 M NaOH to dissociate the paraformaldehyde into formaldehyde. Store this solution at room temperature.*
- Tissue can be fixed by vascular perfusion (15 min), followed by immersion in a fixative solution.
- Cells in suspension are fixed in suspension and pelleted by centrifugation at each step. When adherent cells are grown in plastic material, they are fixed *in situ*.

The dialdehyde, e.g., glutaraldehyde, limits the accessibility of antibodies in tissue, but gives the best preservation of morphological structures. It could be used in low concentration (<0.5%) in combination with paraformaldehyde (see Protocol 2): these solutions yield better tissue preservation and lower background staining. Paraformaldehyde penetrates tissue more rapidly than glutaraldehyde and also stabilizes the structures, which can then be fixed more permanently by glutaraldehyde.

PROTOCOL 2: PARAFORMALDEHYDE-GLUTARALDEHYDE FIXATION[1]

- *The fixative solution is a mixture of freshly prepared paraformaldehyde (2 to 4%) and glutaraldehyde (0.05 to 0.1%) in 0.1 M phosphate buffer (pH 7.4).*
- In this type of fixation the phosphate buffer can be replaced by cacodylate buffer (see Appendix).

Other fixative solutions tried out were associated with paraformaldehyde: Zamboni and De Martino describe a mixture comprising picric acid; its rapid penetration gives good preservation of tissue, especially for renal biopsies (see Protocol 3).[2]

PROTOCOL 3: PICRIC ACID AND FORMALDEHYDE FIXATION

- *Add 4 g of paraformaldehyde to 15 ml of a filtered satured aqueous solution of picric acid.*
- *Heat to 60°C and alkalinize with drops of 1 M NaOH to dissociate the paraformaldehyde into formaldehyde.*
- *Cool, filter, and adjust the volume up to 100 ml with 0.1 M phosphate buffer.*
- *This fixative is very stable and can be stored at room temperature for several months.*

Paraformaldehyde and lysine-phosphate-periodate buffer (PLP) have been specially described for immunoelectron microscopy by McLean and Nakane in such a way as to stabilize antigens in tissue without destroying their antigenicity (see Protocol 4):[3]

PROTOCOL 4: PARAFORMALDEHYDE-LYSINE-PHOSPHATE BUFFER FIXATION

Materials
Stock A = 0.1 M lysine-0.05 M sodium phosphate pH 7.4

- Dissolve 1.8 g of l-lysine HCl in 50 ml of water (0.2 *M* lysine HCl).
- Adjust to pH 7.4 with 0.1 *M* disodium phosphate.
- Make up to 100 ml with 0.1 *M* sodium phosphate.

Stock B = 8% paraformaldehyde

- Add 2 ml of 40% paraformaldehyde stock solution to 10 ml of phosphate buffer.

Stock C: aliquot sodium metaperiodate (21.4 mg for 10 ml).

Method

Just before use, mix 3 vol of stock A with 1 vol of stock B and add an aliquot of sodium metaperiodate. The final concentrations are 0.01 M sodium metaperiodate, 0.075 M l-lysine, 0.037 M phosphate buffer, 2% paraformaldehyde, and the pH is approximately 6.2.

In some cases another monoaldehyde, acrolein, has been used to localize neuropeptides.[4] Microwave ovens are now sometimes used to produce a rapid fixation by the rapid movement of molecules during heating. However, this means adjusting power and heat individually for each type of tissue and fixative.

To achieve rapid and uniform diffusion of the fixative in the tissue, fixative perfusion is the method of choice. However, the technique most commonly used after aldehyde fixation consists of quickly freezing the samples to prevent the loss of antigens. The preparation of samples by congelation is treated in Chapter 14.

The immunocytological reaction is glutaraldehyde sensitive. Good morphological results can be obtained with rapid paraformaldehyde fixation (<1 h). Alternatively, tissue may be left for longer periods, up to several weeks. In general, part of the immune reaction will be sacrificed. The choice of the fixative remains the prime necessity for good antigenicity preservation.

B. SAMPLES

Thick tissue sections can be obtained with a cryostat or, for better morphological preservation, with a tissue slicer or vibratome (see Protocol 5). Sections between 30 and 50 μM are then processed in suspension and incubated using the pre-embedding methods described in Section VI.A. For electron microscopic observations, the tissue slicer gives better morphological preservation.

PROTOCOL 5: VIBRATOME SECTIONING

Samples are fixed (see Protocol 1) and washed in buffer.
Depose one drop of cyanolit on a vibratome support.
Place sectioned blocks (~1 cm³) on this drop.
Fix the support in the vibratome jar and fill up with buffer.
Make the sections.

Several parameters have to be determined before sectioning: speed, vibration of cutting, and thickness of the sections depend on the type of tissue and the fixation. The sections are then processed by a pre-embedding method.

Monolayers of cells grown directly in Petri dishes are generally fixed in the presence of a serum-free growth medium. After 15 min the medium is removed and replaced by the fixative solution. All the samples are then processed *in situ* and at this stage the detection of intracellular antigens can be obtained under certain conditions. Aldehyde fixatives lead to the formation of an impermeable barrier in the cell membrane and the cross-linked cytoplasmic proteins. The accessibility of antibodies to their intracellular antigens can be facilitated by mild permeabilization using detergents such as saponin, Triton X100, or digitonin (see Protocol 15). Plastic dishes are incompatible with the classical embedding techniques and need special treatments described in Section III.B (see Protocols 9 and 10).

Cells in suspension can be treated either in suspension or in agarose pellets. In the first case, at each step cells are sedimented by centrifugation, then gently resuspended in the next solution. The final step, that of embedding in resin, is described in Section III.B. In the second case, fixed cells are pelleted in 2% agarose (low melting point) at 37°C. The agarose is then solidified at 4°C. The pellets are cut into small blocks (<1 mm^3) and processed in the same way as small blocks of tissue.

In some cases, other embedding materials have been explored as alternatives to cell preparations, including glutaraldehyde-cross-linked bovine serum albumin (15 to 30%).[5,6] For each antigen determination, three parameters have to be determined in order to obtain the best combination of morphological preservation and immunological detection: the nature of the fixative, the time allowed, and the temperature of the fixation.

III. INFILTRATION IN RESIN

After osmium tetroxide fixation, if this is not incompatible with the resin used, the samples are extensively washed in distilled water or in buffer. Dehydration in graduated alcohols (30°, 50°, 70°, 95°, 2 × 100°) removes all free water from the specimen and replaces it with ethanol, which is then replaced by the embedding resin. These media, which are more or less viscous, are hardened by polymerization.

All the embedding methods require resin infiltration of the samples before the final polymerization step.

A. EPOXY RESINS: ARALDITE, EPON

The reaction between epoxy resins and one or two hardeners is slow at room temperature, but can be sped by the addition of an accelerator. Polymerization occurs at 60°C. The final hardness of the blocks depends on the proportions of the different components. The following protocols (Protocols 6 to 8) describe classical embedding in araldite, epon, and an araldite-epon mixture. With these resins, an infiltration step in a solvent such as propylene oxide is carried out so as to obtain better infiltration than if using ethanol.

PROTOCOL 6: EMBEDDING IN ARALDITE[7]

Materials

Ethanol 100°
Araldite CY 212
Hardener HY 964 (dodecenyl succinic anhydride, DDSA)
Accelerator DY 064 (tridimethylaminomethyl phenol)
Plasticizer, dibutyl phthalate

Method

1. *Ethanol substitution* (1 h)

Araldite CY 212	1 vol
Hardener HY 964	1 vol
Ethanol 100°	2 vol

2. *Resin infiltration* (3 × 6 to 8 h)

Araldite CY 212	1 vol
Hardener HY 964	1 vol

3. *Embedding* (1 h)

Araldite CY 212	10 ml
Hardener HY 964	10 ml
Accelerator DY 064	0.5 ml
Plasticizer	1 ml

The araldite mixture is readily soluble in ethanol or acetone. The four components are mixed and left to polymerize for 2 or 3 days at 48°C or overnight at 60°C to yield a light gold block.

PROTOCOL 7: EMBEDDING IN EPON[8]

Materials

Propylene oxide
Epon 812 or LX 112
Hardener DDSA
Hardener MNA (methyl nadic anhydride)
Accelerator DMP30 (2-4-6-tridimethylaminomethyl phenol)

Method

1. *Stock solutions: Epon A: mix 62 ml Epon with 100 ml DDSA; Epon B: mix 100 ml Epon with 89 ml NMA.*

Epon A and B solutions can be kept 6 months at 4°C. Epon is less viscous than araldite. Embedding is more rapid and different hardnesses can be obtained according to the proportion of epon A and B, and by the addition of 1.5% DMP 30.

	Hard	⇐	⇔	⇒	**Soft**
A (ml)	0	3	5	7	10
B (ml)	10	7	5	3	0

2. *Infiltration in epon*: After ethanol dehydration and propylene oxide (2 × 15 to 30 min), the samples are infiltrated first with a mixture of propylene oxide and epon, and then with pure epon:

Propylene Oxide	Epon	
2 vol	1 vol	1 h or more
1 vol	1 vol	1 h or more
1 vol	2 vol	1 h or more
	Pure epon	1 h or overnight at 4°C

3. *Embedding*: the samples are then placed in gelatin capsules with fresh epon and heated to polymerize at 37°C overnight, followed by 24 h at 60°C or overnight at 60°C.

PROTOCOL 8: EMBEDDING IN ARALDITE-EPON[9]

Materials

Propylene oxide
Epon 812
DDSA
Araldite CY 212
Plasticizer, dibutyl phthalate (DBPht)

Method

1. *Resin preparation:* add successively:
Epon	25 ml
DDSA	55 ml
CY 212	15 ml
DBPht	2 to 4 ml
DMP 30	1.5% of the final volume
2. Infiltration: after ethanol or acetone dehydration, the samples are passed through:
   - Propylene oxide for 2 × 15 min
   - Propylene oxide/resin, 2 vol: 1 vol, 1 h
   - Propylene oxide/resin, 1 vol: 1 vol, 1 h
   - Resin 12 h
   - Fresh resin 1 to 2 h
3. *Embedding:* the samples are placed in gelatin capsules and polymerized either for 2 h at 37°C followed by 3 days at 60°C, or overnight at 40 to 80°C.

B. PARTICULAR EMBEDDING PROTOCOLS

Specific infiltration techniques have been developed to use epon directly on plastic material such as Petri culture dishes, which are incompatible with solvent. Treatment with ethanol/epon mixture is often followed by the depositing of a thin layer of resin polymerized at 37°C overnight and then at 60°C (24 h). Inverted capsules filled with epon are set on selected cells, polymerized at 60°C, and then detached by a thermial shock of liquid nitrogen (see Protocol 9).

PROTOCOL 9: EMBEDDING OF CULTURED CELLS IN ETHANOL/EPON

- *For epon preparation see Protocol 7.*
- *After ethanol dehydration, the resin infiltration is carried out with ethanol, omitting the propylene oxide step:*

Ethanol 100°: Epon	*2 vol:1 vol*	*1 h or more*
	1 vol:1 vol	*1 h or more*
	1 vol:2 vol	*1 h*
Two steps with fresh epon		*1 h*

- *Keep a thin layer of resin and polymerize overnight at 37°C to eliminate all traces of ethanol and continue at 60°C for 24 h.*
- When the resin is polymerized, take the dishes from the oven and put them directly on liquid nitrogen. By the thermal shock, the cell monolayer is released with the thin polymerized film of epon. Then epon-filled capsules are inverted on selected cells and polymerized overnight at 60°C.

Brinkley et al. have described the infiltration of epon in Petri dishes with hydroxypropylmethacrylate (*HPMA*) instead of ethanol;[10] this method had also been described by Tougard and Picart.[11] A thin layer of epon is polymerized at 60°C. Then, selected cells are taken off using a prewarmed punch biopsy.[12] The small dishes of resin obtained are then applied on a prepolymerized epon block using cyanolite glue (see Protocol 10).

PROTOCOL 10: EMBEDDING OF CELLS *IN SITU* IN HPMA-EPON

1. *Dehydration:*

Ethanol 30°	15 min
Ethanol 50°	15 min
Ethanol 70°	15 min
Ethanol 90°	10 min
HPMA 90% in distilled water	3 × 5 min
HPMA 95% in distilled water	15 min
HPMA 97% in distilled water	15 min

2. *Infiltration in epon:*

HPMA/ epon	2 vol:1 vol	15 min
	1 vol:1 vol	15 min
	1 vol:2 vol	30 min

3. *Embedding in epon:* pour a thin layer of epon on the cells; keep overnight at 37°C and 3 days at 60°C.

Cells grown on coverslips or tissue sections processed on glass slides are embedded in resin as described, then removed from the glass during polymerization, which is continued at 60°C overnight.

Cells in suspension are infiltrated with the mixture of propylene oxide-epon only. After centrifugation, the pellet is layered on a capsule filled with fresh resin and centrifuged. Cells form a small pellet in pure resin and are polymerized overnight at 60°C.

C. METHACRYLATES

Methacrylate monomers are colorless liquids with low viscosity and are soluble in ethanol. They are mixed with one or more methacrylates, along with a catalyst and benzoyl peroxide, which removes the inhibitor, hydroquinone, present during storage to prevent polymerization from occurring. The major disadvantages of these embedding resins arise during the polymerization process (disrupted structures, bubbles of gas, etc.), but ultrathin sections are also unstable under electron beams.

PROTOCOL 11: EMBEDDING IN GLYCOL METHACRYLATE[13]

Materials

Glycol methacrylate (GMA)
n-Butylmethacrylate
Luperco

All the infiltration steps are carried out after aldehyde fixation without osmium fixation and ethanol dehydration, at 4°C.

Method

1. *Resin substitution*

GMA 80% in water	20 min
GMA 97%	20 min
Unpolymerized mixture:	20 min
n-Butylmethacrylate	9 ml
Luperco	0.18 g
97% GMA	21 ml

Dissolve the luperco in the butylmethacrylate, using a glass rod, before the addition of the GMA.

2. *Prepolymerized mixture:* In an Erlenmeyer (50 ml) mix to obtain 1 cm maximum at the bottom:

n-Butylmethacrylate	9 ml
Luperco	0.18 g
97% GMA	21 ml

Apply heat to boiling while swirling over a flame or in a 90°C water bath until the mixture becomes viscous. Cool immediately in an ice bath with swirling. This prepolymerized resin can be stored at –20°C.

3. Embedding:
   Incubation at 4°C overnight
   Samples are transferred into gelatin capsules, which are top filled with the prepolymerized mixture and polymerized under UV light, 315 nM at 4°C for 48 h.

D. HYDROPHILIC RESINS

Several hydrophilic embedding media have been investigated with a view to reducing the possible undesirable effects of epoxy processing and enhancing or stabilizing the antigenic sites in tissue.

Lowicryls: Carlemalm et al.[14,15] have developed low-temperature embedding media known as Lowicryl resins, which are hydrophilic resins composed of polar acrylates and methacrylates (K4M and K11M), or hydrophobic resins composed of apolar acrylates and methacrylates (HM20, HM23). Lowicryls are mixtures of low viscosity which polymerize under UV light at temperatures between –30°C and –80°C. They give improved preservation of molecular organization and antigenicity (see Chapter 14).

London Resin White (LR White): this hydrophilic resin has been described by Newman et al.[16] without an accelerator (see Protocol 12). Low concentrations of osmium tetroxide may be used to enhance the morphological structures. For cold curing, one drop of the accelerator should be added per 10 ml of resin. This process is highly exothermic and this mixture polymerizes at room temperature in less than 9 min.

PROTOCOL 12: EMBEDDING IN LR WHITE

1. *Ethanol dehydration*
   Ethanol 30° 1 h or more
   Ethanol 50° 30 min
   Ethanol 70° 30 min (or storage at –20°C)
   Ethanol 100° 2 × 10 min
2. *Resin substitution*

Ethanol 100°/LR White	2 vol/1 vol	1 h
	1 vol/1 vol	1 h
	1 vol/2 vol	1 h
Pure LR White at 4°C		overnight

At each step keep the tubes very well closed to avoid oxidation.

3. *Embedding without accelerator*

Samples are incubated in a fresh batch of resin for 1 h.

Gelatin capsules are filled with LR White, samples are set on the resin and sink down through it.

The polymerization is performed at 50°C.

4. *Embedding with accelerator*

The mixture of LR White and accelerator is prepared immediately before use, polymerization taking place in less than 10 min; all the products and the samples are kept at 4°C.

Mix and vortex immediately for 1 min:

LR White 5 ml

accelerator 5 μl

Beem capsules or easy molds are used:

Set one drop of the mixture in each capsule

Layer samples

Fill up immediately with the mixture

Polymerize at 4°C for 24 to 48 h

Continue at 37°C for 24 h

During polymerization, it is important to avoid the resin from coming into contact with oxygen. The procedure can be carried out using gelatin, easy molds, or beem capsules filled with the resin and tightly closed. Polymerization occurs at 50°C for 24 h. When accelerator is added, the polymerization should be carried out at 4°C (24 h), followed by 1 day at 37°C.

The extremely low viscosity of LR White can be exploited when cells are in suspension. Small drops of concentrated cells fall to the bottom of the Beem capsules, avoiding the centrifugation step.

Immunocytological detection can be performed after embedding in these hydrophilic resins. Ultrathin sections are mounted on nickel grids and directly incubated with the antibodies. This "on-grid staining" is becoming more and more commonly used, given the advantages of colloidal gold conjugates (see Section V.B.2).[17]

IV. IMMUNOGLOBULINS (IgGs)

Immunocytochemical detection is based on the principle that antibodies react with antigens to form a specific complex held together by noncovalent bonds. Several Protocols have been developed to visualize this complex, using antibodies coupled with different markers. Up to now an appreciable number of immunological methods have been applied with success.

In the following part of this presentation, basic information and procedures will be provided and an overview of some useful protocols will be offered. More details can be found in several important well known works.[18,19]

A. MOLECULAR STRUCTURE OF IgGs

Antibodies are the pivotal reagents of all immunological methods whose results will therefore tend to reflect their specificity.

Antibodies belong to the immunoglobulin family, which is designated as the *Ig supergene family.* They are glycoproteins and are distributed in the plasma and various other bodily fluids. They are produced in large quantities by mature B cells. The members of this family present a large homology in their sequence or their structural organization. They differ by their molecular weights, their amino acids, and their oligosaccharide sequences. After electrophoresis they are distributed between the γ and α fractions of normal serum in the following descending order: IgG, IgA, IgM, IgD, and IgE .

An immunoglobulin is a unit consisting of two identical light chains (L) and two identical heavy chains (H) (Figure 1). A number of different regions, composed of about 110 amino acids each, are found along the H and L chains. The immunoglobulin characteristics are supplied by the H chains: the antigenic or structural properties, the class determination (γ, α, ...), and subclass classification (IgG$_1$, IgG$_2$, ...). The H chains give the Ig classes their names and are organized in a single amino-terminal variable region (V) and three or four constant domains (C). The first and the second regions are separated by a hinged domain, which gives the molecule flexibility. The C domain of the H chain determines an antibody's specificity.

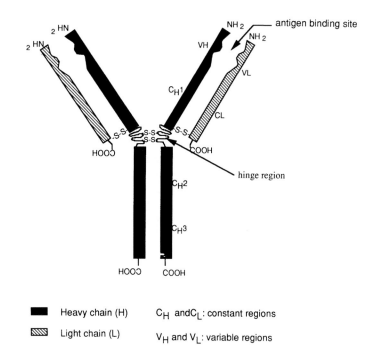

FIGURE 1 Typical structure of immunoglobulin G (IgG). It is composed of two identical light chains (L) and two heavy chains (H). The antigen binding sites are localized between the N-terminal parts of the two chains.

The two L chains are designed as kappa (κ) or lambda (λ) types and distributed specifically in all classes and different species. The κ type in human IgG is two times more expressed than the λ type. The κ chains represent 95% of the mouse IgG, which explains their predominance in mouse monoclonal antibodies. The carboxy terminal part of the L chain is called the C domain, while the amino terminal domain varies from one L chain to another and participates in antigen binding. Many inter- and intrachain disulfide bridges contribute to stability of molecules.

The tetrapeptidic structure of human IgG is also found in other species, with differences that are characteristic of each class: (1) the number of disulfide bridges, (2) their distribution along or between the two chains, and (3) the number of binding interchains, which goes from 2 (IgG_1 or IgG_4) to 15 (IgG_3).

IgG, IgD, and IgE are composed of the tetrapeptidic monomer form alone, with two L chains and two H chains, IgA are monomers or dimers in some animal species, and IgM is of a pentameric form. Despite this diversity, all the antibodies exhibit a common basic structure.

B. IgG FRAGMENTS: Fab, F(ab′)2

IgG is the immunoglobulin fraction that is most frequently used as antibodies. IgG fragments are evenly distributed between the intra- and extravascular compartments. In both humans and mice, they are made up of four subclasses: IgG_1, IgG_2, IgG_3, IgG_4. They mediate a wide range of functions from the neutralization of bacterial toxins to the targeting of bound antigens for destruction through phagocytosis or antibody-dependent cellular cytoxicity, which involves the specific Fcγ receptors.

The structure of IgG (146 kDa, or 170 kDa for IgG_3) shows that the H and L chains are composed of variable (V) and constant (C) domains, which are linked by inter-and intradisulfide bonds (Figure 1). Proteolytic digestion or reductive dissociation have determined the relation structure-function of the IgG:

- *Papain* proteolytic digestion cleaves the immunoglobulin in the inter-heavy-chain disulfide bridges of the hinge region and yields two antigen-binding fragments (Fab) and one crystalline fragment (Fc). The Fab fragment binds the antigen, while the Fc portion mediates the binding with the different components of the complement system: cellular binding (lymphocytes, monocytes, platelets), transplacental passage, or *Staphylococcus* protein A binding (Figure 2).
- *Pepsin* cleaves the gamma chain in the C-terminal side of the inter-heavy-chain disulfide bridges: one bivalent antigen-binding fragment F(ab′)2 is obtained, while the Fc fragment will be destroyed (Figure 3). Antibody formation is a complex process and will not be described here.[18,19]

The use of these Fab or F(ab′)2 fragments results in the reduction of the background due to the nonspecific binding of the Fc fragment in tissue; they can also be directly coupled with a marker, in which case the reduction in the size of the antibodies results in an enhanced intracellular reaction.

C. POLYCLONAL AND MONOCLONAL ANTIBODIES
1. Polyclonal Antibodies

These are produced by a number of different cells and react with a variety of epitopes on the antigen against which they are raised (Figure 4a). Rabbits are most

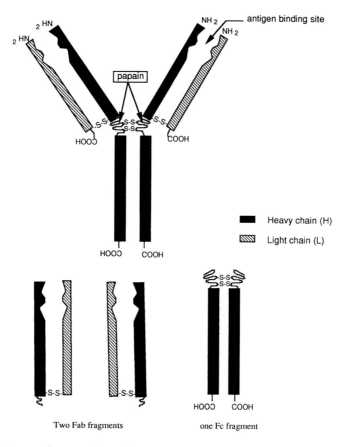

FIGURE 2 Papain cleavage. The results of this enzymatic cleavage are two Fab fragments and one Fc fragment.

frequently used for their production, followed by goats or sheep. Rabbits present some advantages: (1) easy maintenance, (2) rabbit antibodies do not cross-react with human proteins, (3) pools of antibodies obtained from many other animals vary to some extent between batches.

2. Monoclonal Antibodies

These are produced by clones of plasma cells, generally obtained in mice according to the technique described by Köhler and Milstein:[20,21] mice are boosted with the chosen antigen; after the immune response has been achieved, B lymphocytes from spleen are fused with nonsecreting mouse myeloma cells; hybrid cells are selected by the culture medium, cultured, and selected for the desired reactivity. Monoclonal antibodies present numerous advantages over their polyclonal counterparts, the most important being that a given monoclonal antibody clone reacts with a specific epitope on the antigen against which it was raised (Figure 4b). This also gives rise to the major disadvantage: if the epitope is rare it will not be possible to visualize the detection. Antibody sites represent about 15 amino acids in the variable region of Fab fragments; the H and L chains contribute to the formation and the

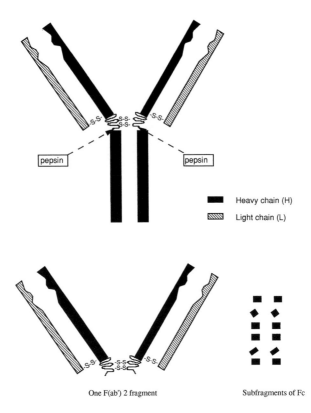

FIGURE 3   Pepsin cleavage. The results of this enzymatic cleavage are one F(ab')2 fragment and destruction of the Fc fragment.

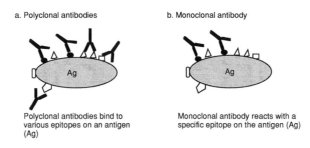

FIGURE 4   Polyclonal and monoclonal antibody binding sites on an antigen (Ag).

hypervariable domains determine the specificity. The corresponding epitope on the antigen may be composed of as little as three amino acids.

Antibody affinity is related to the ability of antibodies to form insoluble immune complexes. Polyclonal antisera are composed of different high- and low-affinity antibodies against antigen epitopes; the lowest affinity antibodies will be lost during washings after the specific incubation of a tissue sample, leaving only the high-affinity antibody complexes. On the other hand, monoclonal antibodies present the same affinity and therefore monoclonal antibodies of high affinity have to be selected in order to avoid their loss during the washing steps.

3. Protein A and Protein G

Protein A is isolated from the walls of *Staphylococcus aureus* and has two possible binding sites for the Fc fragment.[22,23] However, IgGs have only one binding site for Protein A which binds to one Fc in each of several IgG subclasses. Protein A offers the possibility of double labeling (see Section VI.C). Another protein isolated from the group G *Streptococci* also binds to the IgG Fc fragment with some differences between species: Protein G is more useful in mice, rats and goats.[24-26]

V. ANTIBODY TRACERS

The different tracers used in ultrastructural methods are divided into two major groups: the enzymatic tracers with their chromogens and the directly electron-dense tracers such as ferritin and colloidal gold.

A. ENZYMATIC MARKERS

Enzyme-labeled antibodies are used for the localization of a large number of cellular molecules. An enzyme covalently linked to an antibody can be revealed by the appropriate cytochemical reaction and can localize the corresponding antigen-antibody complex in the cells.

Immunoenzymatic staining methods consist in the conversion, by an enzyme reaction, of colorless chromogens into colored end products; horseradish peroxidase (PO) is used as a marker for light as well as for electron microscopy detection. Several procedures have been developed in either one step or two steps, or with amplification bridges.[27,28]

Enzymatic activities are dependent on several variables: substrate and enzyme concentrations, buffers (salts and their concentrations), pH, temperature, and light, as well as several cofactors such as metallic ions (Mg^{++}, Mn^{++}, Co^{++}).

The coupled enzyme should be stable in solution, and possible endogenous activity should be tested for, and where present, eventually inhibited before the staining reaction.

Certain electron-dense tracers (ferritin, colloidal gold, etc.) eliminate this enzymatic reaction and provide a means of direct observation under an electron beam. Colloidal gold staining, now commonly used, is referred to as immunogold staining or IGS.

1. Horseradish Peroxidase

Peroxidase (~40 kDa) is isolated from the root of the hoseradish plant. Its active site consists of an iron-containing heme group (hematin), which can be detected in the same way as other heme goups (red cells) present in the tissue. Peroxidase activity occurs in two phases: (1) the formation of an enzyme-substrate complex in the presence of an electron donor, the chomogen, which becomes colored after oxidation, and (2) peroxidase forms a complex with hydrogen peroxide, which breaks down into water and oxygen. The end products are insoluble and their color is dependent of the chomogen at the photonic level; at the electron microscopic level the intensity of the end products is enhanced by osmium metallic salt deposition on oxidized chomogen and appears very electron dense.

Peroxidase can be attached covalently to other proteins,[29] or noncovalently to antibodies in the peroxidase-antiperoxidase complex or PAP.[30]

2. Peroxidase Chromogens

Graham and Karnovsky[31] have described the revelation of peroxidase with *3-3'-diaminobenzidine tetrahydrochloride (DAB)* in the presence of H_2O_2 (see Protocol 13), whose brown end-product is insoluble in alcohol and detectable at the photonic level (if it is amplified with a cobalt or nickel salt, the end products appear dark blue). The oxidized DAB reacts with the osmium salt and may be visible at the electron microscopic level without any counterstaining of the sections.

PROTOCOL 13: PEROXIDASE REVELATION WITH DIAMINOBENZIDINE (DAB)

- Prepare a solution of DAB* in Tris-maleate buffer, pH 7.6.
- Dissolve 0.4 ml/ml of DAB, and divide this solution in two parts.
- Incubate first with this solution for 20 min at 20°C.
- Add hydrogen peroxide** to the second part.
- Remove the DAB and replace with the DAB-H_2O_2 for 5 to 10 min.
- Wash in buffer and in cacodylate buffer.
- Fix in 1% OsO_4 in cacodylate buffer and then process as described for epoxy resin embedding (Protocols 6 or 7).

Some side effects may be observed during preparation such as microbubbles formed by the H_2O_2 during the enzymatic reaction. To minimize this phenomenon, cells or tissues are surfixed in 1% buffered glutaraldehyde before the DAB reaction.

To enhance the DAB reaction, the Tris-HCl buffer can be replaced by Tris-maleate buffer (see appendix), either 10 mM imidazole[32] or metallic cations (nickel, cobalt, ...) can also be added in the buffer.

In all cases, the most successful results are obtained when the enzymatic reaction is realized in the dark and in two steps: (1) DAB between 0.2 to 0.4 mg/ml in buffer is incubated on cells or tissues alone for 30 min, then is removed and replaced, (2) by a fresh DAB solution in the presence of H_2O_2 (1:2000 of final volume).

Endogenous peroxidase has to be checked. To inhibit this reaction without excessive morphological damage, cells or tissue are incubated in $2 \cdot 10^{-2}$ M aminotriazole in buffer before and during DAB revelation.

---

* DAB tablets are commercially available instead of the powder; use in conformation with the manufacturer's instructions.
** The concentration of H_2O_2 corresponds to 1:2000 of final volume.

The compound *4-chloro-1-naphthol (CN)* was described by Nakane.[33] It precipitates at the photonic level as a dark-blue end product which is insoluble in water but is soluble in alcohol and highly sensitive to light. At the electron microscopic level, this product can be visualized only after osmium fixation, as small dense granulations. It is exclusively used on ultrathin frozen sections, which are not dehydrated with ethanol before embedding in methylcellulose (see Protocol 14).

PROTOCOL 14: PEROXIDASE REVELATION WITH 4-CHLORO-1-NAPHTHOL

- Dissolve 4 mg of 4-chloro-1-naphthol in 50 µl of dimethylformamide or alcohol 95°.
- Add 10 ml of Tris-maleate buffer, pH 7.6, and 5 µl of H_2O_2 for 10 min.
- Wash in buffer and in cacodylate buffer.
- Fix in 1% OsO_4 in cacodylate buffer and then process as described for epoxy resin embedding (Protocols 6 or 7).

3. Bridge Methods

Many procedures have been tried as ways of amplifying the sensitivity of the reaction with a large range of detectable peroxidase molecules.

In the "bridge method",[34] after the specific incubation an unlabeled antibody is allowed to form a "bridge" between the first antibody and the antiperoxidase antibodies. The last one and the specific antibodies have to be raised in the same species, and the second one against this species. The peroxidase is finally visualized as described in Protocol 13.

The peroxidase-antiperoxidase method *(PAP)* was described by Sternberger et al.[30] as an improvement on the latter method: a soluble PAP immune complex is prepared before the incubation step, avoiding the possible admixing of nonspecific immunoglobulins with the antiperoxidase antibodies.

Protein A can be substituted for the bridge between the polyclonal IgG and the PAP complex (0.1 to 1 mg/ml for 30 min). The PAP method[30] is said to be at least 20 times more efficient than the direct method described by Avraméas.[27]

4. Biotinylated Antibodies

Biotin can be conjugated to antibodies and revealed through its high affinity for avidin or streptavidin, which can themselves be coupled directly with peroxidase or complexed with antiperoxidase antibodies. Glycoprotein avidin has four binding sites for biotin. It seems that streptavidin, in the absence of oligosaccharide residues present on the avidin, offers a higher binding specificity with the biotin.[35]

This method is often used to amplify the immunological reaction, using avidin or streptavidin as a bridge between specific antibodies and secondary biotinylated antibodies. This is known as the ABC complex and is commercially available (Figure 5).[36] At the electron microscopic level, however, it seems that this method is more effective if colloidal gold is used as a tracer.

PROTOCOL 16: DIRECT METHOD

Ultrathin sections are incubated directly with colloidal gold conjugated anti-bodies. This method can be carried out quickly, as nonspecific reactions are limited; however, not much signal is observed and this method is now rarely used.

- Prepare the incubation buffer (buffer B):
  - *20 mM Tris-buffer pH 7.6*
  - *150 mM NaCl*
  - *0.05% Tween 20*
  - *0.5% Ovalbumin*
- *For the blocking step, use 1% ovalbumin in this buffer with gentle agitation for 10 min.*
- *Incubate the sections on a drop of gold labeled antibodies at the selected dilution in the incubation buffer for 60 min.*
- *Wash the sections with the same buffer, 4 x 10 min*
- *Wash eventually the sections in cacodylate buffer before fixation in 1% glu-taraldehyde in the same buffer, followed by washing in buffer and distilled water, 4 x 5 min*
- *Stain with uranyl acetate and if necessary with lead citrate.*

PROTOCOL 17: INDIRECT METHOD

A large variety of specific antibodies raised in the same or different species can be used in the first incubation step. Gold-labeled secondary antibodies are directed against the IgG of the species which was used to raise the first antibodies. This method gives higher sensitivity and the possibility of multiple antibody detection.

- *Prepare buffer A:*
  - *100 mM Mono- or di-phosphate buffer pH 7.4*
  - *150 mM NaCl*
  - *0.05% Tween 20*
  - *0.5% Ovalbumin*
- *Sections are incubated in blocking solution: 1% ovalbumin in buffer A for 10 min.*
- *Incubate with the specific antibodies at the selected dilution for 60 min.*
- *Wash with the buffer A, 3 × 5 min.*
- *Pass in buffer B (see Protocole 15) for 5 min.*
- *Continue as described in Protocole 15.*

The biggest advantages of colloidal gold as an electron microscopic tracer are its high electron density, the different possible sizes of the particles, and its high resolving power, which is not diminished by masking of the substructural cellular morphology; disadvantages include the impossibility of tissue penetration in pre-embedding methods and the need to use small particles in order to label the maximum number of antigens. On ultrathin frozen sections and on hydrophilic resin sections, colloidal gold particles are probably the best marker.

C. MULTIPLE LABELING

When the antibodies are characterized by *a simple detection method* (dilution, time of incubation) for the fixation and type of tissue studied, the majority of detections lead to a double visualization using colloidal gold particles of different sizes. Several possibilities may be considered: two different antibodies raised in two different species, both antibodies provided from the same species, pre- and post-embedding are used, antigen detection is realized after *in situ* hybridization. Some of these Protocols are described.

Antibodies from different species — A specific incubation is carried using a mixture of the two antibodies at their final dilutions, already determined by a simple labeling. After washing in the incubation buffer, secondary gold-coupled antibodies are added. The size of these two antibodies has to be changed around in a further incubation to ensure that the same labeling is observed.

Antibodies from the same species — When polyclonal antibodies are raised in rabbits, immunolabeling can be obtained using Protein A gold: the first specific antibody incubation is followed by incubation with Protein A gold. The sections are then incubated with protein A in buffer in order to be sure that the IgG of the first immunoreaction is saturated with Protein A, so that it cannot interfere in the second incubation. The second specific antibodies are incubated in the same way as the fist one and visualized using a second Protein A-gold solution, made distinct by the diameter size of the gold particles used for the first step. The combination of 5 nM and 15 nM gold preparations gives good results. It is important to use the smaller gold particles in the first incubation. Bendayan has demonstrated, when monoclonal antibodies are used, that the protein A/G gives the same sensitivity;[49] it seems, however, that the combination of two different species gives more reproducible results.

Multi-labeling on the same sections — In some cases, the localization of two or three antibodies is carried out using a combination of the pre-embedding method using peroxidase conjugated antibodies and immunogold staining on the ultrathin sections. The classical multigold labeling is known as the double-face method.[49] The incubation of either one or two antibodies on each face of the grid makes possible the detection of different antigens on the same section. This method appears attractive and elegant, but it is not very easy to realize and the gold particles may not be distinct enough to be analyzed.

Simultaneous detection of nucleic acid and antigen using Protein A — This application of the double immunogold labeling can associate either two different antibodies or an antibody corresponding to the visualization of the probe and Protein A-gold (see Protocol 18).

PROTOCOL 18: SIMULTANEOUS DETECTION OF A NUCLEIC ACID AND AN ANTIGEN USING PROTEIN A[6]

The following reagents can be used:

Unlabeled rabbit anti-hybrid label diluted 1:200
Unlabeled rabbit anti-antigen diluted 1:200
Unlabeled protein (100 μg/ml)
Protein A labeled with 5- and 15-nm colloidal gold particles diluted 1:40
PBS (mono- or di-phosphate buffer 10 mM, NaCl 150 mM, pH 7)
Bovine serum albumin (BSA) 5% in PBS
PBS added of 10 mM glycine
Glutaraldehyde 1% in PBS

All the sections are incubated with gentle agitation:

- Wash the grids on PBS, 4 × 2 min
- Incubate the grids on 5% BSA for 10 min
- Wash the grids on PBS-glycine for 2 min
- Incubate the grids on unlabeled rabbit anti-biotin for 60 min
- Wash the grids on PBS-glycine, 6 × 1 min
- Incubate the grids on Protein A labeled with 5-nm gold particles for 30 min
- Wash the grids on PBS-glycine, 6 × 1 min
- Incubate the grids on unlabeled Protein A for 5 min
- Wash the grids on PBS-glycine, 6 × 1 min
- Incubate the grids on unlabeled rabbit anti-antigen for 60 min
- Wash the grids on PBS-glycine, 6 × 1 min
- Incubate the grids on Protein A labeled with 15-nm gold particles for 30 min
- Wash the grids on PBS-glycine, 2 × 1 min
- Wash the grids on PBS, 4 × 1 min
- Fix in 1% glutaraldehyde in PBS for 5 min
- Wash the grids on PBS, 2 × 1 min
- Wash the grids on water, 4 × 1 min
- counterstain the sections.

The localization of two antigens in the same tissue or cells can be easily realized using immunogold staining. The wide range of specific antibodies combined with the availability of gold-coupled secondary antibodies, with different sizes of gold particles, provides an opportunity to detect numerous antigens in the same structure and to find out the nature of their relationships.

VII. CONTROLS

All visualizations of antigens require several control reactions to confirm the specificity of the observations. The different steps of the reactions have to be verified and their overall results serve to validate the immunological results.

The antibody omission give the simplest control: the incubation in buffer alone instead of the specific antibody or the secondary labeled antibody gives the possible background, which may be due to endogenous enzymes such as peroxidase or biotin.

Another incubation will confirm the specificity of the reaction: no signal is observed after incubation with *normal serum* (i.e., frozen samples aliquoted before animal immunization). To enhance the validity of the results, *different specific antibodies* may be incubated on serial sections and a positive signal will confirm their localization.

The absorption test is the control most widely used: when the antibody at the determined dilution is incubated with its antigen molecule, the subsequent incubation with the tissue does not give a specific signal. Using other antigens, at a higher concentration than homologous antigen, the subsequent incubation with the tissue gives a specific signal. The specificities of the antibody correspond to only *one part of the molecule*: the receptor molecules with their cytoplasmic or extracellular segments offer a very good example. Their localization can be followed at different physiological stages with antibodies raised against one or the other segment.

After these controls, the characterization of the signal can be analyzed by the different techniques of detection. Their comparison assumes their subcellular localization and gives precious information in connection with their physiological implications.

VIII. RESULTS

In order to visualize ligand-receptor complexes by immunocytology at the electron microscopic level there are two possibilities: (1) detection of ligand immunoreactivity, or (2) detection of receptor immunoreactivity.

The density of receptor proteins within the cell is always greater than that of a ligand, but although antibodies against ligands are now available, this is not always the case with receptor antibodies. Thus, it has been easier, up to now, to visualize ligands than the receptors. However, it is necessary to demonstrate that a ligand has not been synthesized in the cell where it was detected. *In situ* hybridization gives the possibility of showing whether or not the mRNA coding for this molecule is expressed in the particular cell involved (see Chapters 14, 15).

A. VISUALIZATION OF A LIGAND

Since the density of these proteins is always low, more sensitive methods must be used. Combinations of immunocytological detection and an ultrathin frozen section method, which have been shown to constitute a more sensitive approach,[50] have permitted the detection of a number of these peptide ligands (see Section IX) as well as steroid ligands.[51] For example, in rat pituitary gland neuropeptide-like thyroliberin (TRH) (Figure 6),[52] growth hormone-releasing factor (GRF) (Figure 7),[53] or vasoactive intestinal peptide (VIP)[54] have been detected in particular cell populations using the PAP method on ultrathin frozen sections. All these peptides are known to be taken up by the cell populations of the anterior pituitary. They bind to membrane receptors and are difficult to visualize in this subcellular

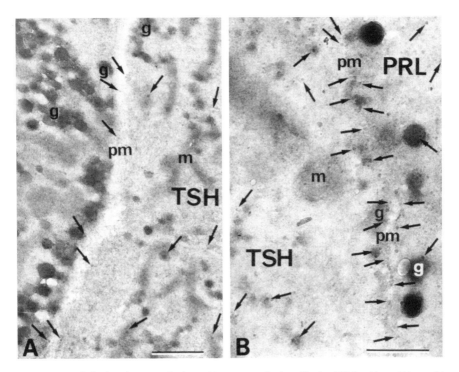

FIGURE 6   Cellular localization of a ligand in target cell: thyroliberin (TRH) without (A) or with (B) injection of the ligand. The receptor for this ligand is a membrane receptor. TRH-like immunoreactivity (PAP method on ultrathin frozen section of pituitary gland) is localized at the plasma membrane level. The dilution of anti-TRH serum: $2 \cdot 10^{-5}$. Some of the TRH-like immunoreactivity granules are identified by arrows. A: localization of endogenous TRH. TRH-like immunoreactivity is scarcely distributed along the plasma membrane (pm) of thyrotrophs (TSH). B: localization of exogenous TRH 15 min after i.v. injection of TRH. TRH-like immunoreactivity is very strong along the plasma membrane of the thyrotroph and lactotroph (PRL); m: mitochondrion; g: secretory granules. Bar = 0.5 μm. (From Morel, G. et al., *Neuroendocrinology*, 41, 312, 1985. With permission.)

compartment in physiological conditions (Figure 6A), but can be observed after peptide injection (Figure 6B). Nevertheless, they can also be observed in the cells (Figure 7), in particular in the nucleus and the cytoplasmic matrix. The absence of immunoreactivity in other cell populations (i.e., adjacent cells) (Figure 7) remains a useful control, along with physiological variations and comparisons with other methods.[55,56]

Similar results have been obtained with steroid ligands (Figure 8)[57-59] or T3.[60] In these cases immunoreactivities were mainly revealed in the nuclear compartment.

B. VISUALIZATION OF A RECEPTOR

If an antibody against a receptor molecule is available, this makes possible its direct detection by an immunocytological reaction. In this way, peptides, T3,[60] Vitamin D3,[61] and steroid receptors[62,63] have been visualized in cells. For example, growth hormone receptors (Figure 9)[64] as well as prolactin receptors[65] are found in

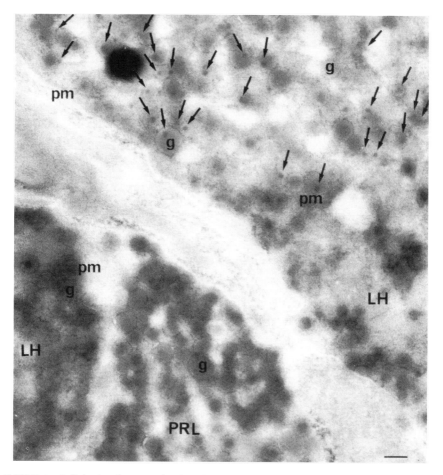

FIGURE 7   Cellular localization of a ligand in a target cell: growth hormone-releasing factor (GRF). Anti-GRF serum (455-6, dilution $2 \cdot 10^{-4}$) is used in the PAP method on ultrathin frozen section of pituitary gland. The GRF-like immunoreactivity (arrow) is localized in the somatotroph only. No reaction is observed in lactotrophs (PRL) and gonadotrophs (LH). G: secretory granules; pm: plasma membrane. Bar = 0.5 µm. (From Morel, G. et al., *Neuroendocrinology*, 38, 123, 1984. With permission.)

rat pituitary gland. These receptors are present in particular endocrine cell populations in the case of growth hormone receptors,[64] and in all endocrine cell populations, in the case of prolactin receptors.[65] Labeling density varies between cell populations. In other types of tissue, prolactin receptors have been described in hepatocytes.[66] Their subcellular localization is identical for these two types of cytokine receptors, i.e., on the plasma membrane, in the cytoplasmic matrix, on the nuclear envelope, and in the nuclear matrix.[64-66]

For steroid receptors (e.g., Vitamin D3 receptor,[61] or estrogen[62] and progesterone[63] receptors) their subcellular localization is the same as for their biochemical detection, i.e., mainly in the nuclear matrix where they bind to DNA, and in the cytoplasmic matrix where they are synthesized (Figure 10).

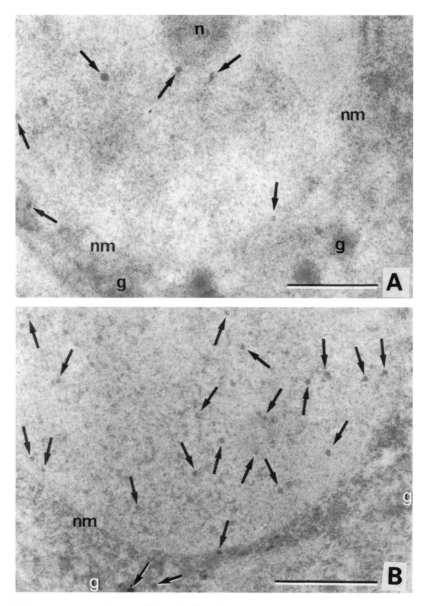

FIGURE 8 Detection of a steroid ligand molecule: testosterone. Immunocytochemical detection (PAP method on ultrathin frozen section) of testosterone in anterior pituitary gland. Immunoreactivity (arrow) is detected mainly in the nucleus of untreated rat (A), or injected rat (B); g: secretory granule; n: nucleolus; nm: nuclear membrane. Bar = 1 μm. (From Morel, G. et al., *Cell. Tissue Res.*, 235, 159, 1984. With permission.)

C. QUANTIFICATION

In order to show that the signal detected is different from the background and nonspecific binding of antibodies and to reveal the cell(s) or the cell compartment(s) that contain(s) the ligand or the receptor, quantitative analysis is performed.

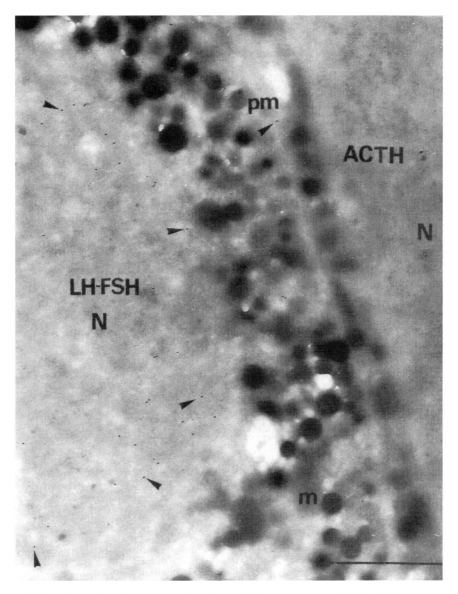

FIGURE 9   Detection of a receptor molecule: growth hormone receptor (GHR). GHR immunore-activity (arrowhead) (immunogold method on ultrathin frozen section of pituitary gland, MAb 263 anti-GHR 0.1 μg/μl, 10-nM gold particles) is visualized in pituitary gland in gonadotroph (LH-FSH) but not in corticotroph (ACTH); pm: plasma membrane; N: nucleus; m: mitochondrion; sg: secretory granule. Bar = 1 μm. (From Mertani, H.C. et al., *Neuroendocrinology, 59*, 483, 1994. With permission.)

To determine the background, the signal is measured over sections (i.e., negative cell), over the membrane on the grid, or over resin. This nonspecific staining is compared to the signal over the positive cells. For example, in the pituitary gland the nonspecific signal for immunoreactivity of somatostatin (a neuropeptide which inhibits

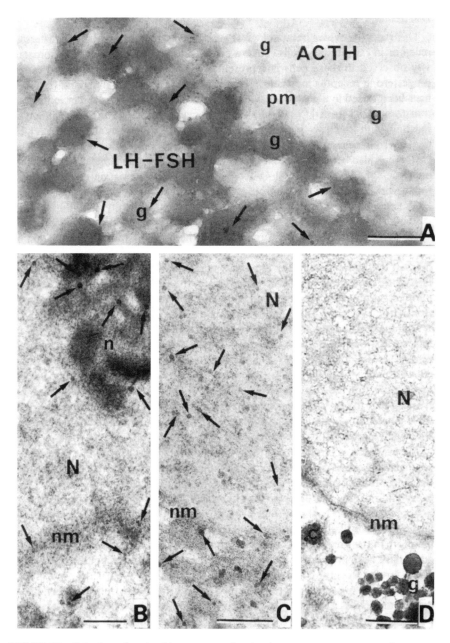

FIGURE 10   Detection of a steroid receptor molecule: the progesterone receptor. Immunocytochemical detection (PAP method on ultrathin frozen section) of progesterone receptor in anterior pituitary gland. Immunoreactivity (arrow) is detected in cytoplasm where receptor is synthesized (A) of gonadotroph (LH-FSH) but not of corticotroph (ACTH), and mainly in the nucleus (N) of untreated rat (B), or injected rat (C). No signal is observed in control experiment. (D); pm: plasma membrane; g: secretory granule; n: nucleolus; nm: nuclear membrane. Bar = 0.5 μm. (From Morel, G. et al., *Exp. Cell Res.*, 155, 283, 1984. With permission.)

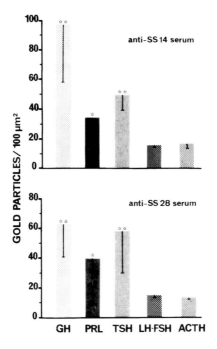

FIGURE 11   Quantification of gold particles resulting from somatostatin-like immunoreactivity (anti-SS14 and anti-SS28 sera) in different cell populations of anterior pituitary; GH: somatotroph, PRL: lactotroph, TSH: thyrotroph, LH-FSH: gonadotroph, ACTH: corticotroph). Gold particles observed in LH-FSH and ACTH represent background, ☆☆ significantly ($p < 0.01$) different from background; ☆ significantly ($p < 0.05$) different from background. (From Mesguich, P. et al., *Cell. Tissue Res.*, 252, 419, 1988. With permission.)

the release of growth hormone, prolactin, and thyrotropin-stimulating hormone in the pituitary gland[67] depends on the marker used: background staining is about 1 to 2 PAP complexes per cell and about 5 times more abundant than this with gold particles.[67] In this way, somatostatin immunoreactivity (two forms of somatostatin are detected in their pituitary target cells: SS14 and SS28) has been localized in the different endocrine pituitary cell populations (Figure 11). The method of quantification used for a given signal (i.e., gold particles or PAP complexes) is the determination of the number of these particles per area (e.g., 100 mm²), expressed by counting at least 30 cells for each cell population. Similar results are obtained using an anti-growth hormone receptor serum and the immunogold method (Figure 12).

At the subcellular level, the percentage of gold particles and the signal for somatostatin 14 and somatostatin 28 immunoreactivity have been determined for the cellular organelles of somatotrophs (cells which released growth hormone) (Table 1).

Dilution of the primary antibody can also be used to estimate the intensity of labeling. In this way, the disappearance of the immunocytological reaction is systematically studied as for a titration. The results are expressed as the maximal dilution factor of the antiserum (MDA),[69] still giving rise to a visible immunocytological reaction. The value of MDA thus obtained corresponds to a titration of an antiserum. Therefore it cannot serve as an antigen assay. This approach can, however, be used to appreciate the relative variations in the concentration of an antigen in a target cell

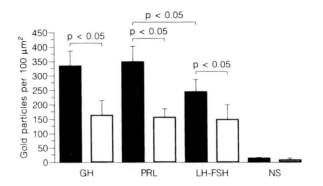

FIGURE 12 Quantification of immunocytological signal resulting from growth hormone receptor-like immunoreactivity (see figure 9). Immunoreactivity (mean number ± SEM of gold particles per 100 μm^2) was quantified in the nucleus (black scattergram) and the cytoplasm (white scattergram). GH: somatotroph, PRL: lactotroph, LH-FSH: gonadotroph, NS: nonspecific signal. (From Mertani, H.C. et al., *Neuroendocrinology*, 59, 483, 1994. With permission.)

TABLE 1   Percentage of Gold Particles for Cellular Organelles of Somatotrophs Using Anti-SS14 and Anti-SS28 Sera (SS14-LI: Somatostatin 14-Like Immunoreactivity; SS28-LI: Somatostatin 28-Like Immunoreactivity)

Cellular Compartments	SS14-LI (%)	SS28-LI (%)
Plasma membrane	7.9	7.5
Golgi apparatus	0.0	0.3
Lysosomes	0.3	0.0
Cytoplasmic matrix	35.2	37.4
Mitochondria	0.8	0.6
Rough endoplasmic reticulum	4.4	4.3
Secretory granules	22.9	25.9
Nucleus	28.5	24.0

population. For example, Figure 13 shows the effect of the time elapsed after thyroliberin injection on the MDA values found in the target cells.

IX. APPLICATIONS

An understanding of molecular communication between cells requires the presence of a surface receptor, the target molecule, which is able to bind the ligand. Different molecular structures characterize different types of receptors, as the number of spanning domains (1 to 7); then they are classified into corresponding families. However, the mechanisms are ubiquitous: a ligand binds to its receptor and the ligand-receptor complex induces intracellular signaling. Some receptors share a

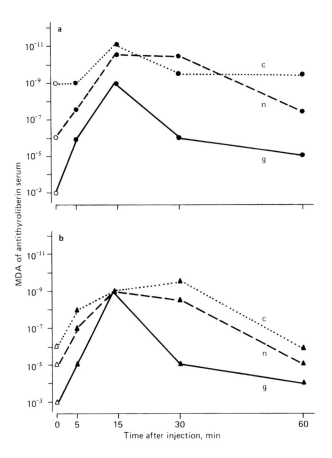

FIGURE 13 Effect of time elapsed after thyroliberin (TRH) injection on the MDA values (anti-TRH serum). The MDA has been determined for the thyrotrophs (a) and the lactotrophs (b) and for each cell compartment: cytoplasmic matrix (c), nucleus (n), and secretory granules (g). (From Morel, G. et al., *Neuroendocrinology*, 41, 312, 1985. With permission.)

protein-kinase motif on their intracytoplasmic domain or are associated with cytoplasmic tyrosine kinase (JAK, Tyk, etc.), which involves a phosphorylation cascade and STAT or non-STAT cytoplasmic-protein complexes. Through this kinase-initiated cascade, receptors are often directly translocated to the nucleus, bind to DNA, and activate the transcription of their target genes. Moreover, activation of different signaling pathways, which depends on the magnitude and rate of activation, leads to interaction between different signal trafficking and intracellular molecules.

Several immunocytological approaches can be tried out as a means of determining the localization and distribution of receptors in the cells: anti-receptor antibodies, labeled ligands, or anti-ligand antibodies, as well as all the possible combinations of these different antibodies.

Anti-receptor antibodies and labeled ligands are tried out first, as a way of following the ligand through the various physiological signaling events,[70] and then localizing the receptor itself or the ligand-receptor complexes in the subcellular structures. In the last decade, a large number of publications have demonstrated the utility of immunolocytological detection at the electron microscopic level. Peroxidase labeling used in pre-embedding methods has been superseded by immunogold staining on tissue sections. The best visualization results have been obtained with colloidal gold as a tracer, associated with hydrophilic embedding resins. This method has brought about significant improvement in localization in areas not previously described, e.g., the central nervous system: the receptor localization has not proven easy, either by biochemical methods or by the absence of high-specific-affinity ligands. However, the expansion in the number of monoclonal antibodies available has increased the possibilities for receptor visualization. They are also frequently used to detect colocalization of a receptor and a ligand, or even different epitopes on the same molecule.

In the last few years, the increasing availability of antireceptor antibodies has produced general improvements in the localization of numerous receptors such as the glutamate receptor,[71-75] the GABAA receptor,[76-79] the glycine receptor,[80,81] and the IGF$_1$ receptor.[82]

Pre-embedding or PAP methods,[83-86] which were at one time very widespread, have now been superseded by post-embedding immunogold detection. However, the highest sensitivity will be observed using cryo-ultrathin sections,[87-90] avoiding the dehydration and resin infiltration procedures. In some cases, tissue fixation can destroy all antigenic activity: many fixatives are cross-linking agents and react with cell surface receptors; low concentrations of glutaraldehyde in the place of paraformaldehyde have permitted the study of the distribution of muscarinic acetylcholine receptors in the rat brain.

All the purified and characterized antibodies and especially the monoclonal antibodies, or antibodies which are directed against restricted regions of the receptor, e.g., in its intra- or extracellular domain, are subjected to a recent immunocytological exploration. These applications are particularly useful for the new development of the nuclear translocation of the ligands and their receptors. The understanding of this phenomenon remains unclear, but receptors, ligands, or their complexes appear to be involved in the signaling of transduction events. The rate of translocation seems to be different for receptors and hormones or their complexes to the nucleus. All the possible determinations using immunocytological methods at the electron microscopic level will potentially give more information on the intracellular signaling pathway.

A large number of peptide ligands and peptide ligand receptors have now been reported in the literature as being localized in the nucleus or as undergoing nuclear translocation (see Table 2).

Ligands have also been extensively used in vesicular trafficking, either directly conjugated with a tracer or detected using secondary labeled antibodies: the visualization of the LDL receptor in coated pits and coated vesicles[70] traffic by different approaches (biochemical or immunocytological) has also revealed other routes of endocytosis such as the caveolae pathway, in which several microdomains of the cellular membrane have been visualized by using immunogold staining.[143-145] Indeed, these caveolae play an important role in several signaling mechanisms as shown by the number of publication.[146-150]

One other important application of this immunocytological approach seems to be that of multiple detection. On the same sections, as well as on ultrathin frozen

TABLE 2 Nuclear Localization of Peptide-Ligands and Peptide-Receptors

	Receptors	Ligands
ANF		91
Angiotensin	92,93	94
Angiotensin II		95
Arginine vasopressin		96
Calcitonin		97
Corticotropin RH		98,99
EGF	100,101,103,104	100,102,104
FGF	105	105–107
Gonadotropin	108	
GnRH	109	69,110
GH	64,111–113	112,114,115
GHRH		53,116
Insulin	117-122	120,123-125
Interferon β		126
Interferon γ	126	126
Interleukin 1	127,128	
Interleukin 1 α		129
NGF	125,130	125,130,131
Neuropeptide Y		132,133
Oxytocin		134
PDGF	125	125,135
Prolactin	65,66,136–138	138–141
Somatostatin		67,68
Substance P		142
TRH		52,133
VIP		54

sections as on resin sections, a lot of different combinations between monoclonal anti-receptor antibodies, or anti-receptor antibodies and labeled ligands, or even anti-receptor and anti-ligand may be associated. Two monoclonal antibodies directed against two different epitopes of an antigen can be detected with anti-mouse immunoglobulins, then with protein-G, using two different sizes of gold particles.[151,152]

Classically, anti-receptor and anti-ligand antibodies raised in different species are associated and visualized using complementary anti-species antibodies coupled with different sizes of gold particles.[51,64,64,66] In the same way, the association of a gold-labeled ligand and anti-receptor antibodies improves the intracellular trafficking in cultivated cells.[153,154]

All the various combinations can be used for the comparison of antibodies — one unknown antibody vs. a well-characterized one. (S-Endo 1)(In the exploration of the CD present on the cell surface, this type of comparison appears as routine determination in flux cytometry).[153,154]

Immunoelectron microscopy also offers a novel development in the understanding of the intracellular vesicular trafficking. The COPI-coated vesicles in the Golgi system share on their surfaces receptors as v-SNAREs, which determine the anterograde or retrograde direction of their destination, especially in Golgi trafficking. Import and export between two cellular compartments have to be balanced in order to maintain size and volume of these subcellular fractions. If v-SNARE addresses a vesicle to its target membrane, this v-SNARE has to be recycled and its partner (t-SNARE) identified in the vesicular membrane. Using immunogold staining, a v-SNARE has been localized in the Golgi cisternae by Nahagama et al.[155]

APPENDIX

Sodium phosphate buffer (0.1 *M*)

Stock A: 0.1 *M* monosodium phosphate: 1.56% $NaH_2PO_4 \cdot 2H_2O$
Stock B: 0.1 *M* disodium phosphate: 3.58% $Na_2HPO_4 \cdot 12H_2O$

To obtain 100 ml of phosphate buffer pH 7.4, mix 20 ml of A and 80 ml of B.

Cacodylate buffer (0.2 *M*)

Stock A: 0.2 *M* sodium cacodylate: 4.28% $Na(CH_3)_2\ AsO_2\ 3H_2O$
Stock B: 0.2 N HCl

To obtain 0.2 *M* buffer at pH 7.4, to 25 ml A add 1.4 ml B and complete to 100 ml with distilled water. Check the pH.
Comment: prepare cacodylate solutions under a hood.

Tris-maleate buffer (0.05 *M*, pH 7.6)

Stock A: Tris-aminomethane (0.2 *M*)-maleic acid (0.2 *M*)
Stock B: 0.2 *M* NaOH

To obtain 0.05 *M* buffer at pH 7.6; mix 25 ml A and 2.9 ml B and complete to 100 ml with distilled water. Check the pH.

REFERENCES

1. Karnowsky, M.J., A formaldehyde-glutaraldehyde fixative of high osmolality for use in electron microscopy, *J. Cell Biol.*, 27, 137A, 1965.
2. Zamboni, L., and De Martino, C., Buffered picric acid-formaldehyde: a new rapid fixative for electron microscopy, *J. Cell Biol.*, 35, 148A, 1967.
3. McLean, I.W., and Nakane, P.K., Periodate-lysine-paraformaldehyde fixative. A new fixative for immunoelectron microscopy, *J. Histochem. Cytochem.*, 22, 1077, 1974.
4. King, J.C., Lechan, R.M., Kugel, G., and Anthony, E.L.P., Acrolein: A fixative for immunocytochemical localization of peptides in the central nervous system, *J. Histochem. Cytochem.*, 31, 62, 1983.
5. Kraenhenbuhl, J.P., Racine, L., and Jamieson, J.D., Immunocytochemical localization of secretory proteins in bovine pancreatic exocrine cells, *J. Cell.Biol.*, 72, 406, 1977.
6. Bernadac, A., Bolla, J.M., Lazdunski, C., Inouye M., and Pages, J.M., Precise localization of an overproduced periplasmic protein in Escherichia coli: use of a double immuno-gold labeling. *Biol. Cell*, 61, 141, 1987.

7. Glauert, A.M., and Glauert, R.M.,Araldite as an embedding medium for electron microscopy, *J. Biophys. Biochem. Cytol.*, 7, 27, 1960.
8. Luft, J.H., Improvements in epoxy resin embedding methods, *J. Biophys. Biochem. Cytol.*, 9, 409, 1961.
9. Mollenhauer, H.H., Plastic embedding mixtures for use in electron microscopy, *Stain Technol.*, 39, 111, 1964.
10. Brinkley, B.R., Murphy, P, and Richardson, L.C., Procedure for embedding *in situ* selected cells cultured *in vitro*, *J.Cell Biol.*, 35, 279, 1967.
11. Tougard, C., and Picart, R., Use of preembedding ultrastructural immunocytochemistry in the localization of a secretory product and membrane proteins in cultured prolactin cells, *Am. J. Anat.*, 175, 161, 1986.
12. Picart, R., and Tixier-Vidal, A., Description d'une méthode permettant la sélection et l'étude ultrastructurale de plages cellulaires dans des monocouches hétérogènes cultivées en flacons de plastique, *J. Microsc.*, 20, 80a, 1974.
13. Leduc, E.H., and Bernhard, W., A reliable and easily sectioned epoxy resin embedding medium, *Anat. Rec.*, 150, 129, 1964.
14. Carlemalm, E., Garavito, R.M., and Villinger, W., Resin developement for electron microscopy and an analysis of embedding at low temperature, *J. Microsc.*, 126, 123, 1982.
15. Carlemalm, E., Villinger, W., Hobot, J.A., Acetarin, J.D., and Kellen-Berger, E., Low temperature embedding with lowicryl resins: two new formulations and some applications, *J. Microsc.*, 140, 55, 1985.
16. Newman, G.R., Jasani, B, and Willimas, E.D., The preservation of ultrastructure and antigenicity, *J. Microsc.*, 127, RP5, 1982.
17. De Mey, J., Moeremans, M., Geuens, G., Nuydens, R., and De Brabander, M., High resolution light and electron microscopic localization of tubulin with the IGS (immunogold staining) method, *Cell Biol. Int. Rep.*, 5, 889, 1981.
18. Roitt, I., Brostoff, J., and Male, D., *Immunology,* Gower Medical Publishing, 1989.
19. Paul, W.E., *Fundamental Immunology,* Raven Press, New York, 1993.
20. Köhler, G., and Millstein, C., Continuous cultures of fused cells secreting antibody of predefined specificity, *Nature,* 256, 495, 1975.
21. Köhler, G., and Millstein, C., Derivation of specific producing tissue culture and tumor cell lines by cell fusion, *Eur. J. Immunol.*, 6, 511, 1976.
22. Forsgren, A., and Sjöquist, J., "Protein A" from *S. aureus*. I. Pseudo-immune reaction with human g-globulin, *J. Immunol.*, 97, 822, 1966.
23. Langone, J.J., Protein A: A tracer for general use in immunoscience, *J. Immunol. Methods*, 24, 269, 1978.
24. Richman, D.D., Cleveland, P.H., Oxman, M.N., and Johnson, K.M., The binding of staphylococcal Protein A by the sera of different animal species, *J. Immunol.*, 128, 2300, 1982.
25. Reiss, K.J., Hansen, M.F., and Bjorck, L., Extraction and characterization of IgG Fc receptors from group C and group G Streptococci, *Mol. Immunol.*, 23, 425, 1986.
26. Bendayan, M., and Garzon, S., Protein G-gold complex: comparative evaluation with Protein A-gold for high-resolution immunocytochemistry, *J. Histochem. Cytochem.*, 36, 597, 1988.
27. Avraméas, S., and Uriel, J., Méthode de marquage d'antigènes et d'anticorps avec des enzymes et son application en immunodiffusion, *C. R. Acad. Sci. Paris*, 262, 2543, 1966.
28. Nakane, P., and Pierce, G.Jr., Enzyme-labeled antibodies preparation and application for the localization of antigens, *J. Histochem. Cytochem.*, 12, 929, 1967.
29. Avraméas, S., and Ternynck, T , Peroxidase labeled antibody and Fab conjugates with enhanced intracellular penetration, *Immunochemistry*, 8, 1175, 1971.
30. Sternberger, L.A., Hardy, P.H., Jr, Cuculis, J.J., and Meyer, H.G., The unlabeled antibody-enzyme method of immunohistochemistry. Preparation and properties of soluble antigen-antibody complexe (horseradish peroxidase-antihorseradish peroxidase) and its use in identification of spirochetes, *J. Histochem. Cytochem.*, 18, 315, 1970.
31. Graham, R.C., and Karnowsky, M.J., The early stages of absorption of injected horseradish peroxidase in the proximal tubules of mouse kidney: ultrastructural cytochemistry by a new technique, *J. Histochem. Cytochem.*, 14, 291, 1966.

32. Strauss, W., Imidazole increases the sensitivity of cytochemical reaction for peroxidase with diaminobenzidine at a neutral pH, *J. Histochem. Cytochem.*, 30, 491, 1982.

33. Nakane, P.K., Simultaneous localization of multiple tissue antigens using the peroxidase labeled antibody methode: a study on pituitary glands of the rat, *J. Histochem. Cytochem.*, 16, 557, 1968.

34. Mason, T.E., Phifer, R.F., Spicer, S.S., Swallow, R.A., and Dreskin, R.B., An immunoglobulin-enzyme bridge method for localizing tissue antigens, *J. Histochem. Cytochem.*, 17, 563, 1969.

35. Costello, S.M., Felix, R.T., and Giese R.W., Enhancement of immune cellular agglutination by use of an avidin-biotin system, *Clin. Chem.*, 25, 1572, 1979.

36. Hsu, S.M., Raine, L., and Fanger, H., Use of avidin-biotin-peroxidase complex (ABC) in immunoperoxidase techniques: a comparison between ABC and unlabeled antibody (PAP) procedures, *J. Histochem. Cytochem.*, 29, 577, 1981.

37. Singer, S.J., and Schick, A.I., The properties of specific stains for electron microscopy prepared by conjugation of antibody with ferritin, *J. Biophys. Biochem. Cytol.*, 9, 519, 1961.

38. Singer, S.J., Preparation of an electron-dense antibody conjugate, *Nature*, 183, 1523, 1959.

39. Davis, W.C., and Silverman, L., Localization of mouse H-2 histocompatibility antigen with ferritin-labeled antibody, *Transplantation*, 6, 635, 1968.

40. Levinthal, J.D, Cerottini, J.C., Ahmad-Zadeh, C., and Wicker, R., The detection of intracellular adenovirus type 12 antigens by indirect immunoferritin technique, *Int. J. Cancer*, 2, 85, 1967.

41. Faulk, W.P., and Taylor, G.M., An immunocolloid method for the electron microscope, *Immunochemistry*, 8, 1081, 1971.

42. Frens, G., Controlled nucleation for the regulation of particle size in monodisperse gold suspension, *Nature Phys. Sci.*, 241, 20, 1973.

43. Horisberger, M., and Rosset, J., Colloidal gold, a useful marker for transmission and scanning electron microscopy, *J. Histochem. Cytochem.*, 25, 295, 1977.

44. Adams, J.C., Heavy metal intensification of DAB-based reaction product, *J. Histochem. Cytochem.*, 29, 775, 1981.

45. Bendayan, M., and Duhr, M.A., Modification of the protein A-gold immunocytochemical technique for the enhancement of its efficiency, *J. Histochem. Cytochem.*, 34, 569, 1986.

46. Geuze, H.J., Slot, J.W., Van Der Ley, P.A., and Scheffer, R.T.C., Use of colliodal gold particles in double-labeling immunoelectron microscopy of ultrathin frozen tissue sections, *J. Cell. Biol.*, 89, 653, 1981.

47. Le Guellec, D., Trembleau, A., Pechoux, C., Gossard, F., and Morel, G., Ultrastructural non-radioactive in situ hybridization of GH mRNA in the rat pituitary gland. Pre-embedding vs. ultrathin frozen sections vs. post-embedding, *J. Histochem. Cytochem.*, 40, 979, 1992.

48. Dimitriadis, G.J., Effects of detergents on antibody-antigen interreaction, *Anal. Biochem.*, 98, 445, 1979.

49. Bendayan, M., Double immunocytochemical labeling applying the protein A-gold technique, *J. Histochem. Cytochem.*, 30, 81, 1982.

50. Hemming, F.J., Mesguich, P., Morel, G., and Dubois, P.M., Cryoultramicrotomy versus plastic embedding: comparative immunocytochemistry of rat anterior pituitary cells, *J. Microsc.*, 131, 26, 1983.

51. Morel, G., Internalization and nuclear localization of peptides hormones. *Biochem. Pharmacol.*, 47, 63, 1994.

52. Morel, G., Gourdji, D., Grousselle, D., Brunet, N., Tixier-Vidal, A., and Dubois, P.M., Immunocytochemical evidence for in vivo internalization of thyroliberin into rat pituitary target cells, *Neuroendocrinology*, 41, 312, 1985.

53. Morel, G., Mesguich, P., Dubois, M.P., and Dubois, P.M., Ultrastructural evidence for endogenous growth hormone-releasing factor (GRF)-like immunoreactivity in the monkey pituitary gland, *Neuroendocrinology*, 38, 123, 1984.

54. Morel, G., Besson, J., Rosselin, G., and Dubois, P.M., Ultrastructural evidence for vasoactive intestinal peptide (VIP) in rat pituitary gland, *Neuroendocrinology*, 34, 85, 1982.

55. Gourdji, D., Tixier-Vidal, A., Morin, A., Pradelles, P., Morgat, J.L., Fromageot, P., and Kerdelhué, B., Binding of a tritiated thyrotropin-releasing factor to a prolactin secreting clonal cell line (GH3), *Exp. Cell Res.*, 82, 39, 1973.

56. Morel, G., Uptake and ultrastructural localization of a [^{125}I] Growth hormone-releasing factor agonist in male rat pituitary gland: evidence for internalization, *Endocrinology*, 129, 1497, 1991.

57. Dubois, P.M., Morel, G., Forest, M.G., and Dubois, M.P., Localization of luteinizing hormone (LH) and testosterone (T) or dihydrotestosterone (DHT) in the gonadotropic cells of anterior pituitary by using ultracryomicrotomy and immunocytochemistry, *Hormon. Metab. Res.*, 10, 250, 1978.

58. Morel, G., Forest, M.G., and Dubois, P.M., Ultrastructural evidence for endogenous testosterone immunoreactivity in rat pituitary gland, *Cell. Tissue Res.*, 235, 159, 1984.

59. Morel, G., Raynaud, J.-P., and Dubois, P. M., Localisation ultrastructurale de l'oestradiol et du moxestrol dans les cellules gonadotropes du rat par immunocytologie après cryoultramicrotomie. Caractérisation de la spécificité hormonale, *Expérientia*, 37, 98, 1981.

60. Ardail, D., Lerme, F., Puymirat, J., and Morel, G., Evidence for the presence of α- and β-related T3 receptors in rat liver mitochondria, *Eur. J. Cell Biol.*, 62, 105, 1993.

61. Boivin, G., Mesguich, P., Pike, J.W., Bouillon, R., Meunier, P., Hausler, M., Dubois, P.M., and Morel, G., Ultrastructural immunocyto-chemical localization of endogenous 1,25 dihydroxyvitamin D_3 and its receptors in osteoblasts and osteocytes from neonatal mice and rats calvaria, *Bone Miner.*, 3, 125, 1987.

62. Morel, G., Dubois, P.M., Benassayag, C., Nunez, E., Radanyi, C., Redeuilh, C., Richard-Foy, H., and Baulieu, E.-E., Ultrastructural evidence of oestradiol receptor by immunochemistry, *Exp. Cell Res.*, 132, 249, 1981.

63. Morel, G., Dubois, P., Gustafsson, J.A., Radojcic, M., Radanyi, C., Renoir, M., Baulieu, E.-E., Ultrastructural evidence of progesterone receptor by immunochemistry, *Exp. Cell Res.*, 155, 283, 1984.

64. Mertani, H.C., Waters, M.J., Jambou, R., Gossard, F., and Morel, G., Growth hormone receptor/binding protein in rat anterior pituitary, *Neuroendocrinology*, 59, 483, 1994.

65. Morel, G., Ouhtit, A., and Kelly, P.A., Prolactin receptor immunoreactivity in rat anterior pituitary, *Neuroendocrinology*, 59, 78, 1994.

66. Ouhtit, A., Ronsin, B., Kelly, P.A., and Morel, G., Ultrastructural expression of prolactin receptor in rat liver, *Biol. Cell*, , 82, 169, 1994.

67. Morel, G., Mesguich, P., Dubois, M.P., and Dubois, P.M., Ultrastructural evidence for endogenous somatostatin-like immunoreactivity in the pituitary gland, *Neuroendocrinology*, 36, 291, 1983.

68. Mesguich, P., Benoit, R., Dubois, P.M., and Morel, G., Somatostatin-28- and somatostatin-14-like immunoreactivities in the rat pituitary gland, *Cell Tissue Res.*, 252, 419, 1988.

69. Morel, G. and Dubois, P.M., Immunocytological evidence for gonadoliberin in rat anterior pituitary gland, *Neuroendocrinology*, 34, 197, 1982.

70. Goldstein, J.L., Brown, M.S., Anderson, R.G.W., Russel, D.W., and Schneider, W.J., Receptor-mediated endocytosis. Concepts emerging from the LDL receptor system, *Annu. Rev. Cell Biol.*, 1, 1, 1985.

71. Spreafico, R., Frassoni, C., Arcelli, P., Battaglia, G., Wenthold, R.J., and De-Biasi, S., Distribution of AMPA selective glutamate receptors in the thalamus of adult rats and during postnatal development. A light and ultrastructural immunocytochemical study, *Brain Res. Dev.*, 82, 231, 1994.

72. Tachibana, M., Wenthold, R.J., Morioka, H., and Petralia, R.S., Light and electron microscopic immunocytochemical localization of AMPA-selective glutamate receptors in the rat spinal cord, *J. Comp. Neurol.*, 344 , 431, 1994 .

73. Ohishi, H., Ogawa-Meguro, R., Shigemoto, R., Kaneko, T., Nakanishi, S., and Mizuno, N., Immunohistochemical localization of metabotropic glutamate receptors, mGluR2 and mGluR3, in rat cerebellar cortex, *Neuron*,13 , 55, 1994.

74. Nomura, A., Shigemoto, R., Nakamura, Y., Okamoto, N., Mizuno, N., and Nak, S., Developmentally regulated postsynaptic localization of a metabotropic glutamate receptor in rat rod bipolar cells, *Cell*, 77, 361, 1994.

75. Meeker, R.B., Swanson, D.J., Greenwood, R.S., and Hayw, J.N., Quantitative mapping of glutamate presynaptic terminals in the supraoptic nucleus and surrounding hypothalamus. *Brain Res.*, 8, 600, 112, 1993.

76. Nusser, Z., Roberts, J.D., Baude, A., Richards, J.G., and Somogyi, P.I., Relative densities of synaptic and extrasynaptic GABAA receptors on cerebellar granule cells as determined by a quantitative immunogold method, *J. Neurosci.*, 15, 2948, 1995.

77. Grunert, U., and Hughes, T.E., Immunohistochemical localization of GABAA receptors in the scotopic pathway of the cat retina, *Cell Tissue Res.*, 274, 267, 1993.

78. Caruncho, H.J., Puia, G., Slobodyansky, E., da-Silva, P.P., and Costa, E., Freeze-fracture immunocytochemical study of the expression of native and recombinant GABAA receptors, *Brain Res.*, 603 , 234, 1993.

79. Hansen, G.H,. Hosli, E., Belhage, B., Schousboe, A., and Hosli, L., Light and electron microscopic localization of GABAA-receptors on cultured cerebellar granule cells and astrocytes using immunohistochemical techniques, *Neurochem. Res.*, 16, 341, 1991.

80. Sassoe-Pognetto, M., Wassle, H., and Grunert, U., Glycinergic synapses in the rod pathway of the rat retina: cone bipolar cells express the alpha 1 subunit of the glycine receptor, *J. Neurosci.*, 14, 5131, 1994.

81. Chiba, T., and Semba, R., Immuno-electronmicroscopic studies on the gamma-aminobutyric acid and glycine receptor in the Intermediolateral nucleus of the thoracic spinal cord of rats and guinea pigs, *J. Auton. Nerv. Syst.*, 36, 173, 1991.

82. Smith, R.M., Garside, W.T., Aghayan, M., Shi, C.Z., Shah, N., Jarett, L., and Heyner, S., Mouse preimplantation embryos exhibit receptor-mediated binding and transcytosis of maternal insulin-like growth factor I, *Biol. Reprod.*, 49, 1, 1993.

83. Littlewood, N.K., Todd, A.J., Spike, R.C., Watt, C., and Shehab, S.A., The types of neuron in spinal dorsal horn which possess neurokinin-1 receptors, *Neuroscience*, 66 , 597, 1995.

84. Czekay, R.P., Orlando, R.A., Woodward, L., Adamson, E.D., and Farquhar, M.G., The expression of megalin (gp330) and LRP diverges during F9 cell differentiation, *J. Cell Sci.*, 108, 1433, 1995.

85. Nakamura, H., Kenmotsu, S., Sakai, H., and Ozawa, H., Localization of CD44, the hyaluronate receptor, on the plasma membrane of osteocytes and osteoclasts in rat tibiae, *Cell Tissue Res.*, 280, 225, 1995.

86. Figueroa, C.D., Gonzalez, C.B., Grigoriev, S., Abd-Alla, S.A., Haasemann, M., Jarnagin, K., and Muller-Esterl, W., Probing for the bradykinin B2 receptor in rat kidney by anti-peptide and anti-ligand antibodies. *J. Histochem. Cytochem.*, 43, 137, 1995.

87. Yamashita, S., Intranuclear localization of hormone-occupied and unoccupied estrogen receptors in the mouse uterus: application of 1 nM immunogold-silver enhancement procedure to ultrathin frozen sections, *J. Electron. Microsc.*, 44, 22, 1995.

88. Kleijmeer, M.J., Ossevoort, M.A., van-Veen, C.J., van-Hellemond, J.J., Neefjes, J.J., Kast, W.M., Melief, C.J., and Geuze, H.J., MHC class II compartments and the kinetics of antigen presentation in activated mouse spleen dendritic cells, *J. Immunol.*, 154 , 5715, 1995.

89. Boivin, G., Anthoine-Terrier, C., and Morel, G., Ultrastructural localization of endogenous hormones and receptors in bone tissue: an immunocytological approach in frozen samples, *Micron.*, 25, 15, 1994.

90. Chaitin, M.H., Wortham, H.S., and Brun-Zinkernagel, A.M., Immunocytochemical localization of CD44 in the mouse retina, *Exp. Eye Res.*, 58, 359, 1994.

91. Morel, G., and Heisler, S., Internalization of endogenous and exogenous atrial natriuretic peptide by target tissues. *Electron Microsc. Rev.*, 1, 221, 1988.

92. Re, R.N., Vizard, D.L. Brown, J., and Bryan, S.E., Angiotensin II receptors in chromatin fragments generated by micrococcal nuclease, *Biochem. Biophys. Res. Commun.*, 119, 220, 1984.

93. Re, R.N., The cellular biology of angiotensin: paracrine, autocrine and intracrine actions in cardiovascular tissues, *J. Mol. Cardiol. V.* (Suppl.), 63, 1989.

94. Moroianu, J., and Riordan, J.F., Nuclear translocation of angiogenin in proliferating endothelial cells is essential to its angiogenic activity, *Proc. Natl. Acad. Sci. U.S.A.*, 91, 1677, 1994.

95. Chabot, J.-G., Gray, D.A., Dubois, P.M., and Morel, G., Presence of angiotensin II in the adult male rat anterior pituitary gland: immunocytochemical study after cryoultramicrotomy, *Exp. Cell Res.*, 180, 189, 1989.

96. Terrier, C., Chabot, J.G., Pautrat, G., Jeandel, L., Gray, D., Lutz-Bucher, B., Zingg, H.H., and Morel, G., Arginine-vasopressin in anterior pituitary cells: in situ hybridization of mRNA and ultrastructural localization of immunoreactivity. *Neuroendocrinology*, 54, 303, 1991.

97. Morel, G., Boivin, G., David, L., Dubois, P.M., and Meunier, P.J., Immunocytochemical evidence for endogenous calcitonin and parathyroid hormone in osteoblasts from the calvaria of neonatal mice. Absence of endogenous and estradiol receptors, *Cell Tissue Res.*, 240, 89, 1985.

98. Morel, G., Hemming, F., Tonon, M.-C., Vaudry, H., Dubois, M.P., Coy, D., and Dubois, P. M., Ultrastructural evidence for corticotropin-releasing factor (CRF)-like immunoreactivity in the rat pituitary gland, *Biol. Cell* , 44, 89, 1982.

99. Morel, G., Enjalbert, A., Proulx, L., Pelletier, G., Barden, N., Gossard, F., and Dubois, P.M., Effect of corticotropin-releasing factor on the release and synthesis of prolactin, *Neuroendocrinology*, 49, 669, 1989.

100. Johson, L.K., Vlodavsky, I., Baxter, J.D., and Gosdopodarowicz, D., Nuclear accumulation of epidermal growth factor in cultured rat pituitary cells, *Nature*, 287, 340, 1980.

101. Murthy, U., Basu, M., Sen-Majumdar, A., and Das, M., Perinuclear location and recycling of epidermal growth factor receptor kinase: immunofluorescent visualization using antibodies directed to kinase and extracellular domains, *J. Cell. Biol.*, 103, 333, 1986.

102. Savion, N.I., Vlodavsky, I., and Gospodarowicz, D., Nuclear accumulation of epidermal growth factor in cultured bovine corneal endothelial and granulosa cells, *J. Biol. Chem.*, 256, 1149, 1981.

103. Rakowicz-Szulcynska, E.M., Rodeck, U., Herlyn, M., and Koprowski, H., Chromatin binding of epidermal growth factor, nerve growth factor and platelet derived growth factor in cells bearing the appropriate surface receptors, *Proc. Natl. Acad. Sci. U.S.A.*, 83, 3728, 1986.

104. Jiang, L.W., and Schindler, M., Nucleocytoplasmic transport is enhanced concomitant with nuclear accumulation of epidermal growth factor (EGF) binding activity in both 3T3-1 and EGF receptor reconstituted NR-6 fibroblasts, *J. Cell Biol.*, 110, 559, 1990.

105. Bouche, G., Gas, N., Prats, H., Baldin, V., Tauber, J.P., Teissie, J., and Amalric, F., Basic fibroblast growth factor enters the nucleolus and stimulates the transcription of ribosomal genes in ABAE cells undergoing G0-G1 transition, *Proc. Natl. Acad. Sci. U.S.A.*, 84, 6770, 1987.

106. Imamura, T., Engleka, K., Zhan, X., Tokita, Y., Forough, R., Roeder, D., Jackson, A., Maier, J.A., Hla, T., and Macaig, T., Recovery of mitogenic activity of a growth factor mutant with a nuclear translocation sequence, *Science*, 249, 1567, 1990.

107. Hawker, J.R., and Granger, H.J., Internalized basic fibroblast growth factor translocates to nuclei of venula endothelial cells, *Am. J. Physiol.*, 262, H1525, 1992.

108. Rajendron, K.G., and Menon, K.M., Evidence for existence of gonadotropin receptors in the nuclei isolated from rat ovary, *Biochem. Biophys. Res. Commun.*, 111, 127, 1983.

109. Millar, R., Rosen, H., Badminton, M., Pasqualini, C., and Kerdlehue, B., Luteinizing hormone releasing binding to purified rat pituitary nuclei, *FEBS Lett.*, 153, 382, 1983.

110. Stenberger, L.A., and Petrali, J.P., Quantitative immunocytochemistry of pituitary receptors for luteinizing hormone releasing hormone, *Cell Tissue Res.*, 162, 141, 1975.

111. Lobie, P.E., Breipohl, W., Lincoln, D.T., Garcia-Aragon, J., and Waters, M.J., Growth hormone receptor expression in the rat gastrointestinal tract, *Endocrinology*, 126, 299, 1990.

112. Lobie, P.E., Barnard, R., and Waters, M.J., The nuclear growth hormone receptor/binding protein. Antigenic and physicochemical characterization, *J. Biol. Chem.*, 266, 22645, 1991.

113. Fraser, R.A., and Harvey, S., Ubiquitous distribution of growth hormone receptors and/or binding proteins in adenohypophyseal tissues, *Endocrinology*, 130, 3593, 1992.

114. Rezvani, I., Maddaiah, V.T., Collipp, P.J., Thomas, J., and Chen., S.Y., Uptake of ^{3}H-human growth hormone into human liver slices in vitro, *Biochem. Med.*, 7, 432, 1973.

115. Bonifacino, J.S., Roguin, L.P., and Paladini, A.C., Formation of complexes between 125I-labeled human or bovine somatotropins and binding proteins in vivo in rat liver and kidney, *Biochem. J.*, 214, 121, 1983.

117. Goldfine, I.D., and Smith, G.J., Binding of insulin to isolated nuclei, *Proc. Natl. Acad. Sci. U.S.A.*, 73, 1427, 1976.

118. Vigneri, R., Goldfine, I.D., Wong, K.Y., Smith, F.J., and Pezzino, V.J., The nuclear envelope: the major site of insulin binding in the rat liver nuclei, *J. Biol. Chem.*, 253, 2098, 1978.

119. Goild, J.A., Insulin binding to isolated nuclei from obese and lean mice, *Biochemistry*, 18, 3674, 1979.

120. Podlecki, D.A., Smith, R.M., Kao, M., Tsai, M., Huecksteadt, T., Brandenburg, D., Lasher, R.S., Jarret, L., and Olefsky, J.M., Nuclear translocation of the insulin receptor — a possible mediator of insulin's long-term effects, *J. Biol. Chem.*, 262, 3362, 1987.

121. Wong, K.Y., Hawley, D., Vigneri, R., and Goldfine, I.D., Comparison of solubilized and purified plasma membrane and nuclear insulin receptors, *Biochemistry*, 27, 375, 1988.

122. Kim, S.J., and Kahn, C.R., Insulin induces rapid accumulation of insulin receptors and increases tyrosine kinase activity in the nucleus of cultured adipocytes, *J. Cell. Physiol.*, 157, 217, 1993.

123. Heguy, A., Baldari, C., Bush, K., Nagele, R., Newton, R.C., Robb, R.J., Horuk, R., Telford, J.L., and Melli, M., Internalization and nuclear localization of interleukin 1 are not sufficient for function, *Cell. Growth Differ.*, 2, 311, 1991.

123. Goldfine, I.D., Jones, A.L., Hradek, G.T., Wong, K.Y., and Mooney, J.S., Entry of insulin into human cultured lymphocytes: electron microscope autoradioagraphic analysis, *Science*, 202, 760, 1978.

124. Smith, R.M., and Jarett, L., Ultrastructural evidence for the accumulation of insulin in nuclei of intact 3T3-L1 adipocytes by insulin receptor mediated process, *Proc. Natl. Acad. Sci. U.S.A.*, 84, 459, 1987.

125. Rakowicz-Szulcynska, E.M., Otwiaska, D., and Kowprowski, H., Plasma membrane mediated nuclear uptake and chromatin binding of insulin in tumour cell lines, *Mol. Carcinog.*, 3, 150, 1990.

126. MacDonald, H.S., Kushnaryov, V.M., Sedmak, J.J., and Grossberg, S.E., Transport of gamma inerferon into the cell nucleus may be mediated by nuclear membrane receptors, *Biochem. Biophys. Res. Commun.*, 138, 254, 1986.

128. Kuno, K., Okamoto, S., Hirose, K., Murakami, S., and Matsushima, K., Structure and function of the intracellular portion of the mouse interleukin 1 receptor (type 1). Determining the essential region for transducing signals to activate the interleukin 8 gene, *J. Biol. Chem.*, 268, 13510, 1993.

129. Grenfell, S., Smithers, N., Miller, K., and Solari, R., Receptor mediated endocytosis and nuclear transport of human interleukin 1a, *Biochem. J.*, 264, 813, 1989.

130. Yankner, B.A., and Shooter, E.M., Nerve growth factor in the nucleus: interactions with receptors on the nuclear membrane, *Proc. Natl. Acad. Sci. U.S.A.*, 76, 1269, 1979.

131. Marchisio, P.C., Naldini, L., and Calissano , P., Intracellular distribution of nerve growth factor in rat phaeochromocytoma PC12 cells: evidence for a perinuclear and intranuclear location, *Proc. Natl. Acad. Sci. U.S.A.*, 77, 1656, 1980.

132. Chabot, J.-G., Enjalbert, A., Pelletier, G., Dubois, P.M., and Morel, G., Evidence for direct action of neuropeptide Y (NPY) in rat pituitary gland, *Neuroendocrinology*, 47, 511, 1988.

133. Morel, G., Leneveu, E., Tonon, M.-C., Pelletier, G., Vaudry, H., and Dubois, P.M., Subcellular localization of thyrotropin-releasing hormone (TRH) and neuropeptide Y (NPY)-like immunore-activity in the neurointermediate lobe of the frog pituitary, *Peptides*, 6, 1085, 1985.

134. Morel, G., Chabot, J.-G., and Dubois, P.M., Ultrastructural evidence for oxytoxin in the rat anterior pituitary gland, *Acta Endocrinol. (Copenh.)*, 117, 307, 1988.

135. Maher, D.W., Lee, B.A., and Donoghue, D.J., The alternatively spliced exon of platelet derived growth factor A chain encodes a nuclear targetting signal, *Mol. Cell. Biol.*, 9, 2251, 1989.

136. Buckley, A.R., Montgomery, D.W., Hendrix, M.J.C., Zukoski, C.F., and Putnam, C.W., Identification of prolactin receptors in hepatic nuclei, *Arch. Biochem. Biophys.*, 2, 96,198, 1992.

137. Lobie, P.E., Garcia-Aragon, J., Wang, B.S., Baumbach, W.R., and Waters, M.J., Cellular localization of the growth hormone binding protein in the rat, *Endocrinology*, 130, 3057, 1992.

138. Rao, Y.P., Olson, M.D., Buckley, D.J., and Buckley, A.R., Nuclear co-localization of prolactin receptor in the rat Nb2 node lymphoma cells, *Endocrinology*, 133, 3062, 1993.

139. Giss, B.J., and Walker, A.M., Mammotroph autoregulation: intracellular fate of internalized prolactin, *Mol. Cell. Endocrinol.*, 42, 259, 1985.

140. Clevenger, C.V., Russel, D.H., Appasamy, P.M., and Prystowsky, M.B., Regulation of interleukin 2 driven T-lymphocyte proliferation by prolactin, *Proc. Natl. Acad. Sci. U.S.A.*, 87, 6460, 1990.

141. Clevenger, C.V., Sillman, A.L., and Prystowsky, M.B., Interleukin-2 driven nuclear translocation of prolactin in cloned T-lymphocytes, *Endocrinology*, 127, 3151, 1990.

142. Morel, G., Chayvialle, J.-A., Kerdelhue, B., and Dubois, P.M., Ultrastructural evidence for endogenous substance P-like immunoreactivity in the rat pituitary gland, *Neuroendocrinology*, 35, 86, 1982.

143. Schnitzer, J.E., McIntosh, D.P., Dvorak, A.M., Liu, J., and Oh, P., Separation of caveolae from associated microdomains of GPI-anchored proteins (see comments), *Science*, 269, 1435, 1995.

144. Cunningham, A.M., Ryugo, D.K., Sharp, A.H., Reed, R.R., Snyder, S.H., and Ronnett, G.-V., Neuronal inositol 1,4,5-trisphosphate receptor localized to the plasma membrane of olfactory cilia, *Neuroscience*, 57, 339, 1993.

145. Fujimoto, T., Nakade, S., Miyawaki, A., Mikoshiba, K., and Ogawa, K., Localization of inositol 1,4,5-trisphosphate receptor-like protein in plasmalemmal caveolae, *J. Cell Biol.*, 119, 1507, 1992.

146. Geuze, H.J., Slot, J.W., VanderLey, P.A., and Scheffer, R.C.T., Use of colloidal gold particles in double labeling immunoelectron microscopy on ultrathin frozen sections, *J. Cell Biol.*, 89, 653, 1981.

147. Rothberg, K.G., Heuser, J.E., Donzell, W.C., Ying, Y.S. Glenney, J.R., and Anderson, R.G.W., Caveolin, a protein component of caveolae membrane coats, *Cell,* 68, 673, 1992.

148. Anderson, R.G.W., Plasmalemmal caveolae and GPI-anchored membrane proteins, *Curr. Opin. Cell Biol.*, 5, 647, 1993.

149. Anderson, R.G.W., Caveolae: where incoming and outcoming messengers meet, *Proc. Natl. Acad. Sci. U.S.A.*, 90, 10909, 1993.

150. Lisanti, M.P., Scherer, P.E., Tang, Z., and Sargiacomo, M., Caveolae, caveolin and caveolin-rich membrane domains: a signaling hypothesis, *Trends Cell Biol.*, 4, 231, 1994.

151. Sierralta, W.D., and Thole, H.H., Immunogold labeling of the cytoplasmic estradiol receptor in resting porcine endometrium, *Cell Tissue Res.*, 270, 1, 1992.

152. Caruncho, H.J., and Costa, E., Double-immunolabeling analysis of GABAA receptor subunits in label-fracture replicas of cultured rat cerebellar granule cells, *Receptors Channels*, 2, 143, 1994.

153. Brisson, C., Archipoff, G., Hartmann, M.-L., Hanau, D., Beretz, A., Freyssinet, J.-M., and Cazenave J.-P., Antibodies to thrombomodulin induce receptor mediated endocytosis in human saphenous vein endothelial cells, *Thromb. Haemostasis*, 68, 737, 1992.

154. George, F., Brisson, C., Poncelet, P., Laurent, J.C., Massot, O., Arnoux, D., Ambrosi, P., Klein-Soyer, C., and Sampol, J., Rapid isolation of human endothelial cells from whole blood using S-ENDO1 monoclonal antibody coupled to immuno-magnetic beads: demonstration of endothelial injury after angioplasty, *Thromb. Haemostasis,* 67, 147,1992.

155. Nahagama, M., Orci, L., Ravazzola, M., Amherdt, M., Lacomis, L., Tempst, P., Rothman, J.E., and Söllner, T.H., A v-SNARE implicated in intra-golgi transport, *J. Cell Biol.*, 133, 507, 1996.

Chapter **13**

LOCALIZATION AND EXPRESSION LEVEL OF mRNA CODING FOR RECEPTOR BY LIGHT MICROSCOPIC *IN SITU* HYBRIDIZATION

Allal Ouhtit
Didier Decimo
Michèle Crumeyrolle-Arias
Gérard Morel

CONTENTS

0-8493-2644-3/97/$0.00+$.50
© 1997 by CRC Press LLC

I. INTRODUCTION

In situ hybridization (ISH) is widely used in various areas of research, mainly for diagnostic purposes. It serves to visualize nucleic acids, both DNA and RNA, and in particular mRNA coding for specific proteins, which then provide information on the possible activation of corresponding genes in cells.

In order to identify a target cell for a specific ligand, an antibody against a receptor or a ligand must be used to visualize the ligand's immunoreactivity within the cell (see Chapters 9 and 10). An alternative method involves visualization of the mRNA coding for this receptor, but first the sequence of this cDNA must be determined. Moreover, the copy number of this mRNA must be adequate for the use of ISH (minimum 5 to 50 copies per cell).

To get a faithful copy of a target gene, or a fragment of the gene, it suffices to locate the complementary nucleotidic sequence, wherever it may be found, either in a native or a recombinant genome. In optimal conditions, this labeled complementary pure copy (probe), can bind only to a specific target sequence, which will then be easily detectable using radioautography (with radioisotope markers) or by a simple immunohistochemical reaction (with antigenic markers).

Compared to other methods (e.g., Northern blot, Southern blot, polymerase chain reaction, etc.), the most important advantage of ISH is accurate visualization and assessement of specific gene expression, not only within the tissue but virtually in individual specific cells.

II. PROBES

A. CHOICE OF PROBE

Recombinant DNA technology now provides a means of obtaining DNA and RNA probes of any desired sequence. Thus, we can choose between single- and double-stranded probes. Because double-stranded DNA, or cDNA, is the longest type of probe (more than 200 bp) it can induce nonspecific hybridization to particular parts of a nucleic acid. It requires a high degree of stringency to remove nonspecific labeling. However, synthesized single-stranded (oligoprobe) or RNA (riboprobe) probes are currently the most commonly used.

Synthesized oligoprobes were chosen for several reasons:

Their quality, quantity, and purity of copy.
They contain about 30 nucleotides; short probes are very fast at getting into the target nucleic acid in the tissue.
They produce a hight rate of hybridization, especially if several different ones are used together.
Their labeling requires a small quantity of DNA, which gives a high level of specific activity.

Depending on the features of the target mRNA, oligoprobes may bind nonspecifically to homologous sequences, but this disadvantage can be avoided by careful investigation of probe sequence homology in the gene bank before selection.

The RNA probes are very good tools for ISH. They can be labeled to high specific activity and the RNA-RNA hybrids are more stable than RNA-DNA or DNA-DNA hybrids. The nonspecific binding obtained by RNA probes can be further reduced by incubation of the sections in ribonucleases. These enzymes digest remaining single-stranded probes and do not affect the specifically bound double-stranded RNA/RNA hybrids. All the mismatches are removed and increase the signal/background ratio. However, the disadvantages of the RNA probe synthesis are the cloning of DNA of interest in an expression vector and the precautions needed to prevent the RNA degradation by RNase (see Protocol 4).

B. CHOICE OF LABEL

Two types of marker are currently used to label probes for ISH: radioisotopes and antigens.

1. Radioactive Labels

Four radioisotopes are avalaible for labeling: tritium (^{3}H), sulfur-35 (^{35}S), phosphorus-33 (^{33}P), and also occasionally, phosphorus-32 (^{32}P) (Table 1).

TABLE 1   Characteristics of Isotopes Used for Labeling

Radioisotopes	Half-life	E_{max} (MeV)	Specific Activity TBq/mmol	Resolution	Exposure Time
^{3}H	12.3 years	0.018	1.07	+++	+++
^{35}S	87 days	0.17	55.5	++	++
^{33}P	25 days	0.223	111	++	+
^{32}P	14 days	1.6	222	+	+

Each of these radioisotopes has its own advantages and disadvantages.

Given characteristics of the radioisotopes described above, ^{35}S seems to be the best compromise, as well as the easiest to use as a nucleotide labeling isotope;[1-2] ^{33}P can also be successfully used.

Hybrids containing a radioisotope are revealed by radioautographic procedure (see Protocols 15 and 16).

Radiolabeled probes can detect nucleic acids at extremely low concentrations. Radiolabeled hybridization is thus a highly sensitive method, and can also be used as a quantitative assay. Nevertheless, the use of radioactivity has some disadvantages: a radio-specialized laboratory and scrupulous radio-protection are required, and radioisotopes have short half-lives.

2. Nonradioactive Labels

A number of markers, including biotin-, digoxygenin-, or fluorescein-substituted nucleotides,[3] or photobiotin, have been commonly used in recent years to replace radioactive probes for the detection of target mRNA, or as a source of supplementary information (e.g., distinction between cytoplasm and nucleus, visualization of labeled cell type without staining). These are useful for several reasons:

Advantages related to the use of nonradioactive reagents, which do not need radio-protection.

Fast revelation procedure, using specific antibodies conjugated with enzymatic or fluorescent reagents, or colloidal gold.

High level of cellular resolution.

Nonradioactive probes can be stored for several months at −20°C.

However, nonradioactive probes also have some disadvantages:

Lower sensitivity of mRNA detection; and

Quantitative analysis, requiring a high degree of stringency in standardized conditions (internal control), which is difficult to obtain with this system.

C. LABELING PROCEDURES

These procedures are dependent on the nature of the probe.

1. cDNA Labeling

Nick translation (see Protocol 1) — This is designed for the efficient incorporation of labeled deoxynucleoside triphosphate (dNTP) into the DNA duplex.[4] The reaction combines the simultaneous action of two enzymes: pancreatic deoxyribonuclease I (DNase I) and E. *coli*-DNA polymerase I (DNA pol I). However, it requires a large amount of DNA (1 μg/reaction) and the reaction temperature is crucial (15°C).

PROTOCOL 1: NICK TRANSLATION LABELING

Reagents

Labeled nucleotide: 0.3 m*M* biotin-dUTP, digoxygenin-dUTP or fluorescein-dUTP, 50 μCi (α^{35}S) dATP or dCTP.
dNTP mixture: 0.2 m*M* each of dNTP in 500 m*M* Tris-HCl (pH 7.8), 50 m*M* MgCl$_2$, 100 m*M* β-mercaptoethanol, 100 mg/ml nuclease-free BSA.

Method

1. Mix:

DNA 1 μg
dNTP mixture 5 μl
Labeled nucleotide X μl
DNA pol I, 4 U/μl - DNase I, 40 pg/μl 5 μl
H$_2$O to 50 μl

2. Incubate for 2 h at 15°C.

Random priming reaction: (see Protocol 2) — After temperature denaturation of the double-stranded DNA, DNA polymerase (Klenow fragment of E. *coli* -DNA pol I) synthesizes a new DNA strand complementary to a template strand using a random mixture of hexanucleotides and nonanucleotides, including labeled nucleotides.[5,6] Synthesis has been found to occur very efficiently with nanogram amounts of DNA.

PROTOCOL 2: RANDOM PRIMED DNA LABELING

Reagents

Labeled nucleotide: 0.3 m*M* biotin-dUTP, digoxygenin-dUTP, or fluorescein-dUTP, or 50 μCi (α^{35}S) dATP or dCTP.

Method

1. Mix the following reagents:

Linearized plasmid		1 µg	
Transcription buffer 10X		2 µl	
	Radioactive	or	Nonradioactive
NTP mixture	1 µl		2 µl
Labeled nucleotide	120 µCi		0.7 mM
0.1 M DTT		1 µl	
Placental RNase inhibitor		1 µl	
DEPC treated water		to 20 µl	
RNA polymerase		1 µl	

2. Incubate for 90 min at 37°C
3. Take 1 µl aliqot for scintillation counting (Tr)* or for gel electrophoresis**
4. Add 1 µl of RNase-free DNase I, and incubate 10 min at 37°C
5. Precipitate the probe by adding 1 µl yeast tRNA, 2 µl sodium acetate, and 2.5 volume 100° ethanol
6. Leave the mixture for 30 min at –70°C, pellet the RNA precipitate by centrifugation (10,000 × g, 30 min), wash in 70° ethanol, and air-dry
7. Dissolve the RNA pellet in 50 µl DEPC-treated water, take 1 µl aliquot for scintillation counting (Ir)***, add 30 µl 0.2 M Na$_2$CO$_3$ and 20 µl 0.2 M NaHCO$_3$ (pH 10.2)
8. Incubate at 60°C for t min, as determined by the formula:

$$t = \frac{Lo - Lf}{k \times Lo \times Lf}$$

where: Lo is the initial probe length (kb)
Lf is the final length (0.15 – 0.2 kb)
k is the hydrolysis rate constant (0.11 kb/min)

9. Stop the hydrolysis reaction by adding 3 µl 3 M sodium acetate and ethanol. Precipitate the RNA as described above. Wash the RNA pellet in 70° ethanol and air dry.
10. Calculate the final volume of 10 mM DTT needed to resuspend the probe to 2 ng/µl using the formula given in the Protocol 6**** and store the probe at –20°C for no more than 1 month.

D. PRECIPITATION

DNA and RNA precipitation is necessary to eliminate the unincorporated radioactive or nonradioactive nucleotides and to concentrate the labeled probe (see Protocol

---

* Total amount of radioactivity.
** The quantity of digoxigenin-labeled RNA synthesized is determined by agarose gel electrophoresis. The amount of RNA is estimated with reference to DNA or RNA of known concentration.
*** Incorporated radioactivity.
**** Tr and Ir are necessary to calculate the quantity of labeled probe produced in the transcription reaction.

5). The nucleotides can also be eliminated with spin column purification (Sephadex G25).

PROTOCOL 5: PRECIPITATION OF NUCLEIC ACIDS

1. Mix either 1 μl (10 μg/μl) tRNA, 1/10 final volume of sodium acetate 3 *M* pH 5.2, or ammonium acetate 7.5 *M* and 3 volumes cold ethanol (–20°C).
2. Store overnight at –20°C, or 1 h at –80°C, so as to allow the nucleic acid precipitate to form.
3. Centrifuge for 30 min at 14,000 × g at 4°C.
4. Discard the supernatant.
5. Wash in ice-cold 70° ethanol.
6. Centrifuge for 15 min at 14,000 × g, at 4°C.
7. Discard the supernatant; speed-vac briefly to remove the remainder of supernatant (1 to 2 min).
8. Dissolve the probe pellet in the desired volume of buffer or pure water.

E. LABELING CONTROLS

The specific activity of the probe is an indication of the success of the labeling reaction, and thus the effectiveness of this probe for ISH (see Protocol 6).

PROTOCOL 6: SPECIFIC ACTIVITY OF THE PROBE

1. Take 1 μl of labeled DNA or RNA solution and put it in separate tubes for measurement of radioactivity:
   - Before probe precipitation (step 1 of Protocol 5), evaluate the Tr (total amount of radioactivity) introduced in the reaction (see Protocols 1 to 4);
   - After probe precipitation (step 8 of Protocol 5), determine the amount of incorporated radioactivity (Ir).
2. Make up with scintillation liquid, and put the tubes in the scintillation counter to measure the quantity of radioactivity expressed in dpm (disintegrations per minute).

DNA probe: calculate the percentage of radioactivity-incorporation, using the Ir/Tr formula (at >40% the labeled probe can be used).

RNA probe: the quantity (Q) of RNA synthesized is determined as follows:

$$Q = \frac{50 \text{ Ir}}{20 \text{ Tr}} \times 4 \times 330 \times \left(\left[\alpha\ ^{35}\text{S UTP}\right] + \left[\text{UTP}\right]\right)$$

where: Ir is the total incorporated radioactivity in the 50 μl of solution, Tr is the total amount of radioactivity in 20 μl of solution (see Protocol 4), 330 is the average molecular weight of one nucleotide, [α^{35}S UTP] is the quantity of labeled nucleotide used in the transcription reaction, (expressed in nmole),

[UTP] is the quantity of unlabeled UTP added in the reaction (0.25 nmol), and Q is expressed in ng of RNA neosynthesized.

The specific activity of a labeled probe is the amount of Ir per μg of labeled probe. The Ir value is transformed from dpm to Ci/mmol or Bq using the following formula:

$$Ci = dpm / 2.2 \times 10^{12} \ (dpm = cpm/\% \ efficiency)$$

III. TISSUE PREPARATION

ISH can be done on paraffin, frozen tissue sections, cultured cells, or whole-mount embryos. Each type of material has its own advantages and disadvantages. Paraffin tissue sections give the best histological quality, but the sensitivity is lower than with frozen sections. With whole-mount embryos, direct three-dimensional detection of mRNA expression is possible in early stage embryos. Successful ISH depends on the preservation of tissue structure and the retention of the mRNA. The tissue must be either fixed by perfusion and/or immersion, or frozen as rapidly as possible.

A. PREPARATION OF SLIDES

To avoid detachment and to preserve the structure of the sections during the different ISH steps, the slides need specific treatment. Methods include the use of 3-aminopropyl triethoxysilane,[2] gelatin/chrome alum,[8] and polylysine solutions. The first of these methods gives excellent results (see Protocol 7).

PROTOCOL 7: TREATMENT OF SLIDES

1. Leave the slides overnight in ethanol 95° containing 1% hydrochloric acid (10 N)
2. Wash in running water for at least 2 h.
3. Wash in distilled water for 5 min, then dry the slides at 180°C for 2 h.
4. Dip in 3-aminopropyl triethoxysilane 2% in acetone for 5 s.
5. Dip in acetone for 2 × 2 min.
6. Dip in distilled water for 2 × 2 min.
7. Dry in an oven at 40°C.
8. Store at 4°C.

B. PREPARATION OF SECTIONS
1. Frozen Tissue

Fresh biopsy tissue should be rinsed in 0.1 M mono-di-phosphate buffer (pH 7.4) and immediately frozen in methyl-2-butane (isopentane) on dry ice or in liquid nitrogen until use or store at –80°C.

Cryotome sections (10 μm) are placed on pretreated slides (see Protocol 7), air-dried for 1 h at room temperature or overnight in the cryostat, and stored at –80 or –20°C.

Preservation of structural morphology by fixation of tissue in paraformaldehyde at 4% (see Protocol 8) gives maximal hybridization efficiency, but a deproteinization step should be carried out in optimal conditions in order to remove proteins (which can mask the target nucleic acid) with the least damage to tissue structure. The best protocol which we have perfected includes pre- and postfixation steps using 4% paraformaldehyde (see Protocol 8). However, either protease treatment (see Protocol 8a) or a warmed formamide denaturation (see Protocol 8b) can be used to improve probe penetration and to expose target mRNA.

PROTOCOL 8: TISSUE SECTION FIXATION AND PRETREATMENT

Reagents

Phosphate-buffered saline (PBS): 8 g NaCl, 0.2 g KCl, 1.15 g $Na_2HPO_4 \cdot 7\ H_2O$, 2 g KH_2PO_4 for 1 l paraformaldehyde.

Method

1. 4% paraformaldehyde: dissolve 4 g paraformaldehyde in 100 ml PBS, at 65°C, add 10 drops 10 N NaOH and cool to 4°C (the solution can be stored at –20°C for some months).
2. 40% paraformaldehyde: dissolve 40 g paraformaldehyde in 100 ml H_2O and 0.5 to 1 ml 10 N NaOH, heat at 65°C, filter and cool to 4°C (the solution can be stored at room temperature for 2 weeks or at –20°C for some months). Dilute to 4% in buffer.

a. *Protease Treatment*
Reagents

Proteinase K: 10 mg/ml H_2O
NaCl: 9%
Tris-CaCl₂ buffer: 20 mM Tris-HCl (pH 7.6), 2 mM CaCl₂

Method

1. Prefixation

Wash dry sections in phosphate buffer for 5 min
Fix in 4% paraformaldehyde in PBS for at least 20 min
Rinse in the same buffer, then in NaCl solution for 5 min each
Incubate in proteinase K 1 μg/ml Tris-CaCl₂ buffer at 37°C for 15 min
Wash sections in PBS for 5 min

2. Post-fixation

Fix sections in 4% paraformaldehyde in PBS for 15 min
Rinse them in the same buffer and then in sodium chloride solution (9%) for 5
    min each
Dehydrate them in graded ethanol (50 to 100°) and dry at room temperature for
    1 h
Store at –20°C until use.

b. Formamide Treatment
Reagent

Acetone
Formamide
20X SSC: 3 M NaCl, 0.3 M sodium citrate (pH 7)

Method

1. Dip cold dry slides in acetone at 4°C for 3 to 5 min and allow to air dry.
2. Fix them in 4% paraformaldehyde in PBS at 4°C for 15 min.
3. Rinse twice in PBS for 5 min each time.
4. Acetylate the frozen section (see Protocol 10).
5. Rinse in 2X SSC for 3 min.
6. Incubate in 50% formamide-1X SSC at 60°C for 10 min.
7. Dehydrate the sections in 50°, 70° ethanol which have been cooled to –20°C
   and then 100° ethanol.
8. Air dry the sections.

2. Paraffin Embedding and Tissue Sections
Paraffin embedding gives excellent preservation of morphology. However. the
fixatives used for ISH must also preserve this morphology, as well as retaining
mRNA and facilitating probe penetration. Cross-linking fixatives such as paraformaldehyde and glutaraldehyde offer the best compromise between these different
points. Before dissection, an animal can be perfused with the fixative. The tissue is
fixed for several hours, with optimal fixative penetration being obtained with slices
of 3 to 5 mm (see Protocol 9).[8]

PROTOCOL 9: PARAFFIN EMBEDDING

Materials and reagents

Oven
Plastic embedding molds
4% Paraformaldehyde (see Protocol 8)
Phosphate buffered saline (PBS): 100 mM mono-diphosphate, 150 mM NaCl

Ethanol
Xylene
Paraffin wax

Method

1. Fix the tissue in 10 volumes of 4% paraformaldehyde overnight at 4°C.
2. Rinse in PBS for 30 min at 4°C.
3. Dehydrate the tissue twice 15 min each time in 50° ethanol, 30 min each in 70° and 90° ethanol and twice, 30 min each time, in 100° ethanol and xylene.
4. Transfer the tissue onto melted paraffin wax at 58°C, and place in an oven for at least 3 h (with two changes of paraffin for small pieces) or overnight for sections more than 3 mm thick.
5. Take the molds out of the oven, let the paraffin solidify, and store indefinitely at room temperature.

Microtome sections 5 to 7 µm thick are mounted on adhesive-coated slides (see Protocol 7) on a heating plaque at 45°C, with 10° ethanol to unpleat the paraffin sections. The slides are air-dried and stored in a box containing silica gel at 4°C.

The sections must be dewaxed and rehydrated in preparation for the pretreatment and hybridization stages. Digestion with a proteolytic enzyme (e.g., proteinase K, pepsin, pronase, etc.) is necessary in order to improve probe penetration and to expose the target mRNA (see Protocol 10). The protease concentration and the incubation time of the protein digestion must be controlled in such a way as to preserve tissue structure. Finally the sections are acetylated to reduce nonspecific binding of the probe to the tissue.

PROTOCOL 10: PRETREATMENT

Reagents

Proteinase K 10 mg/ml
Buffer Tris-CaCl$_2$ (pH 7.5): 50 mM Tris-HCl, 2 mM CaCl$_2$
100 mM triethanolamine adjusted to pH 8 with concentrated HCl
Acetic anhydride

Method

1. Dip the slides in xylene twice, 5 min each time.
2. Dip twice in 100° ethanol, and once in 95°, 85°, 60°, and 30° ethanol, 30 s each time and then for a few seconds in DEPC-treated water.
3. Incubate the slides in 1 µg proteinase K/ml buffer Tris- CaCl$_2$ at 37°C for 30 min.
4. Rinse in water.

   5. Place the slides in 100 mM triethanolamine for 5 min.*
   6. Add 0.25% acetic anhydride.
   7. Incubate for 10 min with agitation.
   8. Wash in water for a few seconds.
   9. Dehydrate with 50°, 70°, 95° and 100° ethanol.
   10. Air dry.

IV. *IN SITU* HYBRIDIZATION

A. GENERAL CONSIDERATIONS

The labeled antisense probe included in the hybridization buffer targets and matches the complementary strand in the sections, leading to hybrid formation, whose stability can be affected by several factors, primarily sodium ion concentration (Na⁺)[9] and temperature, which determine the melting temperature, Tm (the temperature at which 50% of DNA is in single-strand form). Tm also depends on the length of the probe DNA sequence and the percentage of G-C.[10]

$$Tm = 16.6 \log (Na^+) + 0.41 \,(\% \, G + C) + 85$$

The addition of formamide to the hybridization buffer reduces the Tm by 0.65°C/% formamide.[11] The hybridization temperature should be 10 to 15°C below the Tm.[9]

B. HYBRIDIZATION BUFFER FOR TISSUE SECTIONS

The buffer ensures efficient hybridization between the labeled probe and the target nucleic acid. It must contain (see Protocol 11): a buffer with an optimal pH and a saline concentration suitable for the hybridization process, formamide to reduce the hybridization temperature, and a labeled probe at a saturating concentration.

To reduce nonspecific hybridization, Denhardt's solution and tRNA should be added to the hybridization buffer. Dextran sulfate enhances the efficiency of the reaction by increasing the concentration of the probe.[12] In the case of a ³⁵S-radioactive probe 10 mM DTT should be added.

The concentration of the labeled probe should be determined, since it depends on the type (double-stranded DNA, oligonucleotide, or riboprobe) and labeling of the probe used (radioactive or nonradioactive). It is generally 2 pmol radioactive oligoprobe/ml or 10 pmol nonradioactive oligoprobe/ml, 100 ng radioactive labeled RNA probe/ml or 1 μg of hapten labeled RNA/ml hybridization buffer. Different concentrations should be tested in order to determine the saturating concentration of probe after washing and revelation steps (see Protocols 12 to 16).

---

* Steps 5–8 are not necessary if the ISH is performed with a digoxigenin probe.

TABLE 2 Hybridization Buffer

Reagents	Storage Concentration	Final Concentration	Role of Reagents
SCC	20X	2X to 4X	Maintains salt concentration and pH
Deionized formamide	100%	50%	Decreases Tm, allowing stringent conditions which give better preservation of cell morphology
Denhardt's solution	50X	1X	The presence of such polymers can reduce nonspecific hybridization
Dextran sulfate	50%	10%	Enhances the reaction efficiency by increasing probe concentration (efficiency volume)
tRNA	10 mg/ml	250 µg/ml	Binds to the sites where nonspecific hybridization can occur
DTT	1 M	10 mM	Decreases the oxidation of sulfurated probes
Pure water		to 1 ml	Completes volume
Labeled probe	See above		Matches specifically to mRNA target

PROTOCOL 11: PREPARATION OF REAGENTS FOR HYBRIDIZATION BUFFER

- **50X Denhardt's solution:** 1% polyvinylpyrrolidone, 1% Ficoll, 1% BSA (store in aliquot at –20°C).
- **20X SSC:** 3 M NaCl, 0.3 M sodium citrate (pH 7).
- **Deionized formamide:** mix 50 ml formamide with 5 g mixed-bed ion exchange resin. Stir for 30 min at room temperature. Filter twice through Whatman number 1 filter paper. Dispense into 500 µl aliquots and store at –20°C. If it thaws, it must not be used.
- **50% Dextran sulfate:** dissolve 50 mg dextran sulfate in 65 µl distilled water. Stir gently with a Pasteur pipette and centrifuge.
- **tRNA** (10 mg/ml) is stored in aliquot at –20°C.
- **DTT** (dithiotreitol) (1 M): dissolve 3.09 g DTT in 20 ml 0.01 M sodium acetate (pH 5.2). Sterilize by filtration. Dispense into 500 µl aliquot and store at –20°C.

The dried and pretreated sections are covered by a volume of hybridization buffer which depends on the size of the tissue section (>1 µl/10 mm²). The slides are incubated under a coverglass overnight at hybridization temperature (at least 37°C for DNA probes to 52°C for RNA probes) in a humid box containing 5X SSC.

C. *IN SITU* HYBRIDIZATION FOR CULTURED CELLS

Cells can be directly grown on sterile slides or coverslips used for ISH or deposited by cytocentrifugation, smearing, or sectioning after freezing. The slides are precoated to avoid cell detachment during the ISH procedure. The coating reagent must be compatible with cell growth. Collagen, Denhardt's solution, gelatin/chrome alum, and polylysine can be used to coat the slides. The cells are washed in PBS, and the slides are dipped in acetone for 5 min, allowed to air-dry, and stored at −80°C. The ISH pretreatment of the cells is performed as for frozen tissue sections.

D. WHOLE-MOUNT *IN SITU* HYBRIDIZATION

Whole-mount ISH is principally carried out on whole embryos. This method permits a three-dimensional visualization of gene expression during development in early stages of embryogenesis.[13,14] Hapten-labeled complementary RNA are used in this procedure followed by a chromogenic reaction with anti-hapten antibodies conjugated to enzyme reaction (i.e., alkaline phosphatase, peroxidase, etc.). Here, we describe whole-mount ISH on mouse embryos (6 to 13 days postcoitum embryos) (see Protocols 12 and 18). The ISH is performed principally with digoxigenin- or fluoresceine-labeled RNA probe and not with biotin-labeled probe to avoid the background which can be generated by the presence of endogenous biotin in embryos.

Embryos need to be dissected free of maternal tissues and extra-embryonic membranes. The hybridization protocol is virtually the same as for tissue sections: fixation, mild protease treatment to increase mRNA accessibility, hybridization, washings, and probe revelation. Some precautions should be taken to avoid damages and/or loss of the embryos, such as to keep some liquid over the tissue during the different ISH steps. Both incubation and washing steps must be carried out with gentle agitation on a rocking tray.

PROTOCOL 12: TREATMENT OF EMBRYOS AND HYBRIDIZATION

Reagents

PBS (see Protocol 8)
PBT (PBS containing 0.1% Tween 20)
4% Parafomaldehyde (see Protocol 8) in PBS
2.5% Glutaraldehyde in PBS
25°, 50°, and 75° methanol in PBT
100° methanol
6% Hydrogen peroxide in PBT
Proteinase K 10 mg/ml in PBT
Glycine 2 mg/ml in PBT
20X SSC (see Protocol 11)
Prehybridization buffer (see Protocol 11): 50% deionized formamide (see Protocol 11), 5X SSC, heparin 50 mg/ml, 1% SDS, 100 µg/ml tRNA
Hybridization buffer: prehybridization buffer containing 1 µg/ml digoxigenin RNA probe as prepared in Protocol 4

Method

1. Dissect the embryos in PBS and fix in 4% paraformaldehyde PBS at 4°C for 2 h to overnight.
2. Wash in PBT three times for 5 min each.
3. Dehydrate in 25°, 50°, and 75° methanol in PBT, and twice in 100° methanol for 5 min each time.*
4. Rehydrate them in 75°, 50°, and 25° methanol in PBT and twice in PBT for 5 min each time.
5. Incubate in 6% hydrogen peroxide in PBT for 1 h and rinse three times in PBT for 5 min each time.
6. Treat with proteinase K 10 µg/ml in PBT for 5–15 min.**
7. Wash in glycine 2 mg/ml in PBT to block proteinase K activity for 3 min.
8. Rinse in PBT twice for 5 min.
9. Fix in 4% paraformaldehyde-0.2% glutaraldehyde in PBT for 20 min.
10. Wash twice in PBT for 5 min each time.
11. Incubate in prehybridization buffer at 65°C for 1 h.
12. Change the prehybridization buffer, add RNA probe to final concentration of 1 µg/ml in buffer, and incubate the embryos at 65°C overnight.

V. REVELATION

Hybridization signals can be detected in two different ways, depending on the type of label used: hybrids containing a radioactive isotope are revealed by radio-autography,[15-17] and those which contain an antigenic component are revealed by immunohistochemistry.[13,14]

A. RADIOAUTOGRAPHY
1. Washing

This step is crucial in the reduction of the frequency of nonspecific hybrids, in that it eliminates any probe molecules which are weakly or nonspecifically matched within the tissue, even with high level of stringency. An optimal washing temperature (Tw) also maximizes the number of specific hybrids and minimizes that of nonspecific hybrids. It should be determined as laid out in Section IV. The ionic concentration (SSC) of the washing solution is also important. A low ionic concentration favors the rupture of hydrogen bonds. A decrease in the SSC concentration of the washing medium, therefore, favors the dissociation of molecules.

Double-stranded RNA hybrids are more stable than DNA/RNA or DNA/DNA, and therefore the washing procedure has to be more stringent in order to eliminate nonspecific background binding (see Protocols 13 and 14). Treatment with one or

---

* The embryos can be stored in 75° methanol at −20°C.
** Time is dependent on the embryonic stage (5 min for 6- to 8-day-old embryos and 15 min for older embryos).

two RNases is added to the classical washing steps. The RNases used cut single-stranded RNA and break down all imperfect RNA/RNA hybrids (see Protocol 14).

PROTOCOL 13: WASHING PROCEDURE AFTER HYBRIDIZATION WITH RADIOACTIVE OLIGO- AND DNA-PROBES

1. Dip hybridized sections in SSC solutions, successively:
    5X SSC for 1 min to rinse them and to remove the coverslides
    2X SSC for 1 h at room temperature
    2X SSC for 1 h at the determined washing temperature (see Section IV)
    1X SSC for 30 min at room temperature
    0.5X SSC for 30 min at room temperature
2. Dehydrate in graded ethanol solutions (50 to 100°) and air-dry at room temperature.

PROTOCOL 14: WASHING PROCEDURE AFTER HYBRIDIZATION WITH A RADIOACTIVE RNA PROBE

Reagents

100 µg/µl RNase A
100 U/µl RNase T1
Formamide
20X SSC (see Protocol 11)

Method
The washing solutions must be prewarmed to 55°C or 37°C.

1. Dip the slides at 55°C in 50% formamide-1X SSC, for 30 min.
2. Remove the coverslips with forceps and transfer the slides to 50% formamide-1X SSC, for 90 min.
3. Rinse twice in 2X SSC, 5 min each time at room temperature.
4. Incubate in 2X SSC containing 20 to 50 µg/ml RNase A and 1 to 5 U/ml RNase T1 for 30 min at 37°C.
5. Transfer the slides twice in 50% formamide-SSC 2X at 55°C, 1 h each time.
6. Dip in 0.1X SSC for 15 min at 50°C.
7. Dehydrate the sections in 50°, 75°, and 100° ethanol, 1 min each time, and air-dry.

2. Macroradioautography
The film is placed on washed and hybridized sections to detect the β-radiation corresponding to the given hybrid. The macroradioautogram gives a macroscopic view of the regional localization of the hybridization signal (Figures 1 and 2). The exposure time depends on the specific activity of the probe, the amount of hybridized probe, and the type of radioisotope used (see Table 1).[15-17]

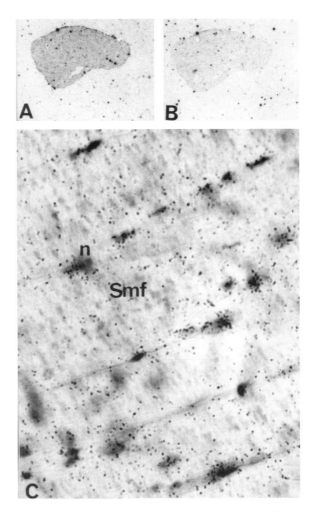

FIGURE 1   Regional and cellular localization of the distinct forms of prolactin receptor (PRL-R) mRNA in skeletal muscle. Hybridization reaction was carried out using a mixture of two different antisens [^{35}S]-oligoprobes, complementary to different regions of the mRNA sequence coding for the intracellular domain of each form of PRL-R. Macroradioautograms of long form (A) and short form (B) mRNAs from adjacent sections of skeletal muscle from male rat. C: silver grains observed at light field (dark grains), corresponding to the long form of mRNA, were concentrated over the cytoplasm of striated muscle fiber (Smf) surrounding the nucleus (n). Original magnification: A and B, ×5; C, ×250.

PROTOCOL 15: MACRORADIOAUTOGRAPHIC PROCEDURE

Materials and Reagents

Film
Autoradiographic cassette

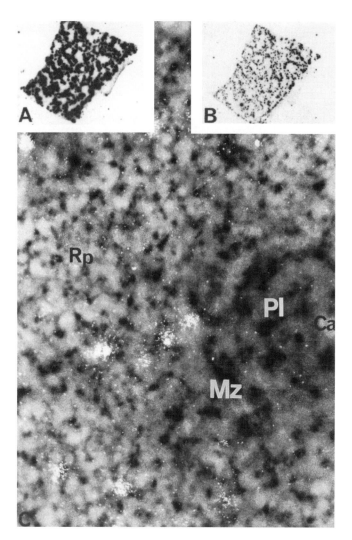

FIGURE 2   Regional and cellular localization of both forms of prolactin receptor (PRL-R) mRNA in spleen. Hybridization reaction was carried out using a mixture of two different antisens [^{35}S]-oligoprobes, complementary to different regions of the mRNA sequence coding for the intracellular domain of each form of PRL-R. Macroradioautograms of long form (A) and short form (B) mRNAs from adjacent sections of spleen from a female rat. C: the density of silver grains (for long mRNA of PRL-R), observed by epipolarization (white grains), is high over cells of the red pulp (Rp) and low in the marginal zone (Mz), periarterial lymphoid (Pl), and around the central artery (Ca) of white pulp. Original magnification: A and B, ×5; C, ×250.

Developer (compatible with film)
Fixative (compatible with film)
Safelight (compatible with film)

1. After the washing steps, place the film on dried sections pre-arranged in light-tight box and expose in a darkroom for at least 24 h.

2. Develop the film by machine or manually, using a developer and a fixative, as follows:

After exposure, place the film in the developer (Kodak, D19) for 2 to 5 min.
Rinse in water.
Put in the fixative (Ilford Hypam) for 5 to 10 min.
Wash in running water.
Rinse in distilled water and dry.

3. Microradioautography
This identifies the hybridization signal in tissues (Figure 3) at the cellular level (Figures 4 and 5) using photographic nuclear emulsion.

PROTOCOL 16: MICRORADIOAUTOGRAPHIC PROCEDURE

Materials and Reagents

Emulsion
Water bath
Developer (compatible with nuclear emulsion)
Fixative (compatible with nuclear emulsion)
Safelight (compatible with nuclear emulsion)

Method
This procedure should be carried out in the dark:

1. Prepare a dipping jar containing distilled water to dilute the nuclear emulsion (1:1 for Kodak NTB2, 2:1 for Amersham LM1, 3:1 for Ilford K5).
2. Place the jar of emulsion and the dipping jar in the water bath (45°C) for 30 min.
3. Fill the dipping jar with emulsion to the required level.
4. Dip a clean slide in the emulsion, shake it gently, and examine it under the safelight. If bubbles are present, gently tap the side of the jar and leave it for a moment.
5. Place the slides carefully in the jar. Drain the excess emulsion onto tissue paper.
6. Leave the dipped slides on a vertical support until dry in order to estimate the thickness and regularity of the emulsion.
7. Put the dried slides in a light-tight box containing a drying agent and keep in a closed box at 4°C for 2 to 30 days (until 7 weeks), depending on the type of the radioisotope used and the signal observed on macroradioautogram
8. Develop the slides in D19 or Dektol developer diluted v/v in the water for 2 min at 18°C.

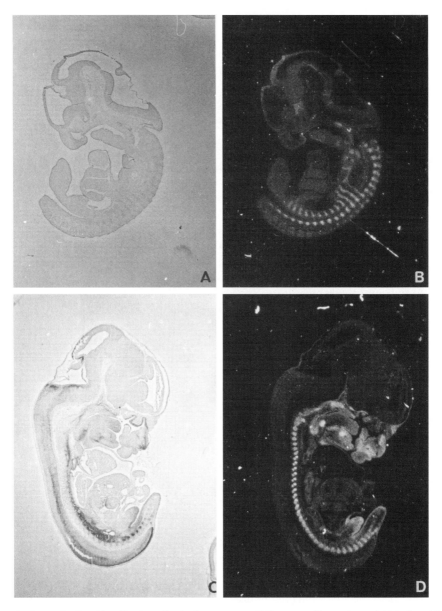

FIGURE 3   Regional localization of retinoic receptor mRNA on paraffin sections of 13.5-day-old mouse embryo. Hybridization reaction was carried out using RNA probes complementary to mRNA sequences coding for the RARγ receptor (A-B) and RXRγ receptor (C-D). A-C and B-D are bright field and dark field observations respectively. In bright field, signal (silver grains) is black and in dark field it appears white. RARγ is only expressed in cartilage and RXRγ is observed in different tissues such as face, tongue, neural tube, genital tubercle, etc. Original magnification: ×15.

9. Wash them briefly in water and then fix them in 30% sodium thiosulfate for at least 2 min.

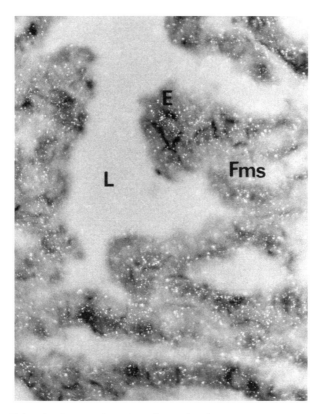

FIGURE 4   Cellular distribution of the long form of prolactin receptor (PRL-R) mRNA in rat prostate gland. Hybridization reaction was carried out using a mixture of two different antisense [35S]-oligoprobes, complementary to different regions of the mRNA sequence coding for the intracellular domain of the long form of PRL-R. After revelation of the hybridization signal, the microradioautogram shows a high density of epipolarized silver grains (white grains), which correspond to the distribution of the long form of PRL-R mRNA over epithelial (E) cells; no specific hybridization is seen over fibromuscular stroma (Fms) or in lumen. Original magnification: ×250.

10. Wash them in water.
11. Stain the sections (see Protocol 17).

PROTOCOL 17: STAINING OF SECTIONS

Reagents

Toluidine blue

Method

1. Prepare 1% solution of toluidine blue in 1% sodium borate.
2. Put the slides in diluted solution of toluidine blue (1:5 in water) for 5 to 10 min at room temperature.

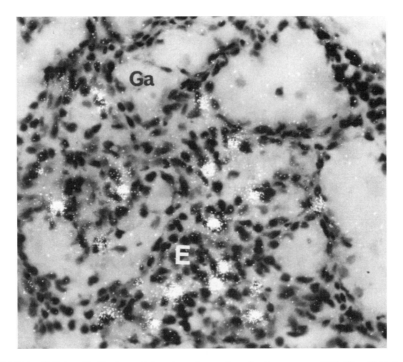

FIGURE 5   Cellular distribution of the short form of prolactin receptor (PRL-R) mRNA in rat mammary gland. Hybridization reaction was carried out using a mixture of two different antisense [^{35}S]-oligoprobes, complementary to different regions of the mRNA sequence coding for the intracellular domain of the short form of PRL-R. The microradioautogram shows a high concentration of silver grains observed by epipolarization (white grains), which correspond to the distribution of the short form of PRL-R mRNA over some PRL-target epithelial cells of glandular alveoli. Original magnification: ×250.

3. Differentiate the staining with 95° ethanol.
4. Dehydrate quickly in 2 × 100° ethanol.
5. Mount on coverslip and leave to dry before examining by light microscopy.*

Other Protocols can also be successfully used.**

B. IMMUNOHISTOCHEMISTRY

ISH with antigen-labeled RNA probes is less sensitive than with radiolabeled riboprobes. For this reason, the ISH washing procedure is less stringent.*** Hybrids are detected by an enzyme-linked immunoassay using an anti-digoxigenin antibody combined with calf intestine alkaline phosphatase**** followed by a color reaction. The endogenous phosphatase activity is inhibited by levamisol and the high pH (>9) of the revelation solution.

---

* Staining should not affect observation (e.g., dark field observation).
** Methyl green, hematoxylin, nuclear fast red, etc.
*** Washing step continues during the revelation of the hybrids.
**** This alkaline phosphatase is not inhibited by levamisol.

PROTOCOL 18: WASHING PROCEDURE AFTER HYBRIDIZATION WITH DIGOXIGENIN-LABELED RNA PROBE AND IMMUNOLOGICAL DETECTION

a. Tissue Sections
Reagents

Anti-digoxigenin alkaline phosphatase (Boehringer-Mannheim, Meylan, France)
Levamisol
TBST buffer: 25 mM Tris-HCl (pH 7.5), 140 mM NaCl, 2.5 mM KCl, 0.1% Tween-20
1% Blocking reagent solution (Boerhinger-Mannheim) or 1% BSA in TBST
NTMT buffer: 100 mM Tris-HCl (pH 9.5), 100 mM NaCl, 50 mM MgCl$_2$, 0.1% Tween-20
Nitroblue tetrazolim salt (NBT, 75 mg/ml in 70% dimethylformamide)
5-Bromo-4-chloro-3-indolyl phosphate (BCIP, 50 mg/ml in 100% dimethylformamide)

Method

1. Dip the slides in 5X SSC, for 5 min at room temperature to detach the coverslip.
2. Incubate in 50% formamide-2X SSC at 50°C for 30 min.
3. Rinse twice in 2X SSC, for 5 min each time.
4. Treat with 10 μg/ml RNase A in 2X SSC, at 37°C for 30 min.
5. Wash in 2X SSC for 10 min at 50°C, two times and in 2X SSC for 10 min at room temperature.
6. Wash twice in TBST buffer, 1 min each time at room temperature.
7. Saturate the sections in 1% BSA-TBST buffer for 30 min at room temperature.
8. Incubate in anti-digoxigenin antibody (1/1000) in 1% BSA-TBST buffer containing 2 mM levamisol for 1 h at room temperature.
9. Wash three times in TBST buffer, 5 min each time at room temperature.
10. Dip in NTMT buffer containing 2 mM levamisol for 5 min at room temperature.
11. Incubate the tissue sections in the staining solution (4.5 μl NBT and 3.5 μl BCIP in 1 ml NTMT containing 2 mM levamisol) in a light-tight box for a few minutes to several hours at room temperature.
12. After sufficient color development has occurred, wash the slides in distilled water at room temperature.
13. Stain the sections in a methyl green solution and mount on glass coverslips at room temperature.

b. Whole-Mount Embryos
Reagents

20X SSC
Formamide

Solution 1: 50% formamide, 5X SSC, 1% SDS
Solution 2: 0.5 M NaCl, 10 mM Tris-HCl (pH 7.5), 0.1% Tween 20
Solution 3: 50% formamide, 2X SSC
100 mg/ml RNase A
Anti-digoxigenin alkaline phosphatase (Boehringer-Mannheim)
Levamisol
TBST: 25 mM Tris-HCl (pH 7.5), 140 mM NaCl, 2.5 mM KCl, 0.1% Tween-20
1% Blocking reagent solution (Boehringer-Mannheim) or 1% BSA in TBST
NTMT: 100 mM Tris-HCl (pH 9.5), 100 mM NaCl, 50 mM MgCl$_2$, 0.1% Tween-
20
Nitroblue tetrazolium salt (NBT, 75 mg/ml in 70% dimethylformamide)
5-Bromo-4-chloro-3-indolyl phosphate (BCIP, 50 mg/ml in 100% dimethylfor-
mamide)

Method

1. Wash the embryos in solution 1, twice for 30 min each at 65°C.
2. Change to 1:1 mixture of Solution 1 and Solution 2 for 10 min at 65°C.
3. Rinse in Solution 2, three times for 5 min each time at room temperature.
4. Treat with RNase A 100 µg/ml in Solution 2, twice for 30 min each at 37°C.
5. Rinse in Solution 2 and after in Solution 3 for 5 min each time at room
   temperature.
6. Incubate in Solution 3, twice for 30 min of each at 65°C.
7. Rinse in TBST three times for 5 min of each.
8. Incubate in 1% blocking reagent in TBST for 1 h at room temperature.
9. Incubate in the same solution containing anti-digoxigenin alkaline phosphatase overnight at 4°C.
10. Wash three times in TBST buffer for 5 min each time and five times in TBST
    for 1 h each at room temperature.
11. Rinse in NTMT buffer containing 2 mM levamisol three times for 10 min
    each at room temperature.
12. Incubate in substrate solution (4.5 µl NBT and 3.5 µl BCIP in 1 ml NTMT
    buffer containing 2 mM levamisol) at room temperature. Color develops in
    30 min to 24 h.*
13. Stop the staining reaction by washing in PBT and clear the embryos in 50%
    glycerol during 1 h and store in 80% glycerol in PBT buffer.

VI. CHECKING PROCEDURES

To validate the signal specificity of the ISH, a number of checks are necessary.
A nonspecific signal could be caused from the label, the probe, the hybridization,

---

* To avoid precipitate formation during long exposure time, the substrate solution must be changed. The
staining reaction can be stopped and started again later. In this case, the embryos are washed in NTMT
and stored in the same solution at 4°C.

or the revelation procedure. For these reasons several points must be checked out in order to confirm the real specific localization of the target mRNA.

A. PROBE SPECIFICITY

The presence of a nonspecific hybridization signal might be due to the nucleotide sequence homologies between the probe used for the target mRNA and other nucleic acids. This problem can be avoided by checking the homology of the selected probe, e.g., using the gene bank system. This specificity can also be confirmed by using a sense oligonucleotide or RNA probe: no signal should be found.

The specificity of the probe can also be checked during the determination of the washing temperature (see Section VII.F). The variation of signal intensity due to increases in temperature also gives an idea of the degree of access of the probe to its specific mRNA target.

B. SPECIFICITY OF THE HYBRIDIZATION PROCEDURE

Numerous points should be taken into account in checking the nonspecific binding of the probe.[15-16]

Positive tissue — When the tissue contains a high level of the target RNA, it can be used as an index of the positivity of the tissue investigated.

Heterogeneous distribution of the hybridization signal — In tissue made up of different types of cells, only the specific areas which express the mRNA studied should be positive. The absence of a signal in other cells or structures is a good check of specificity (Figure 2).

Negative tissue — In the same experimental conditions, tissue which exhibits an absence of the hybridization signal also provides a good check of specificity (Figure 4).[15-19]

Competition reaction — An excess of nonlabeled probe (100 times more concentrated than the labeled probe) added to the labeled probe under the same experimental conditions can decrease, or even completely abolish, the specific hybridization, whereas a heterologous probe applied independently, in similar conditions, does not modify the signal.[15-19]

Heterologous probe — Hybridization with a heterologous probe containing the same amount of G+C, and carried out under similar experimental conditions might exhibit unusual nonspecific binding of nucleic acids.[15-19] This check is carried out with a known positive tissue for the target mRNA.

Enzyme pretreatment — RNase or DNase pretreatment can destroy the signal and/or demonstrate the specificity of RNA or DNA target nucleic acid. The specific ISH signal and not the background disappear with this treatment.

Other techniques — The specificity of the ISH results can be confirmed by other approaches such as Northern blot, RNase protection, RT-PCR, or immunohistology.

C. REVELATION PROCEDURE

The determination of background noise after radioautography can be measured by densitometry either on the films or on the revealed slides (see Section VII).

VII. QUANTIFICATION

The ISH signal can be quantified in standardized conditions in order to assess and compare the level of mRNA expression. Different quantitative procedures are carried out in order to find the approach which gives the most accurate and significant evaluation of mRNA expression. However, these procedures, which require a calibration standard, can be used only with radioactive probes because it is difficult to assess the background in the case of nonradioactive probes.

A. PREPARATION OF CALIBRATION STANDARD

Fresh biopsy tissue from normal rat brain* is ground up and homogenized. The mixture obtained is divided up equally among eight tubes. A known concentration of [^{35}S] deoxy-ATP (or other dNTP) radioactive solution (nCi/mg of tissue) is added to each tube to obtain a concentration gradient (serial concentrations). The resulting pellets are immersed in liquid nitrogen. Cryostat sections (10 μm) are obtained from each frozen fragment, recovered, and arranged (one section from each fragment) on pretreated slides (see Protocol 7). The slides undergo macro- (see Protocol 15) and microradioautographic (see Protocol 16) procedures. This revelation gives a biological standard (slides with eight sections containing known concentrations of radioactivity) which can be used for quantification.[15-19]

B. MACRORADIOAUTOGRAPHIC QUANTIFICATION

After 24 to 48 h exposure (the same time as ISH), the eight macroradioautograms are obtained on film (darkened spots), and the optical density of each spot is measured with a densitometer (Camridge 900, Centre de Quantimetrie, Alcatel, France). Figure 6 shows straight line regression analysis of the mean values of the optical densities of each standard section and its corresponding radioactivity value (expressed as nanocuries per standard tissue section).

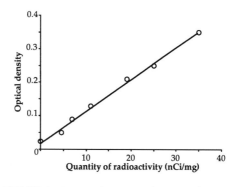

FIGURE 6   Macroradioautographic quantification.

---

* Brain is the tissue generally used but other soft tissues could also be used.

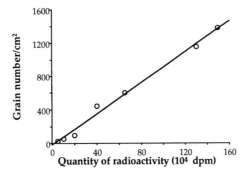

FIGURE 7 Microradioautographic quantification. The signal was analyzed by epipolarized light microscopy.

C. MICRORADIOAUTOGRAPHIC QUANTIFICATION

After microradioautographic revelation (see Protocol 16) and staining (see Protocol 17) procedures, the eight microradioautograms are obtained on slide and the grain density of each section standard is measured by epipolarized light microscopy (Alcatel, Grenoble, France). Figure 7 gives straight line regression analysis of the mean values of the grain density of each standard section and its corresponding radioactivity value (expressed as dpm per standard tissue section).

D. RADIOIMAGER QUANTIFICATION

1. Imaging

The cartography of the labeling of tissue sections can be obtained from silver-grain radioautograms produced by the standard procedure of development of photographic emulsion. The radioautograms can also be digitally generated by radio-imagers (Table 3). These detectors of radioactivity provide images by reconstituting the spatial distribution of the points of emission of the radioactivity. Those that give the best resolution are the μ-imager and the Phosphorimager, which provide results similar to those obtained with macroradioautograms (Figure 8, compare A with B and C). They are sensitive (Table 4), characterized by high speed (Figure 9), and time-saving, providing rapid evaluation of novel experiments, and particularly of exposure time (Figure 9).

2. Radioimager Systems

Radioimagers can be passive or real-time detectors. Passive detectors (e.g., the Phosphorimager) reconstitute the cartography of the labeling procedure, which was previously stored on sensitive screens. Real-time detectors (e.g., the μ-imager and the β-imager) provide real-time images by building up the events induced by radioisotopes, with reconstitution of the points of emission. In this case, the radioautograms come into view progressively on the PC screen.

The phosphorimager — The sensitive surface is a plate composed of fine crystals of BaFBr: Eu^{2+} in an organic binder.[25] This reusable phosphor screen stores the cartography of the radioactivity resulting from the excitation of crystals by

TABLE 3   Imaging Comparative Methods

	Radiosensible Material	Surface of Analysis (mm²)	Radioisotope Detected	Quantification Procedure
Films	Photographic emulsion	Various	^{125}I; ^{32}P; ^{33}P; ^{3}H; ^{35}S; ^{14}C	Relative (standard curves)
Phosphoimager 445 SI	Sensitive screens	200 × 250	^{125}I; ^{32}P; ^{33}P; ^{3}H; ^{35}S; ^{14}C	Absolute
(Molecular Dynamics)[a]	BaFBr:Eu²⁺	350 × 430		(counts/mm²)
β-Imager 2000	Gas mixture	150 × 125	^{125}I; ^{32}P; ^{33}P; ^{3}H; ^{35}S; ^{14}C	Absolute
(Biospace Inst.)	Triethylamide + argon			(counts/mm²)
μ-Imager	Scintillating sheet	13 × 9	^{125}I; ^{32}P; ^{33}P; ^{3}H; ^{35}S; ^{14}C	Absolute
(Biospace Inst.)[b]	SiY2O5(Ce) crystals			(counts/mm²)

[a]   Molecular Dynamics SA Sunnyvale, CA, USA.
[b]   Biospace Instruments, Paris, France.

TABLE 4   Technical Specifications of Radioimagers

	Spatial Resolution (μm)	Linearity of Dose-Response Scale	Units of Measurement (Count/Time/ Unit Surface)	Detection Threshold (dpm/mm²/h)
Phosphoimager 445 SI	^{125}I: 200[20] ^{32}P: 300[21]	5 Log	cp/h/mm²	^{35}S: 1.10[a] ^{32}P: 0.17[a]
β-Imager 2000	^{125}I: 200[22] ^{3}H: 30[23]	6 Log	cp/min/mm²	ND
μ-Imager	^{125}I: 30[20] ^{5}S: 15[24]	4 Log	cp/min/mm²	ND

[a]   Technical note from Molecular Dynamics.

contact with a radioactive source. The latent image produced by the emission of light at 365 nm is reconstituted by scanning with a helium-neon laser at 635 nm.

The β-imager —The ionized particles emitted by the sample trigger an avalanche of secondary particles in a cloud chamber, by the effect of electric fields between parallel electrodes. The gas is a mixture of triethylamide and argon. The ionized particles cause the emission of light spots which, at a suitable level of intensification, are detected by a charged-coupled device (CCD) camera.[23]

The μ-imager — The sensitive surface is a thin scintillator sheet [10 μm thick, Y2SiO₅(Ce)] coupled with an intensified CCD system driven in a specific mode.[26] The detected particle energy is locally converted into light, which is secondarily intensified in order to display the emission point localized by the CCD on a PC screen.[24]

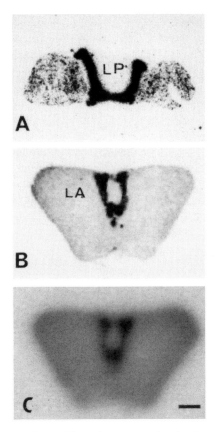

FIGURE 8 Digitized images obtained from the same section by three imaging system. The rat pituitary sections were hybridized using an antisens [35S]-oligoprobe, complementary to a POMC (proopiomelanocortin) mRNA sequence. A: macroradioautoradiogram on film; B: digital radioautogram provided by the μ-imager; C: phosphoimager. Dense signal is present over the intermediate lobe of the gland and is lower over the anterior lobe (LA). The signal over the posterior lobe (LP) is corresponding to the background. Exposure times were 24 h for the film and the μ-imager, and 48 h for the phosphoimager Storm. Bar = 0.5 mm.

3. Quantification

Macroradioautograms and microradioautograms can be quantified using image analyzers relating silver grain densitiy to amounts of radioactivity by means of curves constructed from known standards.

Radioimagers provide direct measurements expressed in counts per minute, e.g., classical β or γ particle detectors with a wide linear dynamic range, in contrast to the photographic emulsion method (Table 4). For precise quantification, the cpm values have to be corrected by determining the efficiency of the detector. Moreover, when all the samples cannot be analyzed at the same time (in cases where the area available for analysis is too small) the cpm measurements have to be corrected for radioactive decay. If the specific activity of the probe is known, the exact quantities bound (mol/mm^2) can be calculated (see Protocol 19).

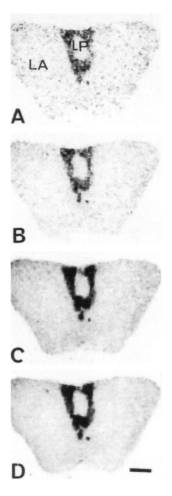

FIGURE 9   Imaging speed. Images were from the μ-imager. Exposure times were A, 42 min; B, 2 h; C, 5 h; D, 15 h; no gain was observed after 5 h. Bar = 0.5 mm.

PROTOCOL 19: QUANTIFICATION USING RADIOIMAGERS

Materials

Radiolabeled tissue sections; radioimager and interface with a PC computer; PC computer and software for imaging and quantification of digital radioautograms; radio-sensitive materials (sensitive screens: Phosphorimager; scintillating sheet: μ-imager; gas mixture: β-imager).

Method

1. Preparation of radiolabeled sections.
2. Exposure of the sections in close contact with the radiosensitive material.

3. Exposure, whose duration depends on the activity of the sample. The accumulation of events generates digital radioautograms in real time (the μ-imager and the β-imager) or after reconstitution of a latent image stored on the sensitive material; the phosphoimager.

4. Delineation of labeled tissue in cell clusters; results are expressed in counts/mm^2 (= total counts).

5. The specific labeling is determined by subtraction of the nonspecific labeling from the total counts.

6. The precise determination of the amount of radioactivity (Ci) bound to specific areas of the sections. This can be obtained after determination of the efficiency of the detector (Ci = dpm/2.2 × 10^{12}, dpm = cpm/% efficiency).

7. The probe binding is expressed in moles after calculation of the probe specific activity (mol = Ci/specific activity in Ci per mol).

E. CHOICE OF QUANTITATIVE METHOD

The ^{35}S biological standard can be used for different experiments, essentially in macroradioautographic quantification, because in microradioautographic grain counting, a standard slide must be prepared for each particular experiment so as to undergo similar conditions of revelation (see Protocols 16 and 17). The latter constitutes a genuine source of background, because the exposure and revelation times are long compared to those produced by macroradioautographic procedures.

Quantitative analysis on film is thus the best approach, because it is easy to carry out and gives the best evaluation of gene expression. On the other hand, the use of this biological calibration standard is limited in time, because of the short half-life of the ^{35}S radioisotope. A commercial standard containing ^{3}H as a radioisotope which can be stored and used for a long time is still available and also gives good results.

Using a standard curve after macroradioautographic analysis (^{3}H standard slide), it is possible to determine the relative amount of radioactivity contained in a hybridized tissue section (exposed under the same film on a standard slide), by comparing optical density values to level of radioactivity of the standard (Figure 6).[15-17] However, the values obtained should be expressed as arbitrary units, since in order to obtain absolute evaluation, a calibration standard would have to be prepared with the labeled mRNA to be analyzed. This is not yet possible, because it is difficult to obtain and handle stable RNA.

F. DETERMINATION OF WASHING TEMPERATURE (Tw)

In order to get a better understanding of this procedure, one of the studies that we carried out was chosen as an example. Quantitative *in situ* hybridization method has indeed been performed and improved to investigate with accuracy the distribution, evaluation, comparison, and regulation of two prolactin mRNA receptors (PRL-R) in various types of rat tissue.[15-17]

Since the liver is known to contain a high level of PRL-R, it was used in this study to determine the washing temperature (Tw) for PRL-R mRNA probes. Fresh liver biopsy tissue was removed from normal adult rats and immediately frozen in liquid nitrogen. Cryostat sections (10 μm) were put through pre- and postfixation procedures (see Protocol 8). They were independently hybridized with two distinct pairs of oligoprobes, which targeted the two PRL-R transcripts (the two oligonucle-otides being complementary to two distinct sites in the same mRNA). They were radiolabeled by tailing the 3' ends with [^{35}S]dATP (3000 Ci/mmol). The hybridization reaction was maintained for 16 h at 37°C or 52°C (see Protocol 11).

Sections were washed sequentially in graded concentrations of SSC (see Protocol 13), then washed at different temperatures (40 to 70°C) in 2X SSC for 1 h each (two sections for each temperature) and dried. After macroradioautographic analysis (see Protocol 15), as shown in Figure 10, the sigmoid representation of regression between mean values of optical densities of radioautograms indicates signal intensities and their corresponding washing temperatures. The melting temperature (Tm), at which 50% of specific hybrids were denatured, was 50°C. The optimal Tw corresponds at the temperature where the signal began to decrease (Figure 10). In these conditions Tw is 40°C.

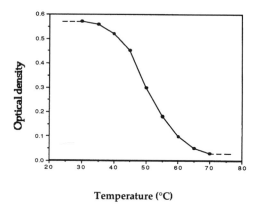

Temperature (°C)

FIGURE 10 Determination of optimal washing temperature (Tw).

Determination of Tw gives the possibility to hybridize at the maximum efficiency and then decrease the nonspecific hybridization in order to obtain the highest signal/background ratio.

VIII. DISCUSSION AND APPLICATION

The different steps in the *in situ* hybridization process should be adapted to the aim of the study and the biological material used (choice of probe type, label, etc.). The fixation (paraformaldehyde) step preserves the quality of the structure and the quantity of the target mRNA. The choice of labeling system is crucial and depends on the amount of the target mRNA present in the tissue investigated. The main advantage of ISH radioautography is its high sensitivity, especially to detect mRNA even when expressed at low level in the tissue. The hybridization reaction requires stringent conditions, especially in the washing step (optimal washing temperature and/or RNase treatment) in order that the hybridized cells should contain the maximum of specific hybrids and the minimum of nonspecific hybrids. The washing temperature is strongly dependent on the type of probe used.

This quantitative assay has been applied for different purposes, mainly diagnostic, e.g., for detection and quantification of hepatitis B virus DNA and human immunodeficiency virus (HIV-1)[27,28] and analysis of metastasis-related genes in human cancers.[29] Furthermore, it is now possible to detect low copy numbers of mRNA while retaining good histologic morphology for the determination of specific gene expression in diseased tissues.[30] ISH was also applied for direct labeling of plant chromosomes.[31]

Quantitative ISH can also be used to compare the cellular distribution and the expression level of different transcripts, either originated from the same or different mRNA. In addition, this can be carried out on adjacent sections of the same tissue or in different tissues.[15-17] On the other hand, mRNA expression level can be quantified in the different stages of either the embryonic development[32] or the reproductive cycle, and also during different physiologic functions.

Simultaneous detections of different labeled probes with nonradioactive and/or radioactive markers can be improved on the same tissue section. These methods give the possibility to observe the expression pattern of different mRNA coding for different receptors in the same tissue or the same cell.[33]

IX. CONCLUSION

The *in situ* hybridization technique is one of the most useful for localizing gene expression, not only at regional level, but also accurately in the specific cells which express this gene within the tissue. However, immunohistochemistry might provide further information to confirm the activation of this gene by analysis of the protein expression level of this gene.

Interestingly, further analysis of the target mRNA at electron microscopic level by ultrastructural *in situ* hybridization (see Chapter 14) and immunocytology (see Chapters 9 and 12) gives not only a supplementary confirmation concerning its presence in the investigated tissue, but also more information on its subcellular distribution.

ACKNOWLEDGMENTS

The authors would like to thank J. Doherty for editing the manuscript and the Alcatel Company for their technical and scientific help concerning the quantification.

REFERENCES

1. Chesselet, M.F., *In situ Hybridization Histochemistry*, CRC Press, Boca Raton, FL, 1990, 208.
2. Rendrop, M., Knapp, B., Winter, H., and Schweizer, J., Aminoalkysilane-treated glass slides as support for *in situ* hybridization of keratin cDNAs to frozen tissue sections under varying fixation and pretreatment conditions, *Histochem. J.,* 18, 271, 1986.

3. Weiss, L. M., and Chen, Y. Y., Effects of different fixatives on detection of nucleic acid from paraffin-embedded tissues by *in situ* hybridization using oligonucleotide probes, *J. Histochem.*, 39, 1237, 1991.

4. Kelly, R. G., Cozarelli, N., Deutsher, M. P., Lehman, I. R., and Kornberg, A., Enzymatic synthesis of deoxyribonucleic acid. *J. Biol. Chem.*, 245, 39, 1970.

5. Feinberg, A. P., and Vogelstein, B., A technique for radiolabeling DNA restriction endonuclease fragments to high specific activity. *Anal. Biochem.*, 132, 6, 1983.

6. Feinberg, A. P., and Vogelstein, B., Addendum: a technique for radiolabeling DNA restriction endonuclease fragments to high specific activity, *Anal. Biochem.*, 137, 266, 1984.

7. Cox, K.H., DeLeon, D.V., Angerer, L.M., and Angerer, R.C., Detection of mrnas in sea urchin embryos by *in situ* hybridization using asymmetric RNA probes, *Dev. Biol.*, 101, 485, 1984.

8. Décimo, D., Georges-Labouesse, E., and Dollé, P., *In situ* hybridization of nucleic acid probes to cellular RNA, *In Gene Probe.* Hames, B.D., Higgins, S.J., Eds., IRL Press, Oxford, 1995.

9. Wetmur, J. G., Ruyechan, W. T., and Douthard, R. J., Denaturation and renaturation of *Penicillium chrysogenum* mycophage double stranded ribonucleic acid in tetraalkylamonium salt solutions, *Biochemistry*, 20, 2999, 1981.

10. Marmur, J., and Doty, P., Determination of the base composition of desoxyribonucleic acid from its thermal denaturation temperature, *J. Mol. Biol.*, 5, 109, 1962.

11. McConaghy, B. L., Laird, C. D., and McCarthy, B. L., Nucleic acid reassociation in formamide, *Biochemistry*, 8, 3289, 1969.

12. Whal, G. M., Stern, M., and Stark, G. R., Efficient transfer of large DNA fragments from agarose gels diazobenzyloxymethyl paper and rapid hybridization by using dextran sulfate, *Proc. Natl. Acad. Sci. U.S.A.*, 76, 3683, 1979.

13. Wilkinson, D. G., Whole-mount in situ hybridization to vertebrate embryos. In *In Situ Hybridization*, Wilkinson, D.G., Ed., IRL Press, Oxford, 1992.

14. Rosen, B., and Beddington, R. S., Whole-mount *in situ* hybridization in mouse embryo: gene expression in three dimensions, *Trends Genet.*, 9, 162, 1993.

15. Ouhtit, A., Morel, G., Kelly, P. A., Visualization of gene expression of short and long forms of prolactin receptor in rat, *Endocrinology*, 133, 135, 1993.

16. Ouhtit, A., Kelly, P. A., Morel, G., Visualization of gene expression of short and long forms of prolactin receptor in rat digestive tissues, *Am. J. Phys.*, 266, G807, 1994.

17. Morel, G., Ouhtit, A., Kelly, P. A., Expression of short and long forms of prolactin receptor in the rat. In *Advances in Contraceptive Delivery Systems*, Hafez, E. S. E., Ed., Reproductive Health Center, Kiawah, SC, 1994, 50.

18. Cau, P., In *Techniques en Microscopie Quantitative:* Stereologie, autoradiographie et immunocytochimie quantitatives, Editions INSERM, Paris, 1990, 262.

19. Williams, M. A., Preparation of electron microscope autoradiographs. In, *Autoradiography and Immunocytochemistry*, Glauert, A. M., Ed., North-Holland, Amsterdam, 1977, 77.

20. Crumeyrolle-Arias, M., Latouche, J., Laniece, P., Charon, Y., Tricoire, H., Valentin, L., Roux, P., Mirambeau, G., and Haour, F., "*In situ*" characterization of GnRH receptors: use of two radioimagers and comparison with quantitative autoradiography, *J. Receptor Res.*, 14, 251, 1994.

21. Johnston, R. F., Pickett, S. C., and Barker, D. L., Autoradiography using storage phosphor technology, *Electrophoresis*, 11, 355, 1990.

22. Crumeyrolle-Arias, M., Jaffarian-Tehrani, M., Cardona, A., Edelman, L., Roux, P., Laniece, P., Charon, Y., and Haour, F., Radioimagers as alternative to film autoradiography for *in situ* quantitative analysis of 125I ligand receptor binding and pharmacological studies, *Histochem. J.*, (in Press).

23. Charpack, G., Dominic, W., and Zaganidis, N., Optical imaging of the spatial distribution of b-particles emerging from surfaces, *Proc. Natl. Acad. Sci. U.S.A.*, 86, 1741, 1989.

24. Laniece, P., Charon, Y., Dumas, S., Mastrippolito, R., Pinot, L., Tricoire, H. J., and Valentin, L., HRRI: a high resolution radioimager for fast, direct quantification in *in situ* hybridization experiments, *BioTechniques*, 17, 338, 1994.

25. Amemiya, Y., and Miyahara, J., Imaging plate illuminates many fields, *Nature*, 336, 89, 1988.

26. Charon, Y., Laniece, P., Bendali, M., Gaillard, J. M., Leblanc, M., Mastrippolito, R., Tricoire, H., and Valentin, L., A self triggered intensified CCD (STIC), *Nucl. Instr. Meth. Phys. Res.*, A310, 379, 1991.

27. Aspinal, S., Steele, A. D., Peenze, I., and Mphahlele, M. J., Detection and quantification of hepatitis B virus DNA: comparison of two commercial hybridization assays with polymerase chain reaction, *J. Viral Hepat.*, 2, 107, 1995.

28. Smith, P. D., Fox, C. H., Masur, H., Winter, H. S., and Alling, D. W., Quantitative analysis of mononuclear cells expressing human immunodeficiency virus type 1 RNA in esophageal mucosa, *J. Exp. Med.*, 180, 1541, 1994.

29. Kitadai, Y., Bucana, C. D., Ellis, L. M., Anzai, H., Tahara, E., and Fidler, I. J., *In situ* mRNA hybridization technique for analysis of metastasis-related genes in human colon carcinoma cells, *Am. J. Pathol.*, 147, 1238, 1995.

30. Szakacs, J. G., and Livingston, S. K., mRNA *in situ* hybridization using biotinylated oligonucle-otide probes: implication for the diagnostic laboratory, *Ann. Clin. Lab. Sci.*, 24, 324, 1994.

31. Reader, S. M., Abbo, S., Purdie, K. A., King, I. P., and Miller, T. E., Direct labeling of plant chromosomes by rapid *in situ* hybridization, *Trends Genet.*, 10, 265, 1994.

32. Srivastava, L. K., Morency, M. A., and Mishra, R. K., Ontogeny of dopamine D2 receptor mRNA in rat brain, *Eur. J. Pharmacol.*, 225, 143, 1992.

Chapter **14**

VISUALIZATION OF mRNA RECEPTOR BY ULTRASTRUCTURAL *IN SITU* HYBRIDIZATION

Béatrice Grandclément
Annie Cavalier
Gérard Morel

CONTENTS

0-8493-2644-3/97/$0.00+$.50

I. INTRODUCTION

In situ hybridization, as a mean of detecting DNA or RNA molecules in tissue sections was initially developed by Gall and Pardue[1] and John et al.[2] Since then, this technique has been widely used in studies of gene mapping and gene expression in various types of mammalian and nonmammalian tissues. Most of these studies have been done at the light microscopic level after hybridization of nucleic acids with radioactive or nonradioactive probes.[3]

Some workers have tried to extend this technique to the electron microscopic level in order to obtain better resolution in the localization of the probe, and thereby gain more accurate information about the subcellular localization of the nucleic acid and the morphology of the cells that express the nucleic acid. The technical feasibility of visualizing DNA by electron microscopy of ultrathin sections after *in situ* hybridization was first demonstrated with oocytes of *Xenopus* tadpoles by Jacob et al.,[4] who used tritiated RNA as a probe. A number of reports describe ultrastructural *in situ* hybridization; the majority of these reports are restricted to pre-embedding procedures.[5,6] Pre-embedding methods (performance of *in situ* hybridization on fixed material before embedding) allow a minimum degree of deviation from established hybridization protocols at the light microscopic level. The disadvantages of this approach are limited:

depth of penetration of probe molecules and gold markers, loss of ultrastructural membrane detail, and potential loss of label during the embedding procedure.

Postembedding techniques, i.e., on-grid hybridization of ultrathin sections, make it possible high-resolution hybridization studies, and also have the advantage that the interiors of cells and organelles are exposed; but probe penetration, even with thin sections of resin-embedded material, can pose a problem.[5,6] Cryoultramicrotomy is often considered to be one of the most sensitive methods in immunocytochemistry, because it allows better antigen retention, and also better antibody access to tissue antigens, than plastic-embedding methods.[7] Moreover, the use of ultrathin frozen sections has the advantage of excluding all the steps of embedding in a hydrophilic or epoxy resin, which could eventually lead to the denaturation of the nucleic acid. In this case, the penetration of the sections by the probe is facilitated, the sensitivity of the method increased, and the preparation of the specimens easier. The main disadvantage of *in situ* hybridization on ultrathin frozen sections is the loss of ultrastructural morphology, and also to some extent the lability of the target nucleic acid, which may be lost during processing. In a previous study, we compared the three main methods used to localize mRNA at the ultrastructural level,[8] showing that all these methods can be used to visualize mRNA. However, the results were better using ultrathin frozen sections and better morphology was obtained using a pre-embedding method, the best compromise being hydrophilic resin-embedded tissue sections.

In subsequent studies, different protocols were developed according to the type of nucleic acid to be detected. The protocols for electron microscopy are similar to those used for the light microscopic visualization of hybrids, except that they specify a particular approach to the tissue and labeling procedure. This requirement is inconsistent with an optimal hybridization procedure, and a compromise has to be found between acceptable morphology and efficient hybridization.

II. TISSUE

All types of tissue, cells — in monolayers harvested after enzymatic treatment or in suspension, either cultivated or not, and pelleted after washing — and clinical biopsies can be used. Optimal sample size is about 1 mm^3. Preparation of specimens for *in situ* hybridization, whether on ultrathin frozen sections or on hydrophilic resin-embedded tissue sections, is the crucial step for ultrastructural preservation.

In the cryosection approach, fixation is the first step. The specimens are then frozen and cryosectioned. Hybridization and all subsequent steps are carried out on grids. Finally, before observation, the specimens are stained and embedded in methylcellulose.

In Lowicryl-embedded tissue sections, fixation is always the first step. The embedding procedure remains a crucial step. Sectioning is easier than in the case of cryoultramicrotomy. Hybridization on grids concerns only the surface of the sections. Staining and observation are done classically.

A. FIXATION

Good fixation of the tissue is particularly important in electron microscopic studies of whole cells, since the acquisition of biological information is related to

- *Cooling system, nitrogen gradient, LM 10 (Air Liquide, Grenoble, France), programmable cooling system (Minicool, Air Liquide)*

Method

1. *Immerse the specimen in saturated or 2.3 M sucrose solution for 1 min*
2. *Place the sample on a cryoultramicrotome support*
3. *Place it in the cooling system*
4. *Lower the temperature slowly to –4°C*
5. *Lower the temperature quickly from –4°C to –130°C or –196°C*
6. *Store the tissue at –196°C.*

2. Cryoultramicrotomy

If freezing the tissue did not result in the formation of large ice crystals or the destruction of tissue or cell structure, tissue sectioning or cryoultramicrotomy step is easy to perform. The most widely used method has been described by Tokuyasu, who identified the thawing of the sections and their transfer from the knife to the grids on a drop of 2.3 M sucrose, the "dry sectioning method", as a critical phase in the procedure *(Protocol 5).*[28]

PROTOCOL 5: DRY SECTIONING METHOD

1. *Place the gold green sections on the knife*
2. *Place a drop of saturated sucrose on the sections on the knife*
3. *Transfer sections onto a carbon-coated film on a grid (Protocol 6)*
4. *Wash in 2X SSC before hybridization.*

PROTOCOL 6: PREPARATION OF COATED GRIDS

Equipment required

A tall beaker
A dropper
An acetate film
200- or 400-mesh nickel grids
1% Collodion solution in isoamyl acetate.

Method

1. *Place one drop of collodion solution on the water in the beaker: the collodion film floats to the surface of the water*
2. *Air-dry*
3. *Place the grids on the collodion film*
4. *Put an acetate film over the collodion film and grids, then remove*

5. *Air-dry*
6. *Deposit a 10-nm layer of carbon in a vacuum by evaporation.*

The sectioning parameters include: knife quality, angle of knife, specimen and knife temperature, and sectioning speed. A diamond knife can be used for making ultrathin frozen sections; great care must be taken in making glass knives. The method described by Griffiths et al. with a 45° knife angle gives excellent results,[29] but in fact 6° seems to be the optimal angle. The temperature of the specimen must be below the recrystallization temperature: $< -100°C$ if the cryoprotectant used was sucrose 0.4 M. The knife temperature must be about $-100°C$. And, the sectioning speed is generally around 5 mm/s, but can be reduced for soft tissue. In some cases, such as that of bone, sectioning is done manually, at a temperature $< -130°C$.[30,31]

Finally, grids with sections and sucrose can be stored at 4°C for a long period before hybridization.

C. LOWICRYL-EMBEDDED TISSUE
1. Embedding

The protocol described (see *Protocol 7*) is based on standard procedure with slight modifications. The polymerization mixture of Lowicryl K4M is composed of polar acrylates and methacrylates and has the ability to infiltrate tissues at low temperature. Some cautions are necessary to use this resin: gloves should be worn since contact with Lowicryl may induce dermatitis, and Lowicryl is toxic, experiments have to be performed under a ventilator hood or within a special system.

PROTOCOL 7: EMBEDDING IN LOWICRYL

Material and Reagents:

- *Polymerization chamber, a simple chamber may be constructed using a box covered with aluminum foil, but commercial chambers are available.*
- *Lowicryl K4M:*

  Crosslinker A: 2.70 ml
  Monomer B: 17.30 ml
  Initiator C: 0.1 g.

Procedure:

1. *Fix the samples in a solution of 4% freshly prepared paraformaldehyde, 0.5% glutaraldehyde in 100 mM phosphate buffer saline for 4 h at 4°C*
2. *Wash in phosphate buffer for 4×20 min*
3. *Dehydrate in a graduated ethanol series, while progressively lowering the temperature. Small pieces need less time for dehydration*

   Ethanol 35° at 0°C for 60 min
   Ethanol 55° at –20°C, for 60 min

Ethanol 70° at –30°C for 1 to 2 h
Ethanol 90° at –30°C for 1 to 2 h

4. *Infiltrate in Lowicryl solution (without agitation). After dehydration, the samples are infiltrated with Lowicryl K4M containing ethanol then transferred to an embedding mixture of pure resin. All Lowicryl-containing solution are stored at –30°C before use. The procedure is performed at –30°C. It should be noted, however that the longer the infiltration time the better are the results.*

Lowicryl K4M solution in ethanol 90° (1 vol + 2 vol), at –30°C, for 1 to 2 h
Lowicryl K4M solution in ethanol 90° (1 vol + 1 vol), at –30°C, for 1 to 4 h
Lowicryl K4M solution in ethanol 90° (2 vol + 1 vol), at –30°C, overnight
Undiluted Lowicryl K4M solution, at –30°C, for 2 to 4 h
Undiluted Lowicryl K4M solution, at –30°C, overnight

5. *Embedding: the samples are transferred to fresh Lowicryl K4M and are simply placed at the tip of a conical capsule (appropriate embedding capsules). Special molds are available for polymerization of resin blocks (LKB easy Mold). Another tray mold is used as an hermetic cap to minimize the contact with air.*

6. *Polymerize under UV, at –35°C, for 5 days*

   • *Indirect illumination is recommended to illuminate the samples from the top, in order to reduce the light intensity at the tissue level. Polymerization will last longer, but without affecting the tissue integrity.*
   • *Distance of samples — UV light: the sample are placed at 10 cm under the UV light. The recommended light intensity for polymerization of Lowicryl K4M at –30°C is given by a 6 W lamp (360 nm). A too-high energy light will interfere with the tissue and will not preserve the nucleic acid integrity.*
   • *Polymerization time of Lowicryl K4M is usually 5 days at –30°C, this time is approximate and can vary according to the quantity of resin and the efficiency of the lamp.*
   • *Polymerization is completed when no drop of unpolymerized Lowicryl K4M is seen at the top of the capsule.*

2. Ultramicrotomy

Ultrathin sections are realized using 45° glass knives or diamond knives and a standard ultramicrotome. The sections are obtained with a lowered water level in the boat at a moderate cutting speed (1 mm/s). Ultrathin sections (80 to 100 nm, respectively, silver and gold) were mounted on a carbon-coated collodion* film on a nickel grid (in order to avoid precipitate on the copper surface during the *in situ* hybridization and revelation procedures). The sections are dried overnight at room temperature, or at 37°C in an oven.

The grids can be stored for several days before hybridization.

---

* The grids must be coated with a film of Formvar, since collodion is soluble in alcohol.

III. PROBE

A. CHOICE OF PROBE

Most of the probes that have been used for *in situ* hybridization up to the present time have been DNA (complementary DNA (cDNA) and oligonucleotide), but some tests have been performed with antisense RNA.[12]

The cDNA used is always limited to the insert. The main advantage of cDNA, as compared to an oligonucleotide probe, is the density of labeling. After labeling, a cDNA probe is cut into 200 to 300 bp labeled fragments, and a target nucleic acid can hybridize several labeled fragments from cDNA. Labeling by nick translation or random priming gives small, labeled nucleic acid fragments.

An oligonucleotide must be synthesized, and this can be done if the sequence of the target nucleic acid is done. The length of the most commonly used oligonucleotides is about 30 mers. Shorter oligonucleotides can give specificity problems, and longer ones are more costly but no more effective. The choice of sequence is the main factor determining specificity. The sequence must be checked against a gene bank, since it is known that some oligonucleotide probes cross-hybridize with other gene sequences. The stability of a hybrid depends on the percentage of G-C bases — 50 to 55% seems to be best. The inconvenience of oligonucleotides compared to cDNA probes, as concerns the density of labeling, can be obviated by the use of several oligonucleotides for a given target nucleic acid.

B. PROBE LABELING
1. Choice of Label

This choice is important, since it determines that of the revelation system after hybridization, and also affects the sensitivity and resolution of hybrid visualization at the electron microscopic level. The use of radioactive nucleotides necessitates radioautographic detection, while that of antigenic-conjugated nucleotides necessitates immunocytological detection.

The advantages of using nonradioactive as opposed to radioactive probes are the same for light and electron microscopy: (1) they do not necessitate long radioautographic exposure time or specific precautions during their handling, and (2) the resolution obtained with these probes is higher than that obtained with radioactive probes. The labeling of the probe can be done either by enzymatic incorporation of labeled nucleotides into the nucleic acid or by chemical attachment of the marker to the nucleic acid. The most successful of these nonradioactive probes use biotin-, digoxigenin-, or fluorescein-conjugated nucleotides. Thus, biotinylated probes have been used in a number of electron microscopic studies.[14,22,32-36] These probes can be obtained by nick translation of DNA and incorporation of a biotin-substituted deoxy-UTP (bio-dUTP), or by *in vitro* synthesis of a cRNA in the presence of bio-UTP or fluorescein-UTP.[37]

In any case, if the rapidity of revelation of immunocytology, compared to radioautography, is an important factor of choice, some disadvantages of this method should be mentioned. These involve the revelation stage where the presence of endogenous antigens such as biotin in the tissue makes checking procedures necessary, and also that of the quantification of the gold particles. Furthermore, a low level of nucleotide labeling can lead to low sensitivity.

2. Method

Nick translation *(Protocol 8)* and random priming *(Protocol 9)* are used to label cDNA. The tailing 3′ end labeling method is performed with oligonucleotide probes *(Protocol 10)*. RNA polymerase is used to obtain antisense RNA *(Protocol 11)*.

PROTOCOL 8: NICK TRANSLATION

Equipment and Reagents Required:

- *Water bath at 15°C*
- *0.1 to 1 μg of cDNA*
- *Nick translation buffer (10X) (500 mM Tris-HCl (pH 7.8), 50 mM MgCl$_2$, 0.1 mg/ml nuclease-free BSA)*
- *dNTP mixture (0.2 mM dATP, dGTP, and dCTP)*
- *Labeled nucleotide: 0.3 mM/l biotin 16-dUTP, or digoxigenin 16-dUTP or fluorescein 12-dUTP*
- *Enzymes (DNA pol I, 4 U/ml; DNase I, 40 pg/ml)*
- *Sterilized distilled water, RNase-free.*

Method:

1. *Mix:*

   Buffer 2 μl
   dNTP 3 μl
   Labeled nucleotide X μl
   DNA 1 μg
   Enzyme 1 μl
   Water to 20 μl

2. *Incubate at 15°C for 2 h*
3. *Precipitate (Protocol 12)*
4. *Dilute in water*
5. *Store at –20°C*
6. *Denature before use by heating to 95°C for 5 min.*

PROTOCOL 9: RANDOM PRIMING

Equipment and Reagents Required:

- *Bath at 37°C and 95°C*
- *10 ng to 3 μg of cDNA*
- *dNTP (0.5 mM of dNTP other than the labeled nucleotide)*
- *Reaction mixture with hexanucleotides (10X): 130 mM K$_2$PO$_4$, 6.5 mM MgCl$_2$, 33 μM dTTP, 32 μg/ml BSA (pH 7.4)*

- *Enzyme (Klenow fragment 2U/µl)*
- *Labeled nucleotide: 0.3 mM/l biotin 16-dUTP, or digoxigenin 16-dUTP, or fluorescein 12-dUTP*
- *Sterilized distilled water, RNase-free.*

Between 10 ng to 3 µg of DNA can be labeled in a standard reaction. Larger amounts can be labeled by scaling up all components and volumes.

1. *Denature cDNA by heating in a water bath for 10 min at 95°C and chilling quickly in ice*
2. *Mix*

    - *DNA 10 ng to 3 µg*
    - *Reaction mixture 5 µl*
    - *dNTP mixture 5 µl*
    - *Labeled nucleotide X µl*
    - *Enzyme 5 µl*
    - *Water to 50 µl*

3. *Incubate at 37°C for 30 min*
4. *Precipitate (Protocol 12)*
5. *Dissolve in water*
6. *Store at –20°C*

PROTOCOL 10 - TAILING 3' END

Equipment and Reagents Required:

- *Bath at 37°C*
- *10 to 100 pmol of oligonucleotide*
- *Buffer (5X): 1 M potassium cacodylate, 125 mM Tris-HCl, 1.25 mg/ml BSA, pH 6.6)*
- *$CoCl_2$ 1.5 mM*
- *Enzyme (terminal deoxynucleotidyl transferase [TdT], 25 U/µl)*
- *Labeled nucleotide: 0.3 mM/l biotin-, or fluorescein-, or digoxigenin-dUTP*
- *Sterilized distilled water, RNase-free.*

Method:

1. *Mix:*

    Oligonucleotide 5 µl
    Buffer 10 µl
    $CoCl_2$ 5 µl
    Labeled nucleotide x µl
    Enzyme 1 µl
    Water to 50 µl

2. *Incubate at 37°C for 1 h, then place on ice*
3. *Precipitate (Protocol 12)*
4. *Dissolve in water*
5. *Store at –20°C.*

PROTOCOL 11: RNA POLYMERASE

- The DNA to be transcribed should be cloned into the polylinker site of a transcription vector containing a promoter for SP6, T7, or T3 RNA polymerase.
- "Run-off" or "run-around" transcripts can be synthesized by using linearized and circular template DNA, respectively.
- The labeled UTP contained in the mixture is incorporated into the transcripts at approximately every 20 to 25th nucleotide under the conditions described below.
- Since the nucleotide concentration does not become limiting when used in the following reaction, approximately 10 µg of full-length labeled RNA is transcribed from 1 µg linear template DNA. Larger amounts of labeled RNA can be synthesized by scaling up the reaction components.

Equipment and Reagents Required:

- *Bath at 37°C*
- *1 µg of linearized DNA*
- *Digoxigenin-, or biotin-, or fluorescein-RNA labeling mixture (10 X): 10 mM ATP, 10 mM CTP, 10 mM GTP, 6.5 mM UTP, 3.5 mM digoxigenin-, biotin-, or fluorescein-UTP pH 7.5*
- *Transcription buffer (10X): 400 mM Tris-HCl, pH 8, 60 mM $MgCl_2$, 20 mM spermidine, 100 mM NaCl, 1 U/µl RNase inhibitor*
- *Sterile distilled water, RNase-free*
- *SP6, T7, or T3 RNA polymerase*
- *EDTA 0.2 M, pH 8.*

Method

1. *The linearized DNA to be transcribed should be purified by phenol/chloroform extraction and ethanol precipitation*
2. *Mix:*

- *DNA 1 µg*
- *Labeling mixture (10X) 2 µl*
- *Transcription buffer (10X) 2 µl*
- *Make up to 18 µl with sterile distilled water and add: SP6, T7, or T3 RNA polymerase 2 µl*

3. *Centrifuge briefly and incubate for 2 h at 37°C. Longer incubations do not increase the yield of labeled RNA*

4. *As the amount of labeled RNA transcript is far in excess of the template DNA (by a factor of approximately 10) it is usually not necessary to remove the template DNA by DNase treatment. If desired, however, the template DNA can be removed by direct addition of 20 U DNase I, RNase-free, and incubate for 15 min at 37°C*

5. *With or without prior DNase treatment, add 2 μl 0.2 M EDTA solution, pH8, to stop the polymerase reaction.*

PROTOCOL 12: PROBE PRECIPITATION

• For Oligonucleotides
Reagents Required:

- *tRNA 10 mg/ml*
- *Ethanol 100% at –20°C*
- *Ammonium acetate 7.5 M or sodium acetate 3 M*

Method:

1. *Add to 1 V of reaction reagent:*

    2 μl of tRNA
    3 V of ethanol
    1/10 V of ammonium acetate 7.5 M, or sodium acetate 3 M

2. *Mix*
3. *Incubate at –20°C overnight or at –80°C for 1 h*
4. *Centrifuge for at least 30 min at 14,000 × g*
5. *Dry*
6. *Dissolve in water*
7. *Store at –20°C.*

• For DNA:
Reagents Required:

- *LiCl 4 M*
- *Ethanol prechilled to –20°C*
- *Ethanol 70% at –20°C*
- *DEPC-treated water.*

Method:

1. *Precipitate the labeled RNA by the addition of 2.5 μl 4 M LiCl and 75 μl prechilled (–20°C) ethanol. Mix well*

2. *Leave for at least 30 min at –70°C or 2 h at –20°C*
3. *Centrifuge, wash the pellets with 50 μl cold ethanol, 70% (v/v), dry under vacuum and dissolve for 30 min at 37°C in 100 μl DEPC-treated water, 20U RNase inhibitor can be added to inhibit possible contaminating RNases.*

C. STORAGE OF PROBES

Nonradioactive probes can be stored for up to 1 year, with the exception of RNA probes, which must be used more rapidly (maximum 1 week).

IV. ULTRASTRUCTURAL *IN SITU* HYBRIDIZATION

A. GENERAL CONSIDERATIONS

The hybridization step consists of putting the probe in contact with the ultrathin sections (which contain the target nucleic acid) to form hybrids by complementarity. The presence of an isotope or antigen in hybrids allows their detection by immuno-cytology provided that nonradioactive probes are used. The effects of various components and conditions on the rate of renaturation and the thermal stability of the resulting hybrids are briefly discussed below.

The hybridization procedure is influenced by several factors, namely, the buffer, the probe, the incubation, and the washes. For each of these factors, a number of parameters have to be set precisely:

Buffer:
    Saline concentration
    Formamide concentration
    Macromolecules (tRNA, DNA, Denhardt's solution)
Probe:
    Nature (cDNA, oligonucleotide, cRNA)
    Label (nonradioactive)
    Density of labeling
    Concentration
Incubation:
    Temperature
    Period
Wash:
    Saline concentration
    Temperature
    Time

The content of the buffer must be determined in such a way that hybrids are formed in optimal conditions. Saline (Standard Saline Citrate: SSC) concentration and deionized formamide concentration have similar effects in conventional hybridization. The possibility of decreasing the temperature by hybridizing in higher concentrations of formamide is limited, since this treatment also has a deleterious effect on ultrastructural morphology.[14] Some studies have, however, reported acceptable ultrastructural preservation after incubation with high formamide concentrations.[11,14,22] Macromolecules like tRNA, DNA, and Denhardt's solution are added to reduce nonspecific adsorption between the probe and the tissue section.

The choice of the probe determines the hybridization protocol, while the choice of label defines the revelation and influences probe concentration. Short probes are required in *in situ* hybridization because the probe has to diffuse into the dense matrix of cells. Fragment length also has an effect on thermal stability.

As previously shown, an incubation period of 3 h is adequate for *in situ* hybridization at the electron microscopic level, given that a longer period, as in light microscopy, leads to the destruction of the sections. Concerning the hybridization temperature, 20 to 30°C is the optimal value.

During hybridization, hybrids form between perfectly matched sequences and between imperfectly matched sequences. The extent to which the latter occur can be manipulated to some extent by varying the stringency of the hybridization reaction. To remove the background associated with nonspecific hybrids, sections should be incubated on drops of saline solution at decreasing concentrations (SSC 5× to 0.5 or 0.1), for 10 min to 1 h each, at room temperature to 40°C, in order to preserve specific hybrids and eliminate nonspecific hybrids.

B. PRETREATMENT

The aim of this step is to bring about the contact between the probe and the nucleic acid. Pretreatments must be optimized in most cases. In the absence of an embedding procedure in hydrophilic or epoxy resins, as used in the ultrathin frozen section method, pretreatments are reduced to a minimum (fixation, proteinase K, prehybridization). The role of the proteinase K is to increase accessibility by removing the cellular proteins that surround the target nucleic acid *(Protocol 13)*. The intensity with which the digestion is to be carried out depends on the preparation of the target sequences. HCl can also be used when the signal level is extremely low. Moreover, RNase treatment serves to remove endogenous RNA, and may improve the signal-to-noise ratio in hybridizations to DNA targets. This treatment can also be used as a control in hybridizations with (m)RNA as the target *(Protocol 14)*. It is carried out after proteinase K pretreatment.

There are two techniques that can be used to reduce background noise during *in situ* hybridization: acetylation *(Protocol 15)* and prehybridization. The first of these, which is commonly used in *in situ* hybridization with light microscopy, has a detrimental effect on the tissue at the electron microscopic level.[14] Prehybridization consists of incubating sections on a hybridization buffer for 1 h, then hybridizing without a washing step.[38] Its main inconvenience is the dilution of the probe. It is necessary to incubate the grids in the hybridization buffer.

PROTOCOL 13: PROTEINASE K PRETREATMENT

1. *Wash sections in phosphate buffer for 10 min*
2. *Fix ultrathin frozen sections in 4% paraformaldehyde in phosphate buffer for 10 min*
3. *Wash sections in phosphate buffer for 2× 5 min*

4. *Wash sections in Tris-HCl-CaCl$_2$ 1X buffer (10X Tris-HCl-CaCl$_2$ buffer: Tris-HCl 20 mM, CaCl$_2$ 2 mM; pH 7.4) for 5 min at room temperature*
5. *Incubate in proteinase K solution (1 μg/ml) for 5 to 10 min at room temperature*
6. *Wash in Tris-HCl-CaCl$_2$ 1X buffer for 2× 5 min at room temperature*
7. *Postfix in 4% paraformaldehyde in phosphate buffer for 10 min*
8. *Wash sections in phosphate buffer for 2× 5 min*
9. *Wash sections in 2X SSC for 2× 5 min*
10. *Prehybridize or hybridize.*

Steps 1 to 3 and 7 and 8 are necessary for ultrathin frozen sections only in order to stabilize nonembedded tissue sections.

PROTOCOL 14: RNase PRETREATMENT

1. *Incubate the sections with RNase A (1 mg/ml) in 2X SSC for 1 h*
2. *Wash the sections with 2X SSC for 3× 5 min*
3. *Postfix in 4% paraformaldehyde in phosphate buffer for 10 min*
4. *Wash sections in phosphate buffer for 2× 5 min.*

Steps 3 and 4 are necessary for ultrathin frozen sections only in order to stabilize nonembedded tissue sections.

PROTOCOL 15: ACETYLATION

1. *Incubate the sections with 100 mM triethanolamine (Sigma, L'Isle d'Abeou, France) for 10 min*
2. *Add 25 mM acetic anhydride and leave for 5 min**
3. *Wash sections in phosphate buffer for 5 min*

*The reaction occurs within the first 30 s after addition of acetic anhydride.

C. HYBRIDIZATION ON ULTRATHIN FROZEN SECTIONS
1. Oligonucleotide

The probes used are 30-mer synthetic oligonucleotides synthesized by solid-phase phosphoramidite chemistry. They are labeled by tailing the 3' end *(Protocol 10)* using a nonradioactive nucleotide. For ultrastructural studies nonradioactive markers are used for preference, e.g., biotin-, fluorescein-, or digoxigenin-labeled dUTP. The length of the carbon chain (6 to 21 atoms) does not affect labeling or detection methods.

As an example, to detect GH mRNA or prolactin (PRL) mRNA in the cells of the pituitary gland, where these hormones are synthesized, we used complementary probes from a sequence absent in the mRNA of the other hormone.[8]

No pretreatment was carried out before hybridization on ultrathin frozen sections using oligonucleotides probes. The buffer used contained:

- 30% Deionized formamide
- 20X SSC
- 50X Denhardt's solution
- 250 μg/ml tRNA

The oligonucleotide probe was added to the hybridization buffer at a concentration of 3 to 5 pmol/ml of hybridization buffer for a radioactive labeled probe, or 5 to 30 pmol/ml of hybridization buffer for a nonradioactive labeled probe. For the radioactive labeled probe, dithiotreitol was added to the hybridization buffer at a final concentration of 10 mM. Hybridization incubation was performed between 20° and 40°C, depending on the oligonucleotides used, in a humid chamber containing SSC buffer (5X concentration) for 3 h. The grids were washed at room temperature in SSC buffer as follows:

- 4X SSC for 5 min
- 2X SSC for 3 to 6× 10 min
- 1X SSC for 2× 5 min

The grids were then postfixed with 2% glutaraldehyde or 4% paraformaldehyde in a 0.1 M sodium phosphate buffer (pH 7.4) for 5 min, and washed in the same buffer. For nonradioactive labeled probes, the grids remained in 0.1 M sodium phosphate buffer (pH 7.4) for the immunocytochemical detection of the label.

D. HYBRIDIZATION ON LOWICRYL-EMBEDDED TISSUE SECTIONS
1. cDNA Probe
The two cDNA probes are prepared using inserts purified by gene cleaning and labeled with digoxigenin. The first one corresponds to the mouse type I receptor of interleukin-1 and the second to type II. Type I receptor was cloned from lymphocytes T and type II from β cells. The length of the inserts, respectively, are 535 and 700 bp.

No pretreatment was carried out before hybridization on Lowicryl-embedded tissue sections using cDNA probes. The buffer used was:

- 30% Deionized formamide
- 4X SSC
- 1X Denhardt's solution
- 250 μg/ml tRNA

The cDNA probes were added to the hybridization buffer at the concentration of 1.5 μg/ml of hybridization buffer. Hybridization incubation was performed at room temperature in humid chamber containing SSC buffer (4X concentration) for 3 h. Grids were washed at room temperature in 2X SSC for 3× 10 min, then the grids were incubated in buffer A (100 mM mono-phosphate buffer, 650 mM NaCl, 1% ovalbumin, 0.01% Tween 20, pH 7.4) before immunocytochemical detection of label.

V. IMMUNOCYTOLOGICAL REVELATION

A. GENERAL CONSIDERATIONS

In electron microscopy, the revelation of hybrids labeled with a nucleotide conjugated with biotin, digoxigenin, or fluorescein, considered as antigen molecules, is done by immunocytology. Antigen localization is based on specific high-affinity molecule binding.

An immunocytological reaction may be applied in two ways: either the high-affinity binding molecule may be conjugated to the label and used to stain the antigen directly (the "direct" method), or a primary unconjugated antibody may be applied to the antigen, and the antibody thus bound then stained with a secondary antibody labeled against the primary antibody (the "indirect" method). The primary antibody can be a molecule with a high affinity for the antigen (e.g., streptavidin for biotin). The antibody can bind two antigens with a Fab fragment, while streptavidin can bind four molecules of biotin.[39]

The marker used must be electron dense. Colloidal gold is the most widely used marker, due to the fact that it permits multiple detections (the diameter of these particles is 5, 10, 15, or 20 nm). However, there are certain limitations to this approach: (1) poor penetration of the cryosections by the colloidal gold particles that are routinely used means that antigens below the cut surface are not detected, and (2) there may be dissociation of antibodies adsorbed to the colloidal gold. Recently 1.4-nm gold probes with covalently attached secondary antibodies, in conjunction with silver enhancement procedures, have been shown to be a useful reagent for pre-embedding ultrastructural immunocytochemistry.[40] At present, colloidal gold is normally conjugated to an antibody (primary or secondary), to protein A (used as the secondary antibody), or to streptavidin. It is possible to use an enzymatic marker (peroxidase or others), but these are less electron dense,[41,42] and the resolution is poor.

In other words, biotinylated probes can be detected indirectly with an anti-biotin serum and a second antibody anti-species to the first serum conjugated to peroxidase or to gold particles, or by a protein A complex. Avidin molecules coupled with peroxidase or gold particles have also been used to detect biotin-labeled DNA after hybridization with viral DNA[36] and cellular mRNA.[22] However, nonspecific adhesion of avidin to cells may limit the use of this method.[18] Thus it is preferable to use gold particles as the marker since they give higher cellular and subcellular resolution. This has been demonstrated by Webster et al.[22] on mRNA in rat Schwann cells hybridized with biotinylated cDNA probes and by Hutchison et al.[33] and Manuelidis et al.[34] on mouse satellite DNA hybridized with DNA- or RNA-biotinylated probes. Using a biotinylated-protein A-gold method on *Drosophila* follicle cells, Binder et al.[14] by extrapolation, arrived at a value of around one gold particle per ten rRNA molecules.

As a recently proposed alternative, fluorescein-labeled nucleotides can be used for direct as well as indirect *in situ* hybridization experiments.[43] Fluorescein-dUTP/UTP/ddUTP can be incorporated enzymatically into nucleic acids using standard techniques. With a direct label no immunocytochemical visualization procedure is necessary, and the background is low. The direct method may, however, be less sensitive than indirect methods, such as those using digoxigenin- and biotin-modified probes.

B. SINGLE DETECTIONS
1. The Direct Method

In this case, the reporter molecule is bound to the nucleic acid probe, and so the molecular hybrids between probe and target sequences can be visualized immediately after *in situ* hybridization. For such methods it is essential that the probe-reporter bond should survive the hybridization and washing operations. Most important, the presence of the reporter molecule should not interfere with the hybridization reaction.

If the hybrids contain biotin, two methods are available: an antibody against biotin or streptavidin. If the antigen is digoxigenin or fluorescein, only antibodies can be used.

The fact that this method utilizes only one antibody means that it can be completed quickly, and that nonspecific reactions are limited. However, since staining involves only one labeled antibody, little signal amplification is achieved. This method is now used only rarely.

PROTOCOL 16: IMMUNOCYTOLOGICAL DETECTION WITH LABELED ANTIBODY

1. *Prepare buffer A (100 mM mono-diphosphate buffer, 650 mM NaCl, 1% ovalbumin, 0.05% Tween 20; pH 7.4)*
2. *Wash the grids in buffer A, with gentle agitation, for 10 min*
3. *Incubate the grids on labeled antibody diluted 1/20 to 1/50 in buffer A, with gentle agitation, for 60 min*
4. *Wash the grids on buffer A, with gentle agitation, for 20 min*
5. *Stain the grids.*

PROTOCOL 17: IMMUNOCYTOLOGICAL DETECTION WITH LABELED STREPTAVIDIN

The same as Protocol 16 except that the third step is modified: the grids are incubated on labeled streptavidin diluted 1/20 to 1/30 in buffer with gentle agitation for 60 min.

2. The Indirect Method

For indirect procedures it is essential that the probe should contain an element, introduced chemically or enzymically, that renders it detectable by affinity cytochemistry; hence the term "indirect". Again, the presence of such an element should not seriously interfere with the hybridization reaction, or affect the stability of the resulting hybrid.

In these conditions, an unconjugated primary antibody binds to the antigen. A labeled secondary antibody directed against the primary antibody (now the antigen) is then introduced, followed, if necessary, by a substrate-chromogen solution. If the primary antibody is rabbit or mouse, the secondary antibody must be directed, respectively, against rabbit or mouse immunoglobulins.

This method is more versatile than the direct method, in that a variety of primary antibodies from a given species can be used with the same labeled secondary antibody. The procedure is also several times more sensitive than the direct method: unlabeled molecules (antibodies or streptavidin) bind better than labeled molecules. Other detection methods, e.g., using protein A, can be used with all labels (enzymes, colloidal gold particles, other electron-dense particles). Any of the amplification systems (PAP, biotinylated antibodies, labeled avidin-streptavidin complexes) can be used, but quantification is difficult.

PROTOCOL 18: IMMUNOCYTOLOGICAL DETECTION BY THE INDIRECT METHOD

1. *Prepare buffer A (see protocol 16) and buffer B (20 mM Tris-HCl buffer, 650 mM NaCl, 0.05% Tween 20, 1% ovalbumin; pH 7.6)*
2. *Wash the grids on a drop of buffer A, with gentle agitation, for 10 min*
3. *Incubate the grids on a drop of unlabeled antibody diluted 1/50 to 1/200 in buffer A, with gentle agitation, for 10 min*
4. *Wash the grids on a drop of buffer A, with gentle agitation, for 10 min*
5. *Wash the grids on a drop of buffer B, with gentle agitation, for 10 min*
6. *Incubate the grids on a drop of labeled antibody diluted 1/30 to 1/50 in buffer B, with gentle agitation, for 30 min*
7. *Wash the grids on a drop of buffer B, with gentle agitation, for 30 min*
8. *Stain the grids.*

C. MULTIPLE DETECTIONS

The possibility of labeling probes with two different antigens permits their revelation, either with specific antibodies (direct method) or by the combination of direct and indirect methods, or by indirect methods alone. The revelation is carried out by combining different types of markers, e.g., colloidal gold particles and enzyme, or more commonly, by using two different sizes of colloidal gold particles.[44] The size of the colloidal gold particles must be checked for their Gauss distribution, and the distinction between large and small particles must be clear. It has been reported that small particles increase the level of staining.[45]

1. Simultaneous Detection of Two Nucleic Acids

PROTOCOL 19: SIMULTANEOUS REVELATION OF TWO HYBRIDS LABELED WITH BIOTIN AND DIGOXIGENIN, RESPECTIVELY, BY DIRECT AND INDIRECT IMMUNOLOGICAL REACTIONS

The following labeled reagents are used:

1. *Streptavidin labeled with 5- and 15-nm colloidal gold particles for biotin detection, and unlabeled mouse anti-digoxigenin and anti-mouse serum labeled with 15- and 5-nm colloidal gold particles for digoxigenin detection.*
2. *Unlabeled rabbit anti-biotin and anti-rabbit serum labeled with 15- and 5-nm colloidal gold particles for biotin detection, and mouse anti-digoxigenin labeled with 5- and 15-nm colloidal gold particles for digoxigenin detection.*

The 5- and 15-nm colloidal gold particles were used with streptavidin and antidigoxigenin, respectively, and then inversely so as to avoid problems related to colloidal gold particle size.

1. *Prepare buffers A and B (see Protocols 16 and 18, respectively)*
2. *Wash the grids on a drop of buffer A, with gentle agitation, for 10 min*
3. *Incubate the grids on a drop of unlabeled serum anti-digoxigenin or anti-biotin diluted 1/50 to 1/200 in buffer A*
4. *Wash the grids on a drop of buffer A, with gentle agitation, for 20 min*
5. *Wash the grids on a drop of buffer B, with gentle agitation, for 10 min*
6. *Incubate the grids on a drop of mixture of labeled reagents (streptavidin diluted 1/30 and anti-mouse serum diluted 1/50, or anti-rabbit serum diluted 1/50 and anti-digoxigenin serum diluted 1/20 in buffer B), with gentle agitation, for 30 min*
7. *Wash the grids on a drop of buffer B, with gentle agitation, for 30 min*
8. *Stain the grids.*

2. Simultaneous Detection of Nucleic Acid and Antigen

A combination of the detection of a nucleic acid by *in situ* hybridization and of an antigen by immunocytology is used to characterize the product synthesized by the nucleic acid, or the cell which contains this nucleic acid.[46] This technique is similar to that used for the detection of two nucleic acids by immunocytology, in that the antigen can be detected by antibodies.[44]

If the hybrids are labeled with biotin, there are two possibilities: (1) streptavidin to detect biotin, or (2) an anti-biotin serum. If the hybrids are labeled with digoxigenin or fluorescein, they can be revealed by antibodies. The possibilities for simultaneous detection are the following:

- If the hybrids are labeled with biotin and can be detected by streptavidin or extrAvidin (Sigma), the antigen can be detected by any antibody in a direct immunocytological reaction (see *Protocol 20*).
- If the hybrids are labeled with biotin, digoxigenin, or fluorescein, and can be revealed by an antibody, the antigen can be detected by an antibody obtained in species other than that of the anti-hybrid label serum, in an indirect immunocytological reaction (see *Protocol 21*).

PROTOCOL 20: SIMULTANEOUS DETECTION OF A NUCLEIC ACID LABELED WITH BIOTIN AND AN ANTIGEN BY A DIRECT IMMUNOCYTOLOGICAL REACTION

The following labeled reagents are used:

- *streptavidin or extrAvidin labeled with 5- and 15-nm colloidal gold particles for biotin detection*
- *and an anti-antigen X labeled with 15- and 5-nm colloidal gold particles for antigen detection.*

The 5- and 15-nm colloidal gold particles are used with streptavidin and anti-antigen X, respectively, and then inversely, so as to avoid problems related to colloidal gold particle size.

1. *Prepare buffers A and B (see Protocols 16 and 18, respectively)*
2. *Wash the grids on a drop of buffer A, with gentle agitation, for 10 min*
3. *Wash the grids on a drop of buffer B, with gentle agitation, for 10 min*
4. *Incubate the grids on a drop of a mixture of labeled streptavidin diluted 1/50, and labeled anti-antigen X diluted 1/50 in buffer B, with gentle agitation, for 30 min*
5. *Wash the grids on a drop of buffer B, with gentle agitation, for 10 to 20 min*
6. *Stain the grids.*

PROTOCOL 21: SIMULTANEOUS DETECTION OF A NUCLEIC ACID AND AN ANTIGEN BY INDIRECT IMMUNOLOGICAL REACTION

The following reagents are used:

- *For hybrid detection: unlabeled mouse or goat anti-probe-label, and sheep anti-mouse labeled with 5- and 15-nm colloidal gold particles.*
- *For antigen detection: unlabeled rabbit (or species other than mouse or goat) anti-antigen serum, and anti-rabbit serum labeled with 15- and 5-nm colloidal gold particles.*

The 5- and 15-nm colloidal gold particles are used for hybrid and antigen binding, respectively, and then inversely, so as to avoid problems related to colloidal gold particle size.

1. *Prepare buffers A and B (see Protocols 16 and 18, respectively)*
2. *Wash the grids on a drop of buffer A, with gentle agitation, for 10 min*
3. *Incubate the grids on a drop of the first unlabeled serum (mouse anti-probe label + rabbit anti-antigen X) diluted 1/50 to 1/200 in buffer A*

4. *Wash the grids on a drop of buffer A, with gentle agitation, for 20 min*
5. *Wash the grids on a drop of buffer B, with gentle agitation, for 10 min*
6. *Incubate the grids on a drop of the mixture of labeled reagents anti-mouse serum, and anti-rabbit serum diluted 1/50 in buffer B, with gentle agitation, for 30 min*
7. *Wash the grids on a drop of buffer B, with gentle agitation, for 10 to 20 min*
8. *Stain the grids.*

D. STAINING

The object of this step is to increase the contrast of sections, as for embedding sections (see Chapter 12), but also to avoid the destruction of the structure and ultrastructure during drying. Stierhof et al.[47] and Keller et al.[48] have shown that the ultrastructure of frozen sections of fixed tissue is perfectly preserved after embedding in epoxy resin followed by classical staining.

It is necessary to define optimal conditions of embedding and staining in order to reveal ultrastructures in conditions similar to those of classical embedding methods. However, the main advantage of using freezing is that it preserves all the soluble substances, which are lost during classical dehydration, and embedding must be preserved during the staining step. There are protocols which can be used to preserve some *(Protocol 7)* or all *(Protocol 22)* of the soluble molecules during staining.[29,49,50-53]

The choice of staining method depends on the ultrastructural aspect desired. Embedding in resin shows up ultrastructures similar to those seen in classical electron microscopy observations (see *Protocol 24*), while methylcellulose gives good results with negative or positive-negative staining. In the field of *in situ* hybridization, the use of methylcellulose is the best way to preserve the mRNA present in a cytoplasmic matrix.

1. Ultrathin Frozen Sections

PROTOCOL 22: EMBEDDING IN METHYLCELLULOSE

- *Prepare 0.8% methylcellulose (25 centipoises) in distilled water*
- *Agitate by reversal for 1 month at 4°C*
- *The preparation can be stored indefinitely at –20°C.*

Method:

1. *Fix by contact with a drop of glutaraldehyde 1 to 2% for 10 min*
2. *Wash in phosphate buffer for 3 × 5 min*
3. *Wash in distilled water for 5 min*
4. *Stain on a drop of 4% neutralized uranyl acetate for 10 min (see Protocol 23)*
5. *Incubate sections on a drop of methylcellulose, with or without 10 to 20% of 4% neutralized uranyl acetate, placed on parafilm on ice for 10 min*
6. *Dry the grids until they turn a gold-purple color*
7. *Observe after 1 h.*

PROTOCOL 23: PREPARATION OF 4% NEUTRALIZED URANYL ACETATE

1. *Mix 4% uranyl acetate with 0.3 M oxalic acid (v/v)*
2. *Neutralize to pH 7.5 with ammonia.*

2. Lowicryl-Embedded Tissue Sections

PROTOCOL 24: STAINING IN 2% URANYL ACETATE

1. *Stain on a drop of 2% uranyl acetate (see Protocol 25) for 25 min*
2. *Wash in 4 to 5 drops of distilled water placed on parafilm for 5 min each*
3. *Wash with a jet distilled water pippe*
4. *Dry the grids*

PROTOCOL 25: PREPARATION OF 2% URANYL ACETATE

1. *Solubilize 2% uranyl acetate in water overnight*
2. *Filter on 0.22 μm size*

The protocols of the two methods (*in situ* hybridization on ultrathin frozen sections and on Lowicryl-embedded tissue sections) are summarized in the Figure 1.

VI. CHECKING PROCEDURE

All the elements of the procedure which might introduce false positives into the reaction signal need to be eliminated in order to increase the ratio of the signal/nonspecific signal, and to demonstrate the specificity of the reaction.

Background noise is due to nonspecific adsorption of the probe, and to the revelation procedure. False positives can be caused by the probe, the hybridization reaction, or the revelation procedure. The signal/nonspecific hybridization ratio can be maximized by decreasing the background noise and increasing the formation of specific hybrids. Several points should be checked to confirm the specific localization of the nucleic acid studied:

- The probe
- The hybridization reaction
- The revelation procedure.

The simultaneous detection of mRNA and the protein synthesized by the nucleic acid, using a combination of hybridization and immunocytology, and the correlation of this method with other techniques, constitute strong evidence for the presence of

| **FROZEN TISSUE** | **LOWICRYL - EMBEDDED TISSUE** |

TISSUE PREPARATION
- Fixation - *Protocol 1*
  - Tissue: by vascular perfusion or immersion
  - Cells: in suspension or *in situ* by immersion

CRYOPROTECTION	**EMBEDDING** - *Protocol 7*
• with hyperosmolar sucrose - *Protocol 2*	• dehydration
• with iso-osmolar sucrose - *Protocol 3*	• infiltration
FREEZING - *Protocol 4*	• polymerization
CRYOULTRAMICROTOMY - *Protocol 5*	**ULTRAMICROTOMY**

- Preparation of coated grids - *Protocol 6*

PROBE PREPARATION
- Probe: cDNA, RNA or oligonucleotide
- Label: non-radioactive
- Labeling of the probe - *Protocols 8-12*

PRETREAMENT (optional)
- Permeabilization - *Protocol 13*
- Treatment to prevent background staining - *Protocols 14, 15*

PREHYBRIDIZATION (optional)
Incubation with a prehybridization solution: hybridization buffer minus probe
performed at the same temperature as hybridization

DENATURATION OF PROBE AND TARGET
- Heating
- Simultaneous or separate denaturation of probe and target
  if double-stranded

HYBRIDIZATION
Determination of hybridization conditions:
- hybridization temperature, formamide concentration
- hybridization buffer (salt concentration, pH, ...)
- probe concentration

Components of the solution:

• *Oligonucleotidic probe:*	• *cDNA probe:*
- Formamide	- Formamide
- Sodium Saline Citrate (SSC)	- Sodium Saline Citrate (SSC)
- tRNA	- Heterologous nucleic acids: RNA, DNA
- Denhardt's solution	- Denhardt's solution
- Probe: $\cong$ 30 mers, 3-5 pmol/ml	- Probe: 1.5 µg/ml

POST-HYBRIDIZATION STEPS

| Stringent washes (SSC) | Washes (SSC) |

IMMUNOCYTOLOGICAL REVELATION
- Single detections - *Protocols 16-18*
- Multiple detections:
  - Two nucleic acids - *Protocol 19*
  - Nucleic acid and antigen - *Protocols 20, 21*

CONTRAST

| Staining in 2 % neutral uranyl acetate | Staining in 2 % uranyl acetate - *Protocols 24, 25* |
| Embedding in methylcellulose - *Protocols 22, 23* | |

OBSERVATION
- Controls
  - Nuclease treatment
  - Use of multiple probes
  - Use of heterologous nucleic acids
- Quantification

FIGURE 1 Summary of protocols for ultrastructural *in situ* hybridization methods on ultrathin frozen section and Lowicryl-embedded tissue section.

nucleic acid in tissue but not for subcellular localization of nucleic acid. Variations of this localization after treatment may argue in favor of its specificity.

A. THE PROBE: LABELING CHECK

Before use, the labeled probe must be checked to determine the density of the label introduced into the nucleic acid. For a radioactive labeled probe it is possible to determine specific activity. For nonradioactive labeled probes, the incorporation of biotin or digoxigenin is difficult to check, and requires a large amount of the labeled probe. The level of incorporation is compared to a standard by spotting on a nitrocellulose filter and visualization by immunohistochemical methods. Recently, it has become possibe to check the labeled probe using fluorescein-labeled nucleotides with UV illumination, after the deposition of serial dilutions on a filter.

B. CHECKING THE HYBRIDIZATION PROCEDURE

Nonspecific hybridization is due to nonspecific adsorbtion of the probe in tissue. This can be overcome by using different types of tissue, or by abolition of the hybridization signal.

1. Tissue

The tissue studied can be hybridized with a heterologous probe (sense riboprobe, cDNA, or oligonucleotide probe containing a percentage of G-C similar to the one currently used). The hybridization procedure should be carried out under identical conditions to those in which the test probes were used, checking for regions of the tissue which might exhibit unusually high nonspecific adsorbtion of nucleic acid.

Positive signals can be detected by carrying out identical hybridization procedures in tissues or cell lines known to contain high levels of the nucleic acid studied: this is the homogeneous positive tissue control.

In the same way, with heterogeneous tissue samples only a small number of the cells expressing the mRNA or DNA studied should be positive. The absence of a signal in other cells or structures is a strong indication of specificity: this is the heterogeneous tissue control.[16]

On the other hand, hybridization with the probe, performed in identical conditions to those in which a positive signal is obtained, should give no signal in tissue known not to contain the nucleic acid: this is the negative tissue control.

2. Abolition of Hybridization Signal

Specific hybridization can be abolished by:

- Competition between the labeled probe and an excess of the same probe, unlabeled (100 times more concentrated). A heterologous probe (in similar conditions) does not modify the signal.[9]
- Destruction of the target nucleic acid by enzymatic pretreatment (DNase or RNase).[25] This technique has frequently been used to demonstrate that the signal observed depends on the presence of RNA in the tissue, and is not due to hybridization to DNA or to fortuitous adsorption to unknown cellular constituents.

C. CHECKING THE REVELATION PROCEDURE

The immunocytological revelation method must not introduce a signal which could mask the specific hybridization signal. Either (1) hybridization is carried out without a probe in a hybridization buffer and under conditions identical to those used with a labeled probe, or (2) the revelation procedure alone is carried out.

The presence of nonspecific adsorption of a labeled probe in the tissue can be checked by examining heterologous tissue, in which only the cells expressing the target nucleic acid will give a positive result. The elimination of this nonspecific adsorption is obtained by increasing the duration and stringency of the washing steps.

Endogenous biotin activity is inhibited by competition with streptavidin. Sections are pre-incubated with streptavidin, then saturated with biotin.[54]

D. OTHER METHODS

Immunocytology or double detection techniques applied to adjacent tissue sections are used to detect the receptor protein. Hybridization on membranes also has to be considered. This technique has been central to recent advances in our understanding of the organization and expression of genetic material. Blotting is characterized by the immobilization of sample nucleic acids on a solid support, generally nylon or nitrocellulose membranes. The blotted nucleic acids are then used as targets in subsequent hybridization experiments. The main blotting procedures are Northern and Southern blotting, dot and slot blotting, and colony and plaque screening. Hybridization can also be used for the analysis of PCR-amplified material, cloned sequences, and microbial genomes.

VII. RESULTS

One logical extension of the capacity to localize a single antigen or group of similar antigens at the electron microscopic level is the development of a reliable procedure for the discrimination of two or more distinct, but coexisting or neighboring antigens.

The search for a reliable multiple immunostaining technique was made considerably easier by the introduction of colloidal gold.[55,56] Homogeneously sized gold particle populations are now available and a number of double immunolabeling procedures utilizing two different particle diameters have been described.[57-61] Colloidal gold particles produced in a range of useful sizes have found many applications when bound to secondary antibodies, monoclonal antibodies, and streptavidin, but also to Protein A and G, and lectins. These gold-conjugated probes provide a useful immunocytochemical marker, and the smallest gold particles give permanent labeling, fine definition, and good sensitivity. With gold probes of different sizes, double "on grid" marking, both direct and indirect, becomes possible. Moreover, double labeling of the same section is sometimes feasible. For example, it is essential when antigens are associated with organelles too small to be serially sectioned, or when close relationships between cells or between organelles must be studied in fine detail, or simply when study is based on the colocalization of an mRNA and its protein. Since, it is clear that the majority of, if not all, "regulatory" peptides are synthesized (as part of a precursor molecule), processed (cleaved, glycosylated, etc.), packaged, and stored in secretory granules within endocrine cells or nerve cell bodies. Release, as a result of specific stimulation, may be endocrine, paracrine, or as a neurotransmitter. However, it is equally clear (from biochemical, histochemical, and immunocytochemical studies) that not all the bioactive components potentially encoded within the precursor mRNA are ultimately stored, though they may be expressed in terminally extended (inactive) nonantigenic form. Electron immunocytochemical techniques using antisera recognizing only the products of precursor cleavage enable the demonstration of coexisting bioactive molecules.

For example, single stainings can be performed by *in situ* hybridization and revealed by direct and indirect techniques both on Lowicryl-embedded tissue (Figure 2) and on ultrathin frozen sections (Figures 3 and 4). In order to colocalize a mRNA and its protein, double stainings are carried out (Figure 5).

The sensitivity of this method makes possible the detection of weakly expressed mRNA.[44,46,62] For example, the mRNA encoding for the growth hormone receptor (Figure 3A) and gonadoliberin receptor (Figure 3B) are revealed in pituitary gland in the somatotrophs, gonadotrophs, and lactotrophs for the growth hormone receptor mRNA, and in gonadotrophs and lactotrophs for the gonadoliberin receptor mRNA.[46,62,63] Prolactin receptors mRNA have been detected in hepatocytes (Figure 4).[64]

The ultrastructural localization of receptor mRNA on Lowicryl-embedded tissue sections shows similar results using cDNA probes. For example, the mRNA coding for the receptor of interleukin-1 can be visualized in Leydig cells in the testis (Figure 2).

The double immunogold staining procedure can be used, for example, to investigate the localization of mRNA coding for Atrial Natriuretic Factor receptors in some cells of the anterior pituitary, and also the hormones specifically secreted by them (Figure 5A).[65] In same way, the presence of gonadoliberin lactotrophs must be demonstrated by using double staining in order to identify the anterior cell population which expressed this receptor mRNA (Figure 5B). But detection of gonadoliberin receptor mRNA in gonadotrophs (Figure 3A) are easily correlated with effects of this neuropeptide on gonadotropin secretions. One potential use of double labeling which has so far not been investigated in depth due to technical problems, is the combination of mRNA hybridization cytochemistry with ultrastructural immunocytochemistry. Basically, this involves the localization of mRNA using labeled DNA probes complementary to the mRNA under investigation (cDNA probes) or oligonucleotidic probes, followed by immunostaining of the peptide product, the synthesis of which is directed by the mRNA.

VIII. QUANTIFICATION

The quantitative analysis of gold particles has several applications:

- Technical controls:
    - To evaluate the specificity of labeling by comparing its intensity in different cell compartments, and after hybridization with unrelated probe or after RNase treatment.

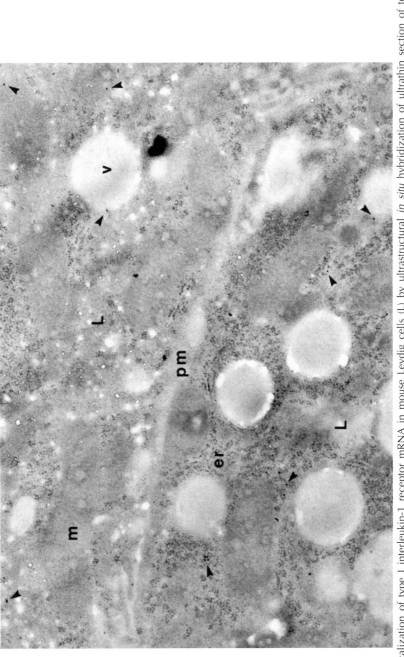

FIGURE 2 Localization of type I interleukin-1 receptor mRNA in mouse Leydig cells (L) by ultrastructural *in situ* hybridization of ultrathin section of testis embedded in Lowicryl K4M. Gold particles of 10 nm (arrowhead), which correspond to hybrids, are detected in cytoplasm near the endoplasmic reticulum (er); pm: plasma membrane with cytoplasmic digitation between Leydig cells; m: mitochondrion with tubular cristae; v: lipidic vesicles. Bar = 0.5 μm.

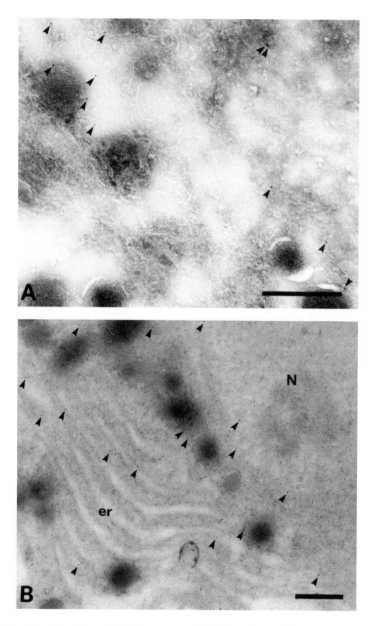

FIGURE 3   Detection of gonadoliberin receptor mRNA (A) and growth hormone receptor mRNA (B) on ultrathin frozen sections of anterior pituitary — 10 nm gold particles (arrowhead) which correspond to hybrids are localized in cytoplasm near the endoplasmic reticulum (er) and in nucleus (N). A: gonadoliberin receptor mRNA is revealed in gonadotrophs characterized by their vesicular endoplasmic reticulum. B: growth hormone receptor mRNA is detected in somatotrophs characterized by their lamellar endoplasmic reticulum. Bar = 0.5 μm.[46,63] (From Mertani, H.C. et al., *Endocrine*, 4(2), 159, 1996. With permission.)

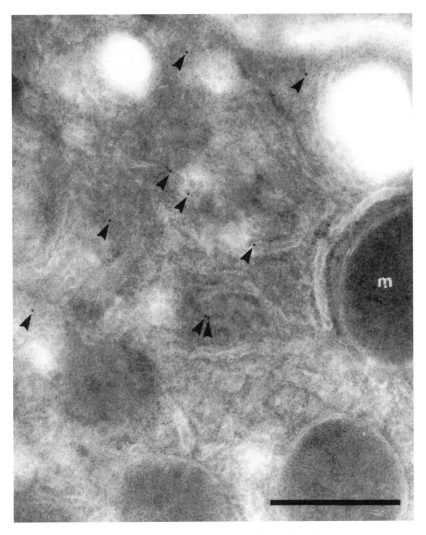

FIGURE 4   Prolactin receptor mRNA in hepatocyte. Gold particles of 10 nm (arrowhead) are visualized in cytoplasmic matrix near the endoplasmic reticulum on ultrathin frozen section of rat liver; m: mitochondrion. Bar = 0.5 μm.[64] (From Ouhtit, A. et al., *Biol. Cell*, 82, 169, 1994. With permission.)

  • To identify the cellular compartment which contains the nucleic acid being studied.

 • Biological analysis:
   • To compare the intensity of the hybridization signal in different physiological situations and variations of gene expression.

The primary purpose of quantification is to determine background, which can be done by comparing the signal over a section to the noise over the grid membrane.

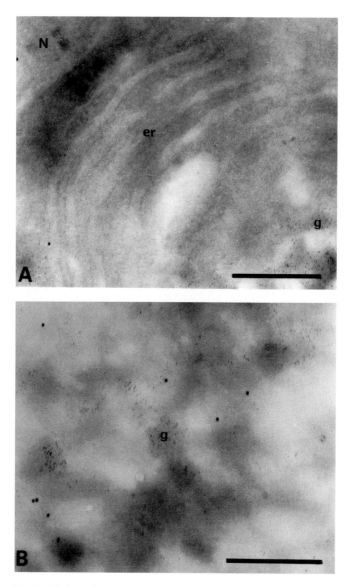

FIGURE 5   Double labeling of receptor mRNA and antigen on ultrathin frozen sections of anterior pituitary. A: Atrial natriuretic petide receptor type A mRNA (15-nm gold particles) are detected in lactotrophic cells characterized by their content of secretory granules (g) (prolactin-like immunoreactivity, 5 nm gold particle). B: Gonadoliberin receptor mRNA (15 nm gold particles) are detected in lactotrophic cells characterized by their content of secretory granules (prolactin-like immunoreactivity, 5 nm gold particle). N: nucleus; er: endoplasmic reticulum. Bar = 0.5 μm.[63,65] (5B from Mertani, H.C. et al., *Endocrine,* 4(2), 159, 1996. With permission.)

However, to determine the nonspecific binding of the probe, it is necessary to identify the cells where the nucleic acid studied is present, and to compare the signal over these cells to the signal over another cell population in the same section. For example, in the pituitary gland, the background for the *in situ* hybridization reaction used to detect GH receptor mRNA with a nonradioactive probe represents about 10% of the signal (Figure 6).

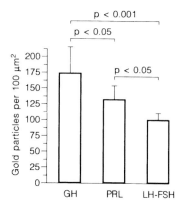

FIGURE 6 Quantification of hybridization signal of growth hormone receptor mRNA on pituitary cell type sections as expressed by mean ± SEM of gold particles per 100 μm². GH: somatotroph; PRL: lactotroph; LH-FSH: gonadotroph.[46] (From Mertani, H.C. et al., *Neuroendocrinology*, 59, 483, 1994. With permission.)

The quantification of *in situ* hybridization with a nonradioactive probe revealed by gold particles is easier since the resolution is better, which means that the percentage of gold particles can be obtained for each cellular compartment individually by dividing the direct count of gold particles by the total number of gold particles. The density of the gold particles can then be used for comparison and analysis. For example, the detection of the mRNA encoding for the GH receptor in different pituitary cell populations has shown that only three cell populations are labeled: somatotrophs, lactotrophs, and gonadotrophs (Figure 6).[46] As an example, the concentration of gold particles in each type of cell after *in situ* hybridization was determined for three animals. Each cell type (n = 60) was photographed in order to count the gold particles per unit area of cells sectioned through the nucleus, to achieve a random selection in a given grid square. All data were expressed as a mean ± SEM (Figure 6). The significance of the data was determined using Student's *t* test.

The density of labeling can also be measured by using an image analysis system, which measures the surface area of different cell compartments such as the mitochondria, nucleus, and cytoplasm. This value is determined by subtracting the nucleus and mitochondria surfaces from the total surface of the negative electron micrographs. In order to recognize the structures, micrographs are taken at a minimal magnification of ×11,500. The boundaries of the nucleus and mitochondria are drawn on the screen, and the measuring is done automatically. For each area, the number of gold particles is counted manually on the corresponding negative electron micrographs. The labeling density is expressed as the mean number of gold particles per square micrometer standard deviation of different samples, and statistical analysis is performed utilizing Student's *t* test.

In other respects, the quantification of the chemical revelation of nonradioactive probes, like that of enzymatic markers, remains difficult at the ultrastructural level.

Quantification and statistical analysis are the only ways to demonstrate nonconventional localization of nucleic acids.[16,46,62,66]

IX. CONCLUSION

The preservation of nucleic acids in frozen tissue is now well demonstrated, as are the advantages of the *in situ* hybridization technique. This makes it possible to localize nucleic acids by *in situ* hybridization on ultrathin frozen sections and on Lowicryl-embedded tissue sections (Table 1), and to use cDNA and oligonucleotidic probes labeled with nonradioactive nucleotides. The signal can be quantified in similar conditions to those used in other methods.

TABLE 1   Chief Characteristics of Ultrastructural Nonradioactive *In Situ* Hybridization Methods on Sections

Method	Frozen Sections	Postembedding
Storage	After freezing	After embedding
Pretreatment	No	No
Probe concentration	4 pmol/ml	100 pmol/ml
Detection system	Colloidal gold	Colloidal gold
Ultrastructural preservation	Low	Medium
Resolution	High	High
Sensitivity	High	Medium

There is no methodological difference between the use of a long probe (cDNA) and that of an oligonucleotidic probe. Labeling is denser with cDNA, but quantification is more difficult. In any case, it is always possible to amplify the signal by using a mixture of oligonucleotidic probes.

Concerning the choice of label, nonradioactive techniques seem to be the most advantageous. The loss of sensitivity is made up for by the short exposure period and the possibility of using radioautographic revelation.

The main disadvantage of this technique, as applied to ultrathin frozen sections, is the poor preservation of cellular ultrastructures. Its main advantage remains its sensitivity.[8] Other positive features are (1) the ability to store frozen specimens as well as ultrathin sections, (2) the rapidity of obtaining ultrathin sections, and thus the hybridization results, and (3) the possibility of simultaneous detections.

The use of *in situ* hybridization techniques at the electron microscopic level, in association with the functional techniques of molecular biology, should provide a useful tool for studying the localization of nucleic acids in cells. Such morphological information should make a useful contribution to the understanding of cell biology.

REFERENCES

1. Gall, J.G. and Pardue, M., Formation and detection of RNA-DNA hybrid molecules in cytological preparations, *Proc. Natl. Acad. Sci. U.S.A.*, 63, 378, 1969.
2. John, H.A., Burnstiel, M.L., and Jones, K.W., RNA-DNA hybrids at the cytological level, *Nature*, 223, 582, 1969.
3. Pardue, M.L., *In situ* hybridization, in *Nucleic Acid Hybridization*, Hames, B. D. and Higgins, S. J., Eds., IRL Press, Oxford, 1985, chap 8.

4. Jacob, J., Todd, K., Burnstiel, M.L., and Bird, A., Molecular hybridization of ^{3}H-labelled ribosomal RNA with DNA in ultrathin sections prepared from electron microscopy, *Biochem. Biophys. Acta*, 228, 761, 1971.

5. Beals, T.F., Ultrastructural *in situ* hybridization, *Ultrastruct. Pathol.*, 16, 87, 1992.

6. Stirling, J.W., Immuno- and affinity probes for electron microscopy: a review of labeling and preparation, *J. Histochem. Cytochem.*, 38, 145, 1990.

7. Van den Pol, A., Ellisman, M., and Deerinck, T., Plasma membrane localization of proteins with gold immunocytochemistry, in *Colloidal Gold: Principles, Methods and Application*, Hayat, T., Ed., Jovanovich, New York, 1989, 451.

8. Le Guellec, D., Trembleau, A., Pechoux, C., and Morel, G., Ultrastructural nonradioactive *in situ* hybridization of GH mRNA in rat pituitary gland: pre-embedding vs. ultra-thin frozen sections *vs* post-embedding, *J. Histochem. Cytochem.*, 40, 979, 1992.

9. Trembleau, A., Calas, A., and Fèvre-Montange, M., Ultrastructural localization of oxytocin mRNA in the rat hypothalamus by *in situ* hybridization using a synthetic oligonucleotide, *Mol. Brain Res.*, 8, 37, 1990.

10. Guitteny, A.F. and Bloch, B., Ultrastructural detection of the vasopressin messenger RNA in the normal and Brattleboro rat, *Histochemistry*, 92, 277, 1989.

11. Tong, Y., Zhao, H., Simard, J., Labrie, F., and Pelletier, G., Electron microscopic autoradiographic localization of prolactin mRNA in rat pituitary, *J. Histochem. Cytochem.*, 37, 567, 1989.

12. Le Guellec, D., Frappart, L., and Desprez, P.Y., Ultrastructural localization of mRNA encoding for EGF receptor in human breast cancer line BT20 by *in situ* hybridization, *J. Histochem. Cytochem.*, 39, 1, 1991.

13. Puvion-Dutilleul, F. and Puvion, E., Ultrastructural localization of viral RNA and DNA by *in situ* hybridization of biotinylated DNA probe on sections of Herpes simplex virus type 1 infected cells, *J. Electron Microsc. Tech.*, 18, 336, 1991.

14. Binder, M., Tourmente, S., Roth, J., Rebaud, M., and Gehring, W.J., *In situ* hybridization at the electron microscopic level: localization of transcripts on ultrathin sections of Lowicryl K4M-embedded tissue using biotinylated probes and protein A-gold complexes, *J. Cell Biol.*, 102, 1646, 1986.

15. Singer, R.H., Langevin, G.L., and Lawrence, J.B., Ultrastructural visualization of cytoskeletal in RNAs and their associated proteins using double-label *in situ* hybridization, *J. Cell Biol.*, 108, 2343, 1989.

16. Morel, G., Dihl, F., and Gossard, F., Ultrastructural distribution of GH mRNA and GH intron 1 sequences in rat pituitary gland: effects of GH-releasing factor and somatostatin, *Mol. Cell. Endocrinol.*, 65, 81, 1989.

17. Angerer, L.M., Stoler, M.H., and Angerer, R.C., *In situ* hybridization with RNA probes: an annotated recipe, in *In Situ Hybridization: Applications to Neurobiology*, Valentino, K. L., Eberwine, J. H., and Barchas, J. D., Eds., Oxford University Press, New York, 1987, chap. 3.

18. Singer, R.H., Lawrence, J.B., and Rashtchian, R.N., Toward a rapid *in situ* hybridization methodology using isotopic and nonisotopic probes, in *In Situ Hybridization: Applications to Neurobiology*, Valentino, K.L., Eberwine, J.H., and Barchas, J.D., Eds., Oxford University Press, New York, 1987, chap. 4.

19. Singer, R.H., Lawrence, J.B., Ultrastructural visualization of cytoskeletal messenger RNAs and their associated proteins using double label *in situ* hybridization, *J. Cell Biol.*, 108, 2343, 1989.

20. Steinert, G. and Thomas, C., and Brachet, J., Localization by *in situ* hybridization of amplified ribosomal DNA during *Xenopus laevis* oocyte maturation (a light and electron microscopic study), *Proc. Natl. Acad. Sci. U.S.A.*, 73, 833, 1976.

21. Morel, G., Dubois, P., and Gossard, F., Détection ultrastructurale des ARN messagers codant pour l'hormone de croissance dans l'antéhypophyse du rat par hybridation *in situ*, *C. R. Acad. Sci. Paris*, 302, 479, 1986.

22. Webster, H. de F., Lamperth, L., Favilla, J.T., Lemke, G., Tesin, D., and Manuelidis, L., Use of a biotinylated probe and *in situ* hybridization for light and electron microscopic localization of Po mRNA in myelin-forming Schwann cells, *Biochemistry*, 86, 441, 1987.

23. Beauvillain, J.C., Mitchell, V., Tramu, G., and Mazzuca, M., GABA axon terminals in synaptic contact with enkephalin neurons in hypothalamus of guinea-pig. Demonstration by double immunocytochemistry, *Brain Res.*, 443, 315, 1988.

Chapter **15**

USE OF RT *IN SITU* PCR FOR mRNA RECEPTOR VISUALIZATION

Brice Ronsin
Gérard Morel

CONTENTS

I. INTRODUCTION

In situ hybridization makes possible the specific localization of nucleic acid sequences in cells or tissues. However, this procedure has a limiting factor: *in situ* hybridization gives a signal only if the nucleic acid copy number is high enough. This is the case for certain genes belonging to viruses such as the human immuno-deficiency virus (HIV), the respiratory syncytial virus (RSV), the Epstein-Barr virus (EBV), the human papiloma virus (HPV), and the hepatitis virus, although some hormones and receptor mRNAs are not detectable by conventional *in situ* hybrid-ization. According to many researchers, the lower limit of detection is between 10 and 100 copies per cell. If the copy number is lower than the threshold of detection, another approach must be adopted.

Thanks to the new methods of molecular biology, and particularly to the variety of enzymes and primers, another technique, namely polymerase chain reaction (PCR) invented in 1985, has contributed considerably to the development of the field of virology, as well as human genetics and biology.[1] Since then, numerous applications have been developed along these lines. PCR techniques are extremely sensitive in that they amplify with specific primers. These results can be obtained after extraction of RNA or DNA, which involves the destruction of cells or tissues, so it is impossible to identify at a given moment the cells that express a gene in a given nucleic acid sequence.

In order to eliminate this threshold effect, which constitutes a limiting factor with *in situ* hybridization, a novel approach was developed for detecting low-expressed nucleic acids in tissues or cells by using an amplification method, namely the *in situ* polymerase chain reaction (*in situ* PCR) or the *in situ* reverse transcriptase polymerase chain reaction (RT *in situ* PCR). A combination of *in situ* PCR or RT-PCR amplification in cells or tissues is followed by a hybridization procedure in order to reveal amplified sequences. The principle is summarized in Figure 1. This technique can be used to amplify DNA (by *in situ* PCR) or RNA (by RT *in situ* PCR) using DNA polymerase and reverse transcriptase, specific primers, and deoxynucleotides triphosphates (dNTPs). The amplified products, or "amplimers" are easily detectable in tissue sections and cells due to their abundance. With indirect *in situ* PCR followed by *in situ* hybridization they are revealed directly in the tissues, whereas with direct *in situ* PCR the marker is incorporated in the amplificates during each elongation cycle of the PCR. However, it is then difficult to preserve the morphological structures, because the samples are subjected to numerous temperature changes and it therefore is important to maintain a compromise between amplification and the preservation of morphology.

II. PRINCIPLE OF RT *IN SITU* PCR AND *IN SITU* PCR

These reactions provide a method for amplifying a sequence from one copy to infinity. The PCR or RT-PCR methods used are a short DNA sequence as primer (oligonucleotides of between 20 to 25 mers), and a DNA replication enzyme, e.g., DNA polymerase (Taq polymerase, extracted from *Thermus aquaticus*, resistant to high temperatures),[1-2] and triphosphate deoxyribonucleotides (dNTPs): dATP, dCTP, dGTP, and dTTP. For RT-PCR an enzyme which transcribes RNA into cDNA is used in a first step, e.g., AMV or M-Mulv reverse transcriptase,

A. REVERSE TRANSCRIPTION

This principle is summarized in Figure 1. RNA is transcribed into the complementary DNA (cDNA), the RNA used as the matrix for the reverse transcription reaction. The hybridization or annealing of 3′ primer (complementary to the 3′ extremity of the RNA), followed by the polymerase effect of the reverse transcriptase, transforms the RNA into cDNA by the addition of dNTPs. In this way, reverse transcriptase prepares the PCR amplification reaction which works only with DNA. This step requires the pretreatment of samples. Generally, 1 μg of RNA, extracted from 10,000 cells, can reveal between 1 and 10 copies per cell.

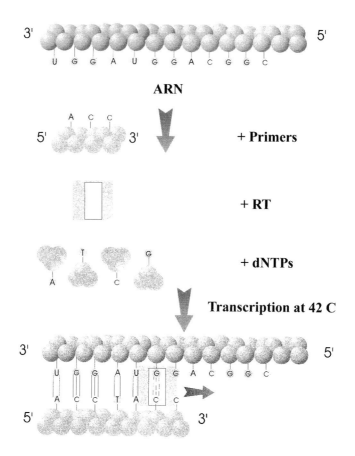

FIGURE 1   Principle of reverse transcription.

B.   POLYMERASE CHAIN REACTION

After the denaturation of the target sequence (generally by heating at 95°C for 5 min), the primers anneal or hybridize to the target sequence. The composition of each primer is determined as a complementary sequence of the 3′ extremities of each strand of the sequence of the matricial DNA. These primers induce the initial annealing. The hybridization temperature is defined by the melting temperature (Tm) determined by the sequences of the oligonucleotide used (in general between 37° and 60°C). This step is followed by the extension of primers in the 5′ → 3′ sense by the addition of deoxyribonucleotide residues of deoxynucleoside triphosphate to the 3′ hydroxyl extremity of the primers by DNA polymerase. Each base added is complementary to the base of the matrix DNA strand. The synthesis is carried out on the two strands of DNA between 65 and 75°C (Figure 2).

For each cycle, denaturation, hybridization, and polymerization are repeated 20 to 30 times so as to produce an exponential amplification of the target DNA. In the liquid phase of the PCR reaction, with double stranded DNA, the number of copies at the end of the reaction is theoretically 2^n (n: cycle number), i.e., between 2^{20} and

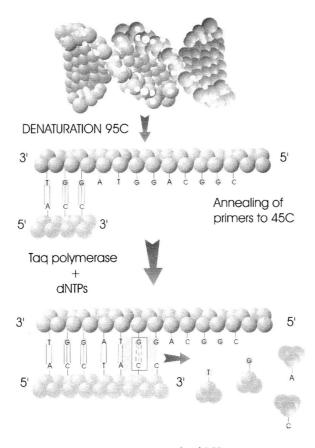

DENATURATION 95C

Annealing of
primers to 45C

Taq polymerase
+
dNTPs

FIGURE 2 Principle of PCR.

2^{30}; but in an *in situ* PCR the reaction is never optimal, and the number of copies is increased by about 20 to 50 times.

PCR allows the amplification of cDNA obtained from RNA, whether cellular mRNA or viral RNA, depending on the type of RNA. Different types of primers are used, e.g., oligo(dT), which can hybridize with the poly A end strand. Hexamers can be used to transform RNAs into cDNAs, and specific primers (A1) are used to transcribe the target RNA into cDNA.

The complementary DNA strand of the cDNA is synthesized during the first step of the PCR, with the second primer (A2) specific to the target sequence. The amplification of the double-stranded cDNA with the two primers can be carried out as previously described.

Depending on the enzyme used, the RT-PCR can be performed in one or two steps: in the one-step process *Thermus thermophilus* (rTth) DNA polymerase is used. This enzyme exibits intrinsic activity, which transforms RNA into cDNA in one step, in the presence of Mn^{2+} and at high temperatures. The polymerase activity can amplify the cDNA.[3] In a two-step process other enzymes, e.g., the Moloney Murine Leukemia Virus (M-MLV) or the the Avian Myeloblastosis Virus (AMV) reverse transcriptase are used in the first step, with DNA polymerase being used in the second step to amplify the cDNA. However, with rTth, a RT-PCR step is not always the best choice, particularly for the detection of weakly expressed RNA.[4]

III. TISSUE PREPARATION

All the different types of material used for *in situ* hybridization: tissue sections, biopsy tissue, or cells, either embedded in paraffin or frozen[2-7] cells fixed on solid support,[8-9] cells in culture,[10] or in suspension[2] can be used as samples for the *in situ* PCR.

For RT *in situ* PCR or *in situ* PCR, as well as for *in situ* hybridization, it is necessary to achieve a compromise betweeen the optimal preservation of structures and the conservation of nucleic acid sequences. The first step is the fixation of the samples, which is followed by an embedding or freezing procedure. The sections are then transferred to pretreated slides, prior to the *in situ* PCR amplification cycles.

A. FIXATION

Immersion or perfusion processes can be used in the *in situ* PCR without significant differences in labeling results. This is one of the most important steps, and it allows good preservation of the morphology as well as maximal amplification. There are two parameters which must be taken into consideration: the type of fixative and the fixation period.[11]

1. Type of Fixative

The fixative must conserve nucleic acids and increase the accessibility of the nucleic acid target to the various compounds of the PCR mixture. The precipitant fixative, e.g., ethanol, acetone, or an ethanol/acetone mixture preserves nucleic acid sequences (see Protocol 1).

PROTOCOL 1: PRECIPITANT FIXATION

1. Fix fresh tissue, biopsy tissue, or cells in an ethanol/acetone mixture 90:10 (v/v), for 20 min to 1 h.
2. Dip the samples in 1 to 4% buffered formaldehyde for 20 min.
3. Rinse two times in the same buffer.

The most common nonprecipitant fixatives are aldehydes, formaldehyde or paraformaldehyde (1 to 4%), added to a small amount (<0.5%) of glutaraldehyde. These are especially useful in RT *in situ* PCR as they limit the degradation of RNA, while avoiding the folding and favoring cDNA synthesis (see Protocol 2).

PROTOCOL 2: ALDEHYDE FIXATION

1. Fix fresh tissue, biopsy, or cells in paraformaldehyde 4% in a 0.1 M phosphate buffer (pH 7), with or without 0.2% glutaraldehyde, for 2 h to 12 h at 4°C.
2. Rinse in the same buffer for the same period.

The majority of researchers agree that a mixture of alcohols (ethanol and methanol) or alcohol/acetone and paraformaldehyde, between 1 and 4% in Sorensen buffer is the best fixative (see Appendix). By contrast, the Carnoy and Bouin fixatives (which contain picric acid) seem to be avoided in the *in situ* PCR reaction because they favor nicks in DNA and RNA.

2. Fixation Period

One of the main problems with *in situ* PCR is first to allow the PCR mixture to reach the target sequences (DNA or RNA), and to keep amplified products from diffusing out of the cells. Fixation time is thus fundamental: if it is too short (5 to 10 min) the structures will not be stabilized and the amplification products will diffuse out of the nuclear and cytoplasmic membranes, and finally out of the cell as such. In this way, they can also diffuse into adjacent negative cells. However, if the fixation step is too long (>12 h) the reticulation of structures by the fixative will prevent the penetration of the PCR mixture, and thus the *in situ* PCR reaction.

However, the fixation process cannot be dissociated from that of proteolytic digestion.[12] A long fixation period, followed by intense proteolytic treatment, may give the nucleic acid targets greater accessibility to the PCR mixture.

B. SECTIONS

In order to obtain sections of fixed or nonfixed materials, the tissue is frozen (see Protocol 3) or embedded in paraffin (see Protocol 4). Unfixed tissue, like cells in suspension or scraped from their support, can be centrifuged. All sections and cells are placed on pretreated slides (Protocol 5).

PROTOCOL 3: FREEZING TISSUE

1. Dip the samples in phosphate buffer (see appendix) for 2 × 15 min at 4°C.
2. Leave the cryoprotected samples in 20-30% saccharose in phosphate buffer at 4°C overnight.
3. Dip the samples in liquid nitrogen.
4. Cut the samples into 10-μm sections on a cryostat.
5. Place the sections on pretreated slides (see Protocol 5).

but the reduction in efficiency observed with glass slides may be due to mechanical factors, i.e., the possible adsorption of Taq polymerase on glass, or may be due to the low thermal conductivity of glass.

PROTOCOL 6: PREPARATION OF SILICONIZED COVERSLIPS

Products
10 N hydrochloric acid
Alcohol 95°
Sterile distilled water
Methanol
Silicone

Protocol

- Leave slides in methanol-alcohol (v/v) overnight
- Wash in running water 2 to 3 h
- Wash in distilled water 1 min
- Dip in methanol 30 s
- Air-dry the slides
- Dip in silicone 2 to 3 min
- Wash in running water and then distilled water 1 min
- Dip in methanol 30 s
- Air dry 1 h at 60°C
- Sterilize 2 h at 180°C
- Store at room temperature

2. Deoxynucleotides Triphosphate (dNTPs: dCTP, dGTP, dATP, dTTP)

These are generally used at a final concentration of 200 μM. A higher concentration could cause a diminution of specificity. Some PCRs can be done using dNTPs labeled with biotin, digoxigenin, fluorescein, or radioisotopes (e.g., ^{35}S or ^{33}P). In this case 40 to 60% of labeled nucleotide is mixed with the same unlabeled nucleotide to obtain a final concentration of 200 μM.

However, it seems that the use of biotin-labeled nucleotides in tissue containing endogenous biotin[25] gives positive results if endogenous biotin is not blocked by specific products. Incorporation of digoxigenin by the Taq polymerase during the extension step can cause a number of problems,[26] and false-positive results may be observed with digoxigenin-labeled nucleotides. One explanation for this may be that DNA fragmented in dying cells acts as a primer in a polymerization process involving an undesirable degree of incorporation of labeled nucleotides.

3. PCR Mixture

This contains Tris/HCl buffer with a basic pH, which is important for the hybridization process. KCl is essential for the activity of the enzyme. Concentration of $MgCl_2$ must always be determined and optimized. It important for the denaturation and hybridization steps, as well as for the specificity of the amplification and activity of the enzyme. A lower concentration of this salt can induce the formation of primer dimers.

4. Enzyme: Taq Polymerase

This is an enzyme of 60 to 70 kDa. Its specific activity lies between 2000 and 5000 U/mg of enzyme. It is generally used in *in situ* PCR at a concentration of 0.5 to 2.5 U/μl (10 times more than in liquid-phase PCR). As in Section IV.A.1, above, this may be due to the possibility of the enzyme being absorbed by glass.

Enzyme activity is temperature dependent. The rate of nucleotide incorporation, is therefore:

>60 nucleotides at 70°C
>24 nucleotides at 55°C
≈1.5 nucleotides at 37°C
≈0.25 nucleotides at 22°C

Many other enzymes have now become available, but their efficiencies must be tested individually for each type of tissue used. They are generally extracted from *Thermus aquaticus,* which resists high temperatures:

>2 h at 92.5°C
≈40 min at 95°C
≈5 to 6 min at 97.5°C

The enzyme may induce errors, but these are unimportant. Some researchers suggest that these amount to only about 17/6692 nucleotides incorporated after 30 cycles i.e., 1 error per 81,000 nucleotides, and its accuracy increases by the diminution of dNTPs and $MgCl_2$ concentrations.

5. Yield of the Reaction

The theoretical reaction yield in liquid PCR is 2^n (n being the cycle number), but in practice this yield reaches almost 70%. With *in situ* PCR, the practical limit to amplification seems to be of the order of 20 to 50.

B. RT *IN SITU* PCR
1. RT Primer

Various types of primers can be used:

- A poly-dT primer can hybridize with the poly-A tail of RNA,
- A hexamer allows a less specific amplification to turn all the RNA into cDNA, but can become a problem with the primers used in the PCR,
- The specific primer for the RNA studied transforms only this RNA into cDNA. The overall specificity of the RT-PCR is thus increased.

2. Enzyme

Reverse transcriptase was first isolated from retroviruses. Two forms of this enzyme are commercially available: the avian form, isolated from the Aviaire Myeloblastosis Virus (AMV), and another form, isolated from the Moloney Murine Leukemia Virus (M-MLV). These enzymes transcribe the RNA into cDNA, but they use only a single strand of DNA as a matrix, and synthesize the complementary strand DNA. The enzyme is generally used at a concentration of 0.5 to 2.5 U/μl of the PCR mixture.

C. REACTION MIXTURE
1. For the RT *In Situ* PCR
The mixture must contain all products required for the hybridization of the primer, i.e., the replication enzyme (RT) and the dNTPs. The specific primer is hybridized to the RNA sequence in the presence of a buffer including $MnCl_2$. This cation gives the RT its specific activity. It is generally used at a concentration of 3 to 5 mM. The incubation is done in a humid chamber at a temperature depending on the enzyme used.

2. For the *In Situ* PCR
The *in situ* PCR uses all the constituents used in primer hybridization (i.e., the polymerase enzyme (Taq) and dNTPs) and extension. The association of the primers with the target nucleic acid must be carried out at a neutral pH and in the presence of 3 to 5 mM of $MgCl_2$. This cation is often retained by cellular constituents.

To avoid nonspecific amplifications, some researchers prefer the hot-start technique, i.e., the introduction of Taq polymerase and primers to the PCR mixture at a temperature higher than the Tm. In some cases no real differences have been observed.

Generally, 20 to 30 μl of this mixture are placed on the sample and covered by a coverslip sealed with nail polish or rubber cement, to avoid any evaporation or drying.

D. MATERIALS
Many thermocyclers are now commercially available, e.g., the Appligene, Hybaid, MJ Research and Perkin-Elmer systems. All of them are similar in their mode of operation: the metal block is heated through electronic induction, and cooled by ventilation. The temperature scales are attained more efficiently with this type of thermocycler, which allows manipulation time to be reduced. Each particular model of thermocycler, however, has its specific advantages.

Appligene, with its Crocodil III, can treat four slides simultaneously; but to perform a liquid PCR, the metal block has to be changed. The advantage of this system is that only one thermocycler is needed to run the two types of PCR.

The Hybaid thermocycler system is modulated in such a way that it can treat four slides at a time, and due to its double satellite block, a liquid PCR can be carried out on eight slides simultaneously. A model capable of accepting 20 slides also exists (Figures 3 and 4).

The Perkin-Elmer thermocycler can manipulate ten slides simultaneously, and has its own anchor clips. Moreover, this system can treat three sections per slide with different mixtures (Figures 5 and 6).

The MJ Research thermocycler can treat 16 slides at the same time, and the metal block allows the isolation of the slides from each other (Figure 7).

E. AMPLIFICATION CYCLES

The slides are placed on the metal block of the thermocycler and put through differents cycles, where the temperatures are defined by the user according to the nature of the sample and the primer sequences.

The number of cycles can vary between 5 and 30. In certain cases the signal reaches a maximum after five cycles then decreases. In such cases, the lowest signal obtained after 10 to 15 cycles corresponds to the loss of amplified product. Generally an amplification cycle consist of the following steps (Figure 8).

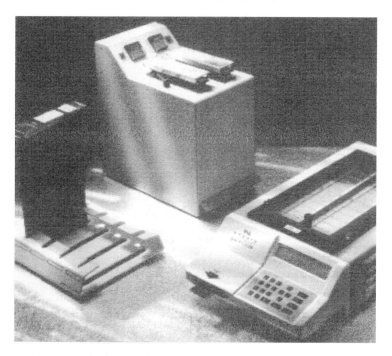

FIGURE 3   Hybaid system for *in situ* PCR: the 20-slide wash system version.

1. Denaturation

Here, the aim is to separate the two strands of DNA in order to allow the primers to hybridize. Generally, denaturation is obtained by heat. A compromise has to be reached between the temperature and the interval during which the tissue is maintained at this temperature: 93 to 94°C for 30 to 60 s or 97°C for 17 to 30 s.

2. Hybridization or Annealing

There are three important points: temperature, time, and specificity. The temperature-time compromise is less important in this case. The hybridization temperature always depends on the Tm but also varies with the nature, size, and concentration of the primers. Generally, the most suitable temperature is 5 to 10°C below the Tm. The annealing of primers occurs between 45 and 60°C.

The specificity of this step increases with the temperature of the annealing of primers, and decreases as the dNTP concentration rises. In these two examples, nonspecific incorporation is observed.

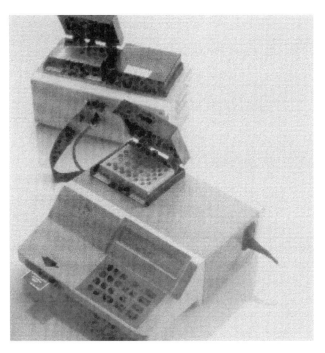

FIGURE 4   Hybaid system for liquid PCR with satellite blocks for *in situ* PCR: version for 2 × 4 slides.

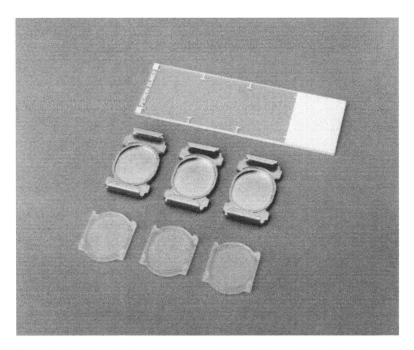

FIGURE 5   Anchor clip system from Perkin-Elmer. Three sections can be tested on the same slide. Photograph used with permission from The Perkin-Elmer Corporation.

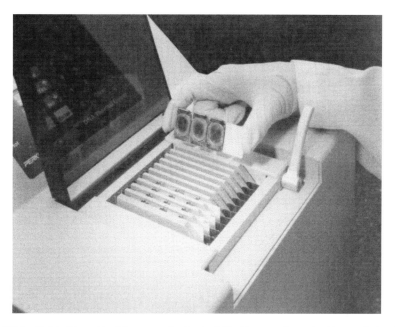

FIGURE 6 Perkin-Elmer 10-slide system for *in situ* PCR. Photograph used with permission from The Perkin-Elmer Corporation.

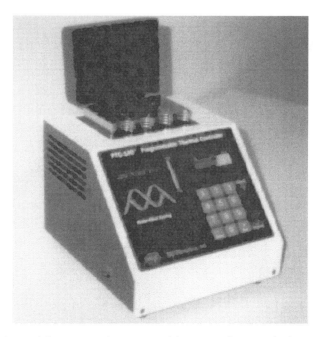

FIGURE 7 The 16-slide MJ Research system Model PTC 100 thermocycler for *in situ* PCR.

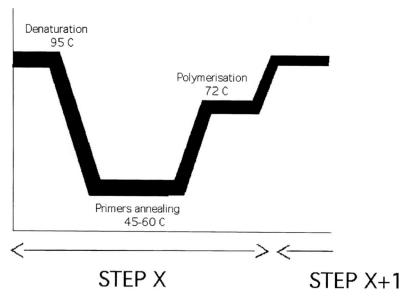

STEP X   STEP X+1

FIGURE 8   Steps for PCR.

3. Elongation

This step depends on the polymerase enzyme activity. It is generally very short, given that polymerase DNA can quickly incorporate a large number. Its optimal temperature is 72°C. The duration of this step depends on the sequence to be amplified, but does not exceed 1 min.

4. Number of Cycles

In liquid PCR, the number of copies increases exponentially with the number of cycles (theoretically 2^n, where n is the cycle number). For *in situ* PCR the increase in the yield of amplified products is arithmetical and must be checked.

V. PRETREATMENT OF SAMPLES

After the steps already described, the samples are prepared for *in situ* RT-PCR. This step is very important. The samples are placed on treated glass slides for the first reaction, in order is optimize the accessibility of the primers and probes to the target of RT *in situ* PCR and PCR (see Protocol 7).

PROTOCOL 7: PRETREATMENT OF SAMPLE

- Hydrated the specimens in 9‰ NaCl for 5 min
- Wash in 0.1 M phosphate buffer (pH 7) for 5 min

- Fix in 4% paraformaldehyde for 10 min
- Wash in 0.1 M phosphate buffer (pH 7), 2 × 3 min
- Wash in Tris-CaCl$_2$ buffer, 5 min at 37°C
- Pretreat with proteinase K 1 μg/ml in Tris-CaCl$_2$ for 15 min at 37°C
- Denature proteinase K, 2 min at 82°C
- Wash in Tris-CaCl$_2$ buffer for 5 min
- Wash in 0.1 M phosphate buffer (pH 7) for 5 min
- Fix in 4% paraformaldehyde for 5 min
- Wash in 0.1 M phosphate buffer (pH 7) for 2 min
- Wash in 9‰ NaCl for 2 min
- Dehydrate in alcohol 95°, 100°, 100° for 1 min each

After the pretreatment, slides carrying the samples are incubated with DNase (although some researchers hold that this step is not necessary) in order to avoid nonspecific amplification of DNA by Taq polymerase. The possible "repair" of endogenous DNA fragments by DNA polymerase, or a non-specific association of primers with other targets,[27] can only be limited by treatment with a DNase solution (see Protocol 8).[28-29]

PROTOCOL 8: DIGESTION OF DNA

- In a sterile tube place:

  - 10 μl of 10 U/μl DNase-Rnase free
  - 10 μl of DNase 10X buffer (see Appendix)
  - 1 μl of RNasin
  - 79 μl of Sterile distilled water

- Distribute 20 to 30 μl of this mixture on sample glass slides
- Outline the section with rubber cement
- Build an incubation chamber with a siliconized coverslip (see Protocol 6)
- Stabilize the coverslip with a second layer of rubber cement
- Incubate overnight (≈16 h) at 37 to 40°C.

This procedure may be necessary even if an RT *in situ* PCR was carried out. At this point two possibilities can be envisaged: direct or indirect RT *in situ* PCR or PCR (Figure 9).

VI. INDIRECT REACTION

The indirect reaction is then carried out. This method is time-consuming compared to the direct reaction, but limits the threshold due to the hybridization step which will reveal the amplified products.

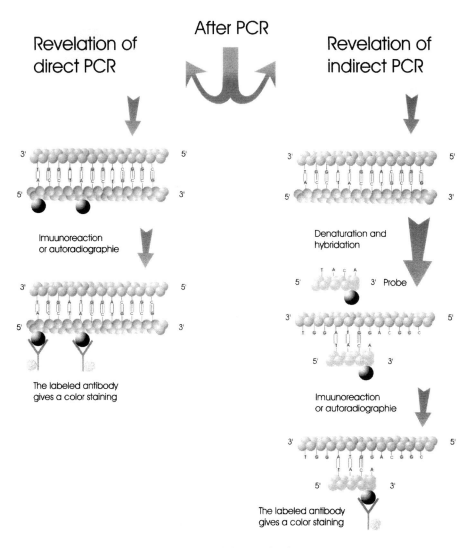

FIGURE 9   Revelation of direct and indirect *in situ* PCR.

A. RT PROCEDURE

The reverse transcriptase procedure requires a primer that anneals to the 3' end of RNA, the four dNTPs, and the reverse transcriptase enzyme (see Protocol 9). Numerous authors have described how the reverse transcriptase reaction is carried out.[5,7,18,23,29]

PROTOCOL 9: REVERSE TRANSCRIPTASE MIXTURE

In a sterile tube, mix:

- 5X RT buffer 20 μl
- 3′ primers ≈1 μM
- dNTPs 1 nmol/μl each
- Reverse transcriptase 1 U/μl
- DEPC-treated water to 100 μl

Method:

- Overlay the sample with 20 to 30 μl of this mixture
- Outline the sections with rubber cement
- Build an incubation chamber with siliconized coverslip (see Protocol 6)*
- Stabilize sections with a second layer of rubber cement or nail polish
- Incubate slides at 42°C for 1 h in a humid chamber
- Remove the coverslips
- Wash in 2X SSC for 10 min at room temperature**

B. PCR PROCEDURE

The sections are then dehydrated once in 95° ethanol and twice in 100° ethanol (2 min each), then air-dried. After this step an *in situ* PCR is carried out (see Protocol 10).

PROTOCOL 10: INDIRECT *IN SITU* POLYMERASE CHAIN REACTION

Prepare:
- 10X PCR buffer 10 μl
- 5 mM MgCl$_2$ 4 to 5 mM optional
- 3′ and 5′ primers ≈1μM
- dNTPs 200 μM each
- Taq polymerase 0.5 to 1 U/μl
- DEPC-treated water to 100 μl
- Distribute 20 to 30 μl of this mixture on samples
- Outline the sections with rubber cement
- Make an incubation chamber with siliconized coverslip (see Protocol 6)
- Stabilize section with a second layer of rubber cement or nail polish
- Put slides on the thermocycler.

VII. DIRECT REACTION

The direct reaction of the RT *in situ* PCR and PCR limit manipulation avoiding the hybridization step. During the amplification steps, the Taq polymerase uses one strand of DNA as a matrix, and with the presence of the primers it synthesizes the

---

* Another possibility would be to use the Perkin-Elmer or Hybaid anchor clips.
** This step is not necessary if rTth is used.

second strand by the incorporation of dNTPs. However, one of these dNTPs is labeled (antigenic or radioactive). Thus the new strands have incorporated the labeled nucleotide and these new strands are labeled. The revelation can be done directly without hybridization. After a classical RT procedure as described above (see Section VI.A) sections are dehydrated once in 95° ethanol and twice in 100° ethanol (2 min each), then air-dried. After this step the direct *in situ* PCR is carried out (see Protocol 11).

PROTOCOL 11: DIRECT *IN SITU* POLYMERASE CHAIN REACTION

Mix:

- 10X PCR buffer 10 μl
- 5mM MgCl$_2$ 4 to 5 mM optional
- 3′ and 5′ primers 1μM
- dNTPs (CTP-GTP-TTP) 200 μM each
- dATP and labeled dATP (1/1) 200 μM of mix
- Taq polymerase 0.5 to 1 U/μl
- DEPC-treated water to 100 μl

Method:

- Cover the samples with 20 to 30 μl of this mixture on sample
- Outline the sections with rubber cement
- Make an incubation chamber with siliconized coverslip (see Protocol 6)
- Stabilize section with a second layer of rubber cement or nail polish (Figure 10).
- Put slides on the thermocycler.

The samples are now ready for the PCR reaction and are placed on the thermocycler. At this point there are two possible ways of limiting the mispriming. When the PCR mixture has been placed over the samples, the primers may hybridize on a nonspecific target. In order to limit this phenomenon, it is necessary to perform a hot start[30-31] in this technique — at least one of the compounds is withheld until the reaction temperature is higher than the annealing temperature of the primers (50 to 80°C, depending on the composition of primers). When an *in situ* PCR is performed on glass slides, the PCR mixture containing all products except the Taq polymerase is placed over the sample. The mixture and sample are heated without coverslips up to the temperature of hybridization, then the Taq is added to the solution at the appropriate concentration. The slides are sealed with the coverslips or by using another system. The hot-start technique is used to increase the specificity and yield of the reaction.[32-34] At the same time, it greatly reduces the number of false positive results caused by mispriming (annealing of the primers to nonspecific sequences). If the start is carried out at room temperature, it increases oligomerization and primer dimerization (resulting in the partial hybridization of primers between each other).

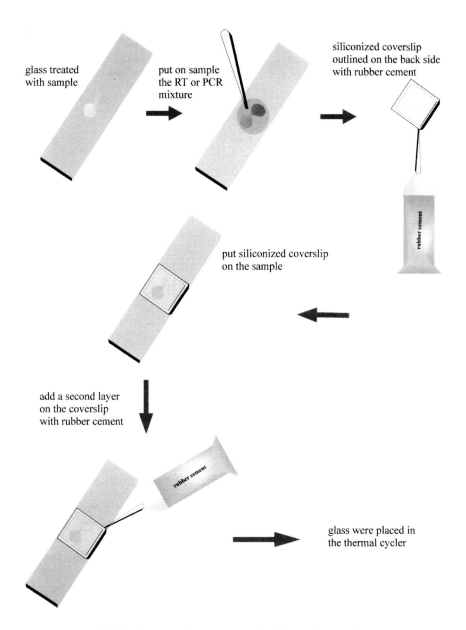

glass treated
with sample

put on sample
the RT or PCR
mixture

siliconized coverslip
outlined on the back side
with rubber cement

rubber cement

put siliconized coverslip
on the sample

add a second layer
on the coverslip
with rubber cement

rubber cement

glass were placed in
the thermal cycler

FIGURE 10   Incubation chamber with siliconized coverslips.

The hot-start technique is thus an important way of reducing nonspecific amplification. According to many researchers the same goal may be achieved by Ampli-
Wax (Perkin Elmer). It increases the sensitivity twice as much as is achieved with
a normal start.[35-36] Hot start can also be performed using Hybaid's coverslips.

If a hot start is not feasible in satisfactory conditions, another solution must be
considered. Indeed, the first assays of *in situ* PCR were made by prolonged denaturation of PCR mix: all the constituents are dropped over the sample and covered

with a coverslip, then put on the thermoblock. At this moment a prolonged denaturation is realized at 92 to 94°C during 3 to 5 min.[37-38] It seems to be a possible way to realize an *in situ* PCR. The second solution is to employ an anti-Taq antibody that hold the Taq polymerase activity until the first cycle at 92 to 94°C, where the antibody is degraded.[39-40]

VIII. CONTROLS

A number of controls are necessary in order to verify the specificity of the *in situ* PCR reaction. These are of two types: first, all the reagents involved in the reaction must be checked, so as to determine if the *in situ* PCR and RT-PCR are feasible; second, checks on the *in situ* PCR reaction itself to show the specificity of the amplification.

A. GENERAL CONTROLS
1. Primers

The size and purity of the primers are controlled by gel electrophoresis in the presence of ethidium bromide (Figure 11). Contamination of the primers by external sources of DNA and/or possible degradation of the nucleic acids that compose the primers must be verified. The primers are generally made in a 10X solution, and aliquoted to limit problems of contamination and degradation.

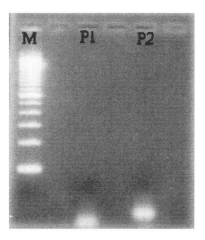

FIGURE 11   Control of single primer by electrophoresis stained with ethidium bromide. M: molecular weight marker (100 mer per band); P1: primer antisens specific of GHR mRNA (21 mer); P2: primer sens specific of GHR mRNA (24 mer).

2. Target Nucleic Acid Samples

With a liquid PCR or RT-PCR, a control is carried out in order to verify whether the amplification reaction is possible in the tissue or cells studied. It needs to extract DNA or RNA from the sample. Total RNA is purified by the guanidium isothiocyanate method.[41-43] Other techniques can only be used if a certain number of positive cells are present.[44] The DNA extraction method uses detergents and proteases to denature the proteins associated with the DNA.

3. Target Nucleic Acid in Histological Samples

In order to test possible degradation of the target in a histological pretreatment (e.g., paraffin embedding), a section must be treated as above in order to extract the nucleic acid. DNA or RNA fragments extracted from a paraffin block are generally no longer than 100 to 400 bases.[45-50] A number of researchers suggest that the guanidium isothiocyanate method should be used with RNA.[49,50] DNA needs another approach, e.g., exposure to microwaves[51] or digestion with proteinase K or a detergent.[52-54] A liquid PCR or RT-PCR is then carried out with the compounds as described for *in situ* PCR and RT-PCR. The amplified products are analyzed by gel electrophoresis. On the agarose gel a single band corresponding to the amplification products indicates that, after embedding or freezing, the DNA or RNA target has been conserved, so *in situ* PCR and RT-PCR are possible.

4. Diffusion

To begin with, the possibility for amplified products to diffuse out of cells must be looked at, since this would affect the results of an *in situ* PCR or RT-PCR. After the reaction, 5 to 10 µl of the PCR mixture are removed for agarose gel electrophoresis (Figure 12). If PCR products have diffused, they will be visible in the gel. This control indicates first that the reaction is functional, and then that the product is no longer present in the cells.[55]

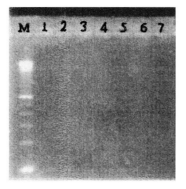

FIGURE 12 Control of diffusion products in the reaction (each band corresponds to 100 bp. Lane 1: after 20 cycles; Lane 2: after 15 cycles; Lane 3: after 10 cycles; Lane 4: after 5 cycles; Lane 5: without Taq; Lane 6: without primer; Lane 7: without reverse transcriptase

An excessively high number cycles can lead to the diffusion of the amplified products, or to an amplification outside of the cells, thus increasing the number of false-positive cells.

B. *IN SITU* PCR AND RT-PCR REACTION

A correct interpretation of *in situ* PCR or RT-PCR results is possible only if the necessary controls have been carried out. These controls must be performed at each step, i.e., during the reaction itself (in the case of direct and indirect reactions) or during the revelation process (in the case of indirect reactions).

1. The RT *In Situ* PCR Reaction

Two controls are necessary: (1) omission of the reverse transcriptase enzyme in the reverse transcription mixture, and (2) digestion of the samples by RNase. In both cases the production of cDNA is impossible. Moreover, some authors recommend that with an RT experiment a DNase digestion of the samples should be carried out, in order to limit DNA amplification during the PCR reaction.[56]

2. The *In Situ* PCR Reaction

The controls carried out in this case depend on whether the reaction is direct or indirect.

Controls for direct *in situ* PCR — Two controls for specificity must be carried out: (1) the omission of the primers, and (2) the omission of the DNA polymerase (Taq). In both cases, amplification is theoretically impossible. However, a control carried out without the primers sometimes gives a signal, due to repair of endogenous DNA by Taq. The enzyme has a polymerase activity in the $5' \rightarrow 3'$ sens, or an exonucleasec activity in the $3' \rightarrow 5'$ sens. The free nucleotides obtained in this way can also be consume a DNA repair reaction.[57]

Controls for indirect *in situ* PCR — The omission of the primers would seem to be the best control of this type, since it prevents any reaction from taking place. Moreover, no signal is detected when irrevelant probes are used to reveal amplification products of *in situ* PCR. To check for possible repair of endogenous DNA, the omission of nucleotides is not necessary. In fact, this control may introduce false-positive results. It must thus be avoided.

3. Controls of Revelation

For the indirect *in situ* PCR, the presence of amplified products is detected by hybridization with an appropriate probe. Thus, all the controls needed for *in situ* hybridization are also necessary for this step (see Chapters 13 and 14).

IX. RESULTS

The results presented here were all obtained by indirect RT *in situ* PCR, revealed by a cocktail of probes. Using digoxigenin as a marker, staining with alkaline phosphatase and using NBT-BCIP (a purple shade) indicates the presence of Growth Hormone Receptor (GHR) mRNA.This receptor has been found in many tissues, including the pituitary gland,[57] where GH is synthesized in the anterior part of the gland. At the ultrastructural level, GHR has been found in three separate cell populations: somatotrophs, lactotrophs, and gonadotrophs (see Chapters 13 and 14).[57] Using classical *in situ* hybridization techniques, a low level of staining can be detected in some cells (Figure 13E). In order to find out whether these negative cells result from negative staining by *in situ* hybridization (GHR expression being lower than the threshold of *in situ* hybridization detection) or from an absence of the mRNA coding for this receptor, we developed the RT *in situ* PCR method. The signal was then found to be higher than with the controls (Figure 14E vs. A to D), indicating that amplification had taken place. Some negative cells, probably thyrotropes and corticotropes, remained of the pituitary gland in the anterior lobe. These results show that after 20 amplification cycles the signal localization was similar for the reaction itself and for the controls (Figure 13), but its intensity was lower in the controls than amplification slides. This indicates that the effects of the growth hormone are initiated by its association with its specific receptor.[58,59] Some *in situ* hybridization and immunocytological studies illustrated the distribution of the growth hormone receptor in the anterior pituitary, and demonstrated that only three out of five cell subtypes contain growth hormone receptor mRNA.[57,58] These results are also found when amplification takes place, due to the fact that some cells are negatively stained.

In these controls, no signal was detected with reverse transcriptase after digestion by RNase and DNase (Figure 14A). The omission of reverse transcriptase gave a low level of staining in some cells (Figure 14B). The same results were found with the controls carried out on the PCR reaction omitting primers and Taq, i.e., a low degree of staining but nonetheless higher than with the omission of reverse transcriptase (Figure 14C and D). In the latter example, synthesized mRNA and cDNA were detectable; only a smaller increase in labeling is observed due to RT reaction. The highest signal obtained in these experiments occurred after 20 cycles. Amplification achieved after 5, 10, and 15 cycles (Figure 13B, C, D), showed a level of staining higher than the controls, but this was not 20 cycles.

A second control was carried out for the pitutiary, consisting of the differential expression of the growth hormone receptor between the three lobes. With or without amplification, the anterior lobe was labeled as has been found in other experiments.

X. CONCLUSIONS

In situ RT-PCR and PCR reactions prove to be procedures sensitive enough to detect a low amount of RNA and DNA in cells (or short turnover times). The future development of such approaches will probably yield further information, and also perspectives on the processes of gene distribution and expression. Cryosections present many advantages, e.g., nondegradation of nucleic acids during the freezing of the sample (unlike the embedded paraffin method), greater sensitivity, and the possibility of using less concentrated compounds. The use of RT *in situ* PCR with frozen sections increases the sensitivity of detection of mRNA targets. New techniques, e.g., thyramin staining amplification and the 3SR reaction, could in the near future open up further possibilities of gene localization, thus limiting the problems involved in the PCR. However, they will probably never provide better amplification than the *in situ* PCR and RT-PCR.

APPENDIX

The RT *in situ* PCR reaction needs a number of reagents for slide pretreatment or tissue preparation; these are

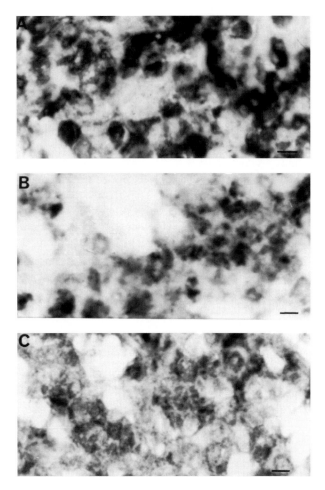

FIGURE 13 Amplification of growth hormone receptor mRNA in anterior pituitary using RT *in situ* PCR showing effect of the number of cycles. Signal was observed in some particular cell populations. A: 20 cycles; B: 15 cycles; C: 10 cycles; D: 5 cycles; E: Control — omission of the reverse transcription step — no amplification can be performed, signal is the result of native growth hormone receptor mRNA present in anterior pituitary. Bar = 5 μm.

1N Hydrochloric Acid:
   With the commercial solution 37% (10 N), diluted at 1/10

1 N NaOH:
   For 100 ml: 4 g NaOH
   Store at room temperature

Proteinase K-1000X:
   Aliquots of 100 μg/100 μl in buffer Tris-CaCl$_2$
   Store at –20°C.

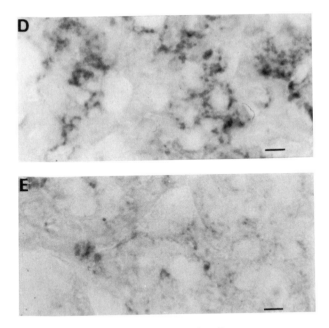

FIGURE 13 (continued)

Tampon Tris-CaCl$_2$ Buffer
   20 mM Tris/HCl, 2 mM CaCl$_2$
   Adjust the pH to 7.6 with HCl

Phosphate Buffer (pH 7.4)
   0.1 M Na$_2$HPO$_4$, 0.1 M NaH$_2$PO$_4$
   Store the solution at room temperature after sterilization

20X SSC Buffer
   3 M NaCl, 0.3 M trisodium citrate
   Adjust the pH to 7 with 1 N NaOH
   Store at room temperature after sterilization

5X Terminal Desoxyribonucleotide Transferase Buffer
   500 mM Potassium cacodylate pH 7.2
   Store at –20°C.

Denhardt's Solution 50X
   1% Bovine serum albumin (BSA)
   1% Polyvinylpyrrolidone (PVP)
   1% Ficoll 400
   Store at –20°C

RNase-free Buffer DNase
   100 mM Tris/HCl pH 7.6

45. Hodges, E. and Smith, J., Isolation of nucleic acid from paraffin embedded tissue for PCR amplification and sequencing of TcR Vbeta genes. *Leukemia Res.*, 19, 183, 1995.
46. Jiang, G., *In Situ* PCR — An overview, *In situ* Polymerase Chain Reaction and Related Technology. Eaton Publishing, Boston, 1-22, 1995.
47. Santa, G. and Schneider, C., RNA extracted from paraffin-embedded human tissues is amenable to analysis by PCR amplification. *Biotechniques*, 11, 3, 1991.
48. Jackson, D.P., Lewis, F.A., Taylor, G.R., Boylston, A.W., and Quirke, P., Tissue extraction of DNA and RNA analysis by the polymerase chain reaction. *J. Clin. Pathol.*, 40, 499, 1990.
49. Ogretmen, B. and Safa, A.R., Mini-preparation of total RNA for RT-PCR from cultured human cells. *Biotechniques*, 19, 374, 1995.
50. Groppe, J.C. and Morse, D.E., Isolation of full-length RNA templates for reverse transcription from tissues rich in Rnase and proteoglycans. *Anal. Biochem.*, 210, 337, 1993.
51. Banerjee, S.K., Makdisi, W.F., Weston, A.P., Mitchell, S.M., and Campbell, D.R., Microwave-based DNA extraction from parrafin-embedded tissue for PCR amplification. *Biotechniques*, 18(5), 768, 1995.
52. Weirich, G., Funk, A., Horpner, I., Heider, U., Noll, S., Pütz, B., Fellbaum, C., and Höfler H., PCR based assays for the detection of monoclonality in non-Hodgkin's lymphoma: application to formalin-fixed paraffin-embedded tissue and decalcified bone marrow samples. *J. Mol. Med.*, 73, 235, 1995.
53. Maher, M., Dowdall, D., Glennon, M., Walshe, S., Cormican, M., Wiesner, P., Gannon, F., and Smith, T., The sensitive detection of fluorescently labelled PCR products using an automated detection system. *Mol. Cell. Probes*, 9, 256, 1995.
54. Wright, D.T. and Manos, M.M., Sample preparation from paraffin-embedded tissues. In *PCR Protocols: A Guide to Methods and Application*. Innis, M.A., Gelfand, D.H., Sninsky, J.J., and White, T.J., Eds., Academic Press, New York, 1990, chap 19.
55. Komminoth, P. and Long, A., *In situ* polymerase chain reaction. An overview of methods, applications and limitations of a new molecular technique. *Virchows Arch. B Cell. Pathol.*, 64, 67, 1993.
56. Mertani, H.C. and Morel, G., *In situ* gene expression of growth hormone (GH) receptor and GH binding protein in adult male rat tissues. *Mol. Cell. Endocrinol.*, 109, 47, 1995.
57. Mertani, H.C., Water, M.J., Jambou, R., Gossard, F., and Morel, G., Growth hormone receptor binding protein in rat anterior pituitary. *Neuroendocrinology*, 59, 483, 1994.
58. Leung, D.W., Growth hormone receptor and serum binding protein: purification, cloning and expression. *Nature*, 330, 537, 1990.
59. Water, M.J., Growth hormone receptor: their structure, location and role. *Acta Paediatr. Scand.*, 336, 60, 1990.

Chapter **16**

X-RAY CRYSTAL STRUCTURE ANALYSIS
OF CYTOKINE RECEPTOR COMPLEXES

_____ Mark R. Walter

CONTENTS

0-8493-2644-3/97/$0.00+$.50

391

I. INTRODUCTION

Many vital cellular communications are transmitted to cells through the formation of cytokine-receptor complexes. Receptor signaling is initiated by dimerization of the receptor upon ligand binding.[1,2] The direct link between receptor binding and biological function has intensified the need for detailed three-dimensional structure information which may be used to understand the signal transduction process and guide the design of clinically useful drugs. The one method that can provide high-resolution three-dimensional structure information on large protein complexes is single crystal X-ray diffraction (Figure 1). The study of macromolecules by X-ray crystallography is a highly integrated and technology-driven science which requires the knowledge of biology, chemistry, physics, and computer science. Recent advances in molecular biology, synchrotron radiation sources, X-ray detector technology, and computer hardware have dramatically increased the speed and ease at which protein structures may be determined. As a result, more and more scientists with diverse backgrounds are beginning their own structural studies or using three-dimensional structure data. Although the focus of this chapter is the crystal structure analysis of cytokine-receptor complexes, once crystals are obtained, the methods outlined for crystal structure determination may be used to determine the structure of any macromolecule (protein, DNA, RNA, virus) or their complexes. The goal of this chapter is to provide a basic overview of the steps required to carry out an X-ray crystallographic experiment and discuss the major results generated for the cytokine-receptor complexes. The Appendix provides a reading list for further study into the subject matter.

II. PROTEIN CRYSTALLIZATION

A. SAMPLE PREPARATION

Obtaining X-ray diffraction quality crystals of a macromolecule is the most problematic step in the structure determination process. The most important variable in the crystallization experiment is the preparation of the molecule being crystallized. Based on results of numerous crystal growth experiments in our lab and the literature,[3] the following criteria must usually be met.

- High purity, >95%
- Soluble at high concentrations
- Conformational homogeneity
- Monodisperse molecular form

A complete structure determination may take anywhere from 5 to 100 mg of protein. The improvement of protein purification equipment and media along with high-yield expression systems have greatly simplified the process of obtaining large quantities of ultrapure protein. For the cytokine-receptor complexes, the individual purified components must be incubated together, followed by purification using size exclusion chromatography.[4-6] Once a purification scheme has been devised and crystals obtained, it is critical that the protocol be rigorously followed since even

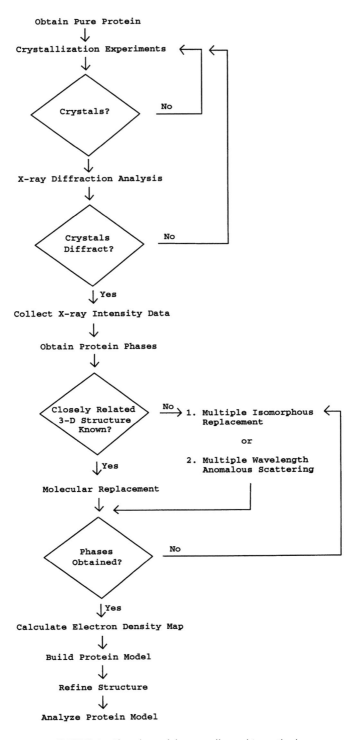

FIGURE 1 Flowchart of the crystallographic method.

small changes (e.g., different salts or buffers) in the sample may prevent crystal formation. In addition to being pure, the sample must be soluble at high concentrations (~4 to 40 mg/ml). Improving the solubility of proteins may be accomplished by simple changes in pH, buffers, and salts, or may require the alteration of the target protein solubility by site-directed mutagenesis.[7,8] Conformational heterogeneity may be an inherent property of the molecule understudy (e.g., multidomain proteins, flexible surface loops), or the result of improperly refolded molecules obtained from solubilized inclusion bodies. Often, flexible molecules or surface loops may be "locked" into one conformation by complex formation with other proteins or small molecules. Improperly folded molecules are often separated by an affinity purification step using an antibody which recognizes the properly folded "active" protein. A final criteria that has been shown to be extremely important for crystallization is the need for a monodisperse protein sample. A mono- or polydisperse protein sample, characterized by one or many aggregation states, may be evaluated by laser light scattering.[9,10] Several studies have now found that samples which exist in multiple aggregation states are not amenable to crystallization. This problem must be overcome empirically using the methods described for solubilizing proteins.

B. THE CRYSTALLIZATION EXPERIMENT

Since it is not possible to predict optimal conditions for protein crystal growth, they must be identified by experimentation. Most crystallization experiments are carried out by vapor diffusion equilibration using the hanging drop method (Figure 2) although several other methods do exist.[11] In the hanging drop procedure, 2 to 4 μl protein drops are suspended over a 1-ml well solution. The well solution is comprised of a precipitating agent that is buffered at a particular pH. Common precipitating agents include ammonium sulfate (30 to 80% saturated), polyethylene glycol 400 to 20,000 (10 to 50% w/v), phosphate (0.5 to 2.5 M), and 2-methyl-2,4-pentanediol (10 to 50%). Typical buffers include citrate, HEPES, MES, TRIS, and phosphate. The experiment may be set up using 24-well Linbro plates (Linbro model 76-033-05) designed for cell culture. The protein drop contains 1 to 2 μl of protein solution and 1 to 2 μl of well solution, which are mixed on a siliconized coverslip and inverted over the well. An airtight chamber is formed by placing a small bead of silicone grease around the circumference of the coverslip prior to inversion. The starting concentration of precipitating agent in the well solution is twice that in the protein drop. The concentration gradient causes the aqueous solution in the protein drop to be slowly transferred through the vapor phase to the more concentrated well solution until the two are in equilibrium. Crystal growth is induced as the protein is concentrated in the drop. Additional variables which affect crystallization are protein concentration, ionic strength, temperature, and other additives such as metal ions and detergents.[12,13] Although the large number of crystallization variables prevent an exhaustive search of all possible conditions, incomplete factorial screening provides a method to search a broad set of different crystallization conditions.[14] Additionally, sparse matrix screening methods provide a biased set of crystallization parameters based on conditions that have previously yielded crystals for other protein samples.[15] These sets of conditions (50 different experiments) are now commercially available (Hampton Research) and have been shown to be quite successful. While these broad screens may not yield usable crystals, they usually provide a starting place for further screening. Crystallization trays should be evaluated every other day using a stereoscopic microscope and the results of each experiment reported. Even if no crystals are obtained, much can be learned by the presence and amount of precipitated protein which can be used to guide further experiments. Experiments should not be discarded until the protein drops have dried up since crystals are sometimes observed only after several months. Once crystals are obtained they may be too small for X-ray diffraction analysis (normally crystals must be at least 100 μm on each edge) and must be optimized using various macro-seeding techniques.[16,17] The conditions for crystallization of the cytokines and cytokine-receptor complexes are shown in Tables 1 and 2.

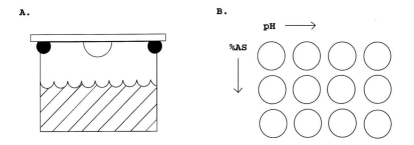

FIGURE 2 Schematic view of a hanging drop experiment (A) and variable screening as described in Section II.B (B). The protein drop is suspended over the well solution (cross-hatched area). The coverslip is sealed with silicone grease (black circles). The screening diagram represents a grid screen crystallization experiment with varying pH across the plate and increasing ammonium sulfate concentrations (%AS) from top to bottom.

C. PROTEIN CRYSTALS AND SYMMETRY

Crystals are by definition an ordered three-dimensional array of molecules; often referred to as the crystal lattice. The exact spatial relationship of each molecule in the array may be described by the crystal's unit cell parameters and spacegroup (Figure 3). The unit cell is defined as the smallest parallelepiped that may be translated in three dimensions to form the entire crystal array. The length of the edges of the unit cell are defined by the parameters a, b, and c (normally reported in Å, 10^{-10}m), while the angles between them are denoted α, β, and γ. Each unit cell is made up of one or more asymmetric units. The asymmetric unit defines the smallest unique volume of the crystal. The spacegroup defines the rotational and translational relationships (symmetry operators) between the asymmetric units of the unit cell. Thus, applying the spacegroup operators to one asymmetric unit will generate the complete unit cell which, when repeated along the three-dimensions specified by α, β, and γ, will form the entire crystal lattice. In practice, the protein content present in the crystal's asymmetric unit must be interpreted to complete a crystal structure analysis. The content of the asymmetric unit varies from crystal to crystal. It may contain one protein molecule, one subunit of a hexameric enzyme, a cytokine receptor complex, or an entire virus particle. Molecular symmetry contained within

TABLE 1   Crystal Structures of the Human Cytokine Receptor Family Ligands

Molecule	Res. (Å)	Unit Cell (Å)	Space Group	Chains in A.U.	R-factor	Crystallization Conditions	Ref.
GH[a]	2.0	79.6 × 58.7 × 48.3, β = 127.2	C2	1	18.5	10% PEG-8000, pH 6.5	Ultsch et. al.[63]
GM-CSF	2.4	47.6 × 59.1 × 126.7	$P2_12_12_1$	2	20.5	21% PEG-8000, pH 6.3	Diederichs et. al.[64]
GM-CSF	2.8	45.9 × 58.7 × 126.7	$P2_12_12_1$	2	25.0	1.8 M PO_4, pH 7.2	Walter et. al.[65]
G-CSF	2.2	91.2 × 110.3 × 49.5	$P2_12_12_1$	3	21.5	8% PEG-8000, pH 5.8	Hill et. al.[66]
M-CSF	2.5	33.5 × 65.3 × 159.6	$P2_12_12_1$	2	19.7	23% PEG-4000, pH 8.5	Pandit et. al.[67]
IL-2	2.5	55.8 × 40.1 × 33.7, α = 90, β = 109.3, γ = 93.2	P1	1	20.2	20% Me_2SO, pH 7.5	Bazan and McKay[68]
IL-4	2.4	91.8 × 46.2	$P4_12_2$	1	23.2	65% A.S., pH 5.8	Walter et. al.[69]
IL-4	2.3	91.8 × 46.4	$P4_12_2$	1	21.8	60% A.S., pH 6.0	Wlodawer et. al.[70]
IL-5	2.4	122.1 × 36.11 × 56.42, β = 98.6	C2	1	22.8	30% PEG-4000, pH 8.5	Milburn et. al.[71]
IL-10	2.0	36.6 × 221.0	$P4_32_2$	1	22.1	35% A.S., pH 7.2	Walter et. al.[72]
IL-10	1.8	70.3 × 70.3	$P3_221$	1	15.6	60% A.S., pH 6.5	Zdanov et. al.[73]
IFN-β	2.2	71.4 × 79.6	$P6_5$	1	19.1	20% A.S., pH 5.3	Senda et. al.[74]
IFN-γ	2.8	123.9, α = 54.8	R32	4	25.0	65% A.S., pH 5.9	Ealick et. al.[75]
LIF	2.0	31.1 × 56.2 × 95.3	$P2_12_12_1$	1	18.6	45% PEG-8000, pH 6.0	Robinson et. al.[76]
CNTF	2.4	62.1 × 208.9 × 53.0	$P2_12_12_1$	4	19.6	7.5% PEG-8000, pH 6.4	McDonald et. al.[77]

Note: A.S. = ammonium sulfate.

A.U. = asymmetric unit.

Res. = resolution.

[a] Human GH mutant

TABLE 2 Complex Crystal Structures of the Cytokine Receptor Superfamily

Ligand: Receptor	Resolution (Å)	Unit Cell (Å) Space Group	Stoichiometry	R-factor (%)	Crystallization Conditions	Ref.
GH:GHR	2.8	145.8 × 68.6 × 76.0 P2₁2₁2	1:2	20.4	40% A.S., pH 5.5	deVos et al.[57]
GH:PLR	2.9	154.2 × 69.8 × 43.4 P2₁2₁2₁	1:1	22.0	15% PEG-3400, pH 6.0	Somers et al.[61]
IFN-γ : IFN-γRα	2.9	145.9 × 180.3 P6₅22	1:2	22.3	1.6 M PO₄, pH 8.8	Walter et al.[58]

Note: A.S. = ammonium sulfate.

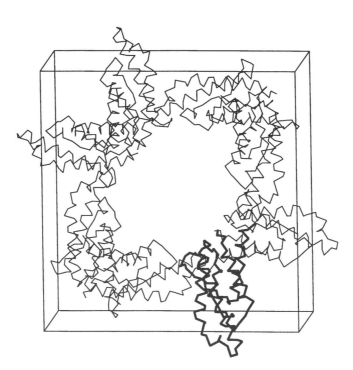

FIGURE 3 Symmetry related molecules and unit cell outline for crystals of IL-4.[69] The unit cell contains eight copies of IL-4 which are represented by α-carbon atoms. One molecule, representing the asymmetric unit, is represented by dark lines.

the asymmetric unit is termed noncrystallographic symmetry since it does not coincide with the symmetry operators of the crystal space group.

The solvent content of protein crystals ranges from about 30 to 90%, with the average being around 50%.[18] Furthermore, crystals are held together by a relatively few number of intermolecular interactions or crystal contacts. As a result, protein crystals are very sensitive to changes in pH, temperature, and ionic strength. For most crystals it is possible to identify a crystal-stabilizing solution which may be used for physical manipulations of the crystals or heavy-atom derivative soaks. The stabilizing solution usually consists of the final crystallization solution that contains a higher concentration of precipitating agent to prevent the crystal from dissolving. Crystals must be transferred into thin-walled glass capillaries or small loops for intensity data collection. Small loops are used when protein crystals are frozen in a stream of nitrogen gas prior to data collection.[19] Crystal freezing techniques are used to eliminate the X-ray-induced decay of the diffraction pattern. Freezing of protein crystals requires the addition of a cryoprotectant to the stabilizing solution. Several common cryoprotectants include ethylene glycol, glycerol, and different molecular weight PEGs. Transferring a crystal from the protein drop to a stabilizing solution may be done with a small needle or with a 0.5 mm glass capillary attached to a syringe.

III. X-RAY DIFFRACTION AND THE PHASE PROBLEM

X-rays, like other waves, may be described by their wavelength, amplitude, and relative phase angle. When passed through a crystal, X-rays interact with the electrons of atoms in the crystal to give rise to a diffraction pattern. Bragg in 1913 showed that the diffracted X-rays may be treated as if they were "reflected" from planes of atoms in the crystal.[20] Thus, a diffracted X-ray may be considered as being reflected from a particular crystal lattice plane defined by the indices *h*, *k*, and *l*. The intensity of each reflection is dependent on the positional coordinates (x,y,z) and atomic numbers of each atom in the crystal. While it is straightforward to measure the intensity of a particular *h,k,l* reflection (which is proportional to the square of the amplitude of the wave), the relative phase angle information is lost. Thus, the job of the crystallographer is to recover the lost phase information and subsequently build a three-dimensional model that represents the resultant electron density. Solving the "crystallographic phase problem" requires a number of mathematical manipulations of the intensity data which is facilitated by high-speed computers. A general outline to the steps required to complete a crystal structure determination are described below.

A.   X-RAY DIFFRACTION EXPERIMENT

The experimental apparatus used to collect X-ray-intensity data sets consists of three main components, an X-ray source, a crystal orienting device, and an X-ray detector, which are shown schematically in Figure 4. The crystal is positioned in the X-ray beam after being mounted in a thin wall glass capillary or small loop. Each *h, k, l* plane is brought into the "reflecting condition" by rotating the crystal through small oscillation ranges (~1.5°). To collect an entire data set, a series of consecutive data images are collected until all of the unique data have been obtained. The number of frames that must be collected is dependent on the symmetry of the crystal lattice.

B.   X-RAY SOURCES

Due to the poor scattering of X-rays by protein crystals, it is important to deliver the largest number of X-ray photons (source brilliance) possible onto the

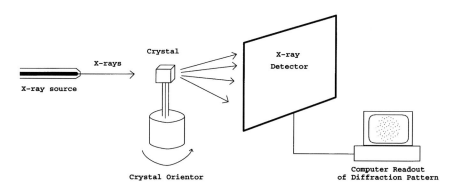

FIGURE 4 Schematic diagram of an X-ray diffraction experiment as described in Section III.A.

crystalline sample. In the laboratory, X-rays are produced by accelerating electrons at high voltages into a metal target. For most protein work, the target material is copper which produces a characteristic wavelength of 1.5418 Å. The intensity of the X-ray beam increases as the current and voltage are increased and is ultimately limited by the rate that heat can be removed from the target. Rotating anode generators are the most common X-ray sources found in crystallography labs. In the rotating anode, heat is removed by forcing cooled water through a disk-shaped copper anode which is spun at ~6000 rpm. The X-rays are directed through a collimator to reduce scatter and provide a X-ray beam with dimensions of approximately 0.3 mm^2. While X-rays produced by rotating anode generators are often adequate for crystal characterization and some data collection, synchrotron radiation provides 10^{3-6} times the brilliance of a laboratory X-ray source. The greater intensity of these sources may be used to obtain higher-resolution data for weakly diffracting crystals or provide a very small, yet intense, X-ray beam to resolve large unit cells. Since synchrotron radiation is comprised of a continuous spectrum of radiation, different wavelengths may be selected for decreasing crystal absorption or for optimizing anomalous scattering which may be directly used to solve the phase problem.

C. X-RAY DETECTORS

The intensities of X-rays diffracted by a crystal are measured by a position-sensitive X-ray detector.[21] Electronic area detectors and image plate detectors are most frequently used for this purpose. After the detector is exposed, it is immediately read out by a computer as a raw intensity data file (Figure 5). Both area detectors and image plates display fairly good X-ray detector characteristics which include high quantum efficiency (sensitivity to X-ray photons), high pixel resolution, low background, large dynamic range, linear and uniform response, and a fast readout time. Using these detectors, complete protein data sets are now collected in 1 to 3 days in the laboratory rather than 1 to 2 weeks using X-ray film or diffractometers. Recently, new detectors based on charged-coupled devices (CCDs) have been developed which are especially suited to rapid data collection (~2 to 6 h) at synchrotron radiation sources.[22]

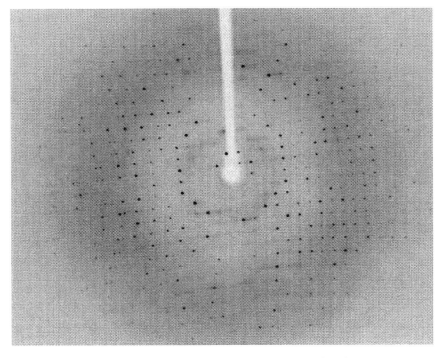

FIGURE 5   Diffraction image collected with an image plate detector.

D. DATA PROCESSING AND REDUCTION

Once raw diffraction images are obtained, the intensity of each reflection must be integrated and converted into a structure factor amplitude (F_{obs}) for subsequent analysis. This is a computer-intensive process which is carried out in several steps. These consist of indexing the diffraction pattern, predicting the position of each reflection on the diffraction image, integrating each reflection, merging and scaling the intensities, and reducing the data into an F_{obs} and an error estimate, σ. Assuming there are no special problems with the collected intensity data, sophisticated computer programs such as DENZO and SCALEPACK allow one to quickly complete these steps to obtain accurate reduced data.[23] Data sets are normally processed at the highest resolution for which diffraction spots are observed. The higher the resolution, the smaller the minimum distance which may be resolved between points in an electron density map. Thus, structures determined at higher resolution (e.g., 2 Å or less) should provide more structural details at higher accuracy than low resolution (~3 Å) structures. Several parameters are used to define the quality of a processed data set. The single most referred to number is the R_{sym} value

$$R_{sym} = \Sigma \mid I_{hkl} - <I_{hkl}> \mid / \Sigma \, I_{hkl}$$

which compares the intensity differences between symmetry-related reflections in an entire data set. Generally, datasets with R_{sym} values less than 10% are useable although R_{sym} values are dependent on data redundancy and resolution. Confidence in the F_{obs} and error estimates may be acquired if the data set contains multiple measurements of the each reflection (multiplicity >3). The multiplicity of a data set may be estimated by comparing the number of observed unique data to the total number of measurements collected. Another useful parameter is data completeness. This value is defined as the number of observed data over the number of possible unique reflections in the data set. The completeness of unique data should be greater than 80% for all resolution ranges.

E. OBTAINING PROTEIN PHASES

The last major hurdle in solving an unknown protein structure is to recover the phase information lost during intensity data collection. There are three main methods of determining protein phases: molecular replacement (MR), multiple isomorphous replacement (MIR), and multiple anomalous dispersion (MAD) techniques. The method of molecular replacement requires that the coordinates of a homologous protein structure are available. Using rotation and translation functions, the homologous protein structure is placed into the unknown protein's unit cell.[24,25] The phases of the known structure are taken to be a good approximation of the unknown structure's phases, which may subsequently be refined. Three particularly useful programs for molecular replacement are AMoRe, MERLOT, and X-PLOR.[26-28]

In contrast to MR methods, MIR and MAD techniques do not require any prior structural models to extract protein phases.[29-31] MIR methods depend on the specific binding of heavy-atom compounds (e.g., K_2PtCl_4 or $HgCl_2$) to proteins. Heavy-atom compounds are soaked for various amounts of time with the crystals, followed by collection of an intensity data set. Each F_{obs} of the native (F_{native})and heavy-atom data set (F_{deriv}) are compared for significant differences which suggest that heavy-atom binding has occurred. This difference in intensity between two data sets is expressed as the mean isomorphous difference (MID)

$$MID = \Sigma \mid F_{native} - F_{deriv} \mid / \Sigma F_{native}$$

which is normally larger than 10% if heavy-atom binding has occurred. Since not every heavy atom binds to every protein, systematic screening for heavy-atom binding must be carried out. Each compound screened (which could be more than 50) requires a new protein crystal for soaking followed by the collection of an intensity data set. The next step in the MIR method is to locate the x,y,z positions of the heavy atoms. Due to poor heavy-atom binding, multiple binding sites, changes in the unit cell dimensions, loss of diffraction due to metal binding, and errors in the data, locating the heavy-atom positions is often very difficult. The most common method of finding heavy-atom sites is using the Patterson function, although other methods have been used with some success.[32] Once heavy-atom sites have been identified, they are refined and subsequently used to calculate protein phases. To obtain useable protein phases normally requires the identification of at least two derivatives to solve the phase problem. Several computer packages, such as the CCP4 suite, have been developed to carry out the data manipulations described above.[33] The protein phases may be improved through a number of procedures which include solvent leveling, histogram matching, and noncrystallographic symmetry averaging.[34-36] The resultant phases may be used to calculate an electron density map.

A final method of protein phasing is the MAD phasing method, which is dependent on the presence of anomalous scattering atoms (such as selenium) in the protein molecule being studied. Anomalous scattering, which results in measurable intensity differences, occurs when the X-ray wavelength matches the absorption edge of the anomalous scatterer. MAD phasing requires the tuneability of synchrotron radiation to accurately optimize the anomalous scattering differences.[37] If an anomalous scatter is not present in the protein, it may be soaked into the crystal or engineered into the protein followed by crystallization. For proteins expressed in *E. coli*, selenium may be incorporated by replacing methionine residues with selenomethionine using a selenomethionine auxotroph.[38] This specific incorporation prevents the need to screen a large number of heavy-atom derivatives. In an analogy to MIR methods, MAD data are collected at several different wavelengths. One data set may be thought of as a native and the other data sets collected at different wavelengths as the derivatives. One advantage of this method is that all data may be collected on the same crystal, which eliminates problems with nonisomorphism. Because even the best anomalous scattering signal is small, great care must be taken to collect accurate intensity data sets. As with MIR methods, the phases may be improved and subsequently used to calculate an electron density map.

IV. CREATION AND REFINEMENT OF PROTEIN MODELS

A. BUILDING PROTEIN MODELS

Under ideal circumstances, building a protein model from its experimentally determined electron density map would be a trivial exercise most easily carried out automatically by a computer. Unfortunately, experimental protein phases usually contain large errors and due to nonisomorphism are rarely available for reflections beyond 2.8 Å. As a result, experimental density maps often contain large breaks as well as poorly defined or missing mainchain and sidechain density. For these reasons, interactive computer graphics programs have been developed that allow a crystallographer to "interpret" poor regions of electron density and subsequently build a suitable protein model. Protein models are usually built in several stages which consist of identification of secondary structural elements (α-helices, β-strands), connection of secondary structure features to obtain the protein connectivity, alignment of the amino acid sequence with the electron density, and building of the full atom protein model. The computer programs used in the laboratory for these tasks are O and Chain.[39,40] Each program makes extensive use of previously determined structural models to assist in the interpretation and refinement of protein structures. Examples of their incorporation into graphics programs include automatic loop and molecule fitting routines of O which are based on the most accurate structures in the protein data bank (PDB).[41,42] Additionally, problem sidechain residues may by fit using the most probable sidechain rotamers which have been tabulated from the PDB.[43] While these programs are very powerful and greatly speed up the model-building process, they are still limited by the quality of the electron density map being used for the model building. The minimal diffraction limit of a crystal needed to fully characterize a protein structure is around 3.5 Å resolution. Lower-resolution maps (5 to 6 Å) usually allow the definition of the protein solvent boundary and sometimes the location of α-helices which are observed as tubes of density. They cannot be used to place sidechain residues.

B. REFINEMENT OF PROTEIN MODELS

To obtain the most accurate protein structure possible, the model derived from the experimental density maps is refined against the experimental data. Refinement programs, such as X-PLOR, make adjustments in the positions of atoms in a protein model that simultaneously minimize the difference between the observed and calculated structure factor amplitudes.[44] The process is complicated by the small number of experimental observations (F_{obs}) to refineable parameters (x, y, z, B) which often leads to inaccurate results. To partially correct for this problem, stereochemically restrained refinement is used. Stereochemical restraints are derived from the accurate bond distances, bond angles, torsion angles, planar groups, and chiral volumes of high-resolution structures of amino acids and peptides. Although simulated annealing techniques have improved the radius of convergence for protein structure refinement, errors in the structure usually must be manually removed.[44] Thus, refinement is a cyclical process which requires a significant amount of analysis of the resulting protein model and electron density map using a computer graphics workstation. The refinement is said to "converge" when a low R-factor is obtained while maintaining good stereochemical properties. The R-factor

$$\text{R-factor} = \Sigma \mid F_{obs} - F_{calc} \mid / \Sigma \, F_{obs}$$

represents the agreement of the observed and calculated structure factor amplitudes. A low R-factor, somewhere around 20%, is one criterion of a correct structure. Cross-validation of the protein structure using the free R-factor described by Brunger has also proven to be a good test for identifying errors in protein models.[45] Other characteristics of an accurate protein model are good agreement with ideal stereochemistry, good mainchain geometry as calculated in a Ramachandran plot, agreement with the database of better-refined proteins (e.g., preferred sidechain conformations, planar peptide bonds), and a good fit of the density to the model (Figure 6).[46,47]

V. CRYSTALLOGRAPHIC RESULTS

A. HELICAL CYTOKINE FOLD

To date, the crystal structures of 12 unique cytokine molecules have been completed (Table 1). The predominant structural feature of the cytokines is an antiparallel four-helix bundle. Each helix in the bundle is labeled A to D from the N- to C-terminus (Figures 7 and 8). The four helices are connected by two long overhand loops (AB-loop, and CD-loop) and one short segment (BC-loop) to form a distinct up-up-down-down topology first described for porcine growth hormone (GH).[48] A

(Ω~35°). The overhand loops contain two short anti-parallel β-strands and the AB-loop passes between the CD-loop and helix D. Crystal structures exhibiting these features include GM-CSF, IL-4 (Figure 7), IL-2, M-CSF, and IL-5. The helices of the LC cytokines are significantly longer than the SC cytokines (~25 amino acids) and pack at angles nearly parallel to one another (Ω~18°). The length of the peptide chain for the LC cytokines range from 160 to 200 amino acids. In contrast to the SC cytokines, the overhand loops of the LC cytokines are devoid of β-strands but contain various nonequivalent helical segments. Also unique to the LC cytokines is the location of the AB-loop, which encircles the exterior face of helix D. The LC cytokine fold is exhibited by the structures of GH (Figure 8), IFN-β, G-CSF, LIF, IL-10, and CNTF.

Although most of the cytokines exist as monomers, four of the molecules (IL-5, M-CSF, IL-10, and IFN-γ) form unique dimers. The M-CSF dimer is stabilized by a disulfide bond between Cys-31 from the twofold-related monomers. Additional molecular interactions are formed between the AB-loop and CD-loop regions of the molecule. For IL-5, IFN-γ, and IL-10, each twofold-related domain of the dimer is formed from the intercalation of two peptide chains (Figure 9). In IL-5, helix D from one peptide chain packs into a cleft formed from helices A, B, and C from the other chain. The same topology is observed for IFN-γ and IL-10 although each peptide chain contains two additional helices that are located in their overhand loop regions (Figure 9). It is interesting that the domain characteristics of IL-5 (SC topology), IL-10 (LC topology), and IFN-γ (SC/LC topology) are each distinct. The angles separating the twofold-related domains of IL-5, IFN-γ, and IL-10 are approximately 180°, 120°, and 90°, respectively. The different domain orientations likely play an important role in positioning the receptor subunits for subsequent signal transduction.

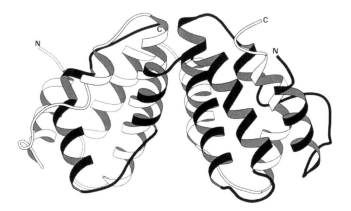

FIGURE 9   Molscript diagram of IFN-γ.[78]

B. CYTOKINE RECEPTOR SUPERFAMILY

The cytokine receptor family includes the receptors for growth hormone (GHR), prolactin (PLR), interleukins (IL-xR) -2, -3, -4, -5, -6, -7, -9, -10, -11, -12, -13, granulocyte colony stimulating factor (G-CSFR), macrophage colony stimulating factor

(M-CSFR), granulocyte-macrophage colony stimulating factor (GM-CSFR), erythro-poietin (EPOR), interferon (IFNR) α, β, γ, ciliary neutrophic factor (CNTFR), leuke-mia inhibitory factor (LIFR), as well as the membrane-bound tether for coagulation protease factor VII and tissue factor (TF).[50-52] This list shows the diversity of the cytokine receptor family, which includes not only the hematopoietic cytokine receptors, but the endocrine and neuropoietic receptors as well. Each receptor contains an extra-cellular ligand binding domain, a single membrane spanning peptide, and a cytoplasmic domain. The common classification of these diverse receptors was based on primary sequence homology among the extracellular regions. In contrast, no sequence corre-lations have been observed for the cytoplasmic domains. Based on unique cysteine sequence patterns, the extracellular domains have been divided into either class 1 or class 2 receptors. The smaller group of class 2 receptors is made up of the receptors for the IFNs, IL-10, and TF.

C. CYTOKINE RECEPTORS

All crystallographic work completed on the cytokine receptors has been accom-plished with the extracellular domains of the receptors rather than the intact molecules. With the exception of TF,[53,54] all cytokine receptor structures (GHR, PLR, IFN-γRα) have been completed as part of a cytokine receptor complex (Table 2). The extracellular regions are comprised of two fibronectin type-III (FBN-III) domains which each contain about 90 amino acids.[55] The N-terminal domain is denoted D1 while the C-terminal domain, which is located near the cell membrane, is labeled D2. The FBN-III domain is a variant of the immunoglobulin constant domain which consists of two β-sheets, formed from β-strands A, B, and E, and the G, F, C, and C′ (Figure 10).[56] The two domains are connected by a short linker segment (~10 residues) which contains one turn of α-helix.

The structures of the class 1 (GHR, PLR) and class 2 receptors (TF, IFN-γRα) are easily distinguished by interdomain angles of 90° and 120°, respectively. Changes in the interdomain angles are accompanied by significant translations (~5 to 8 Å) of the domains with respect to one another. These domain shifts result in considerable differences in the domain interface region of the molecules. The amount of buried surface area in the domain interfaces (~1000 Å^2) of the receptors suggests that the domain orientations are "locked" into prescribed conformations which effect ligand binding. Specifically, the AB-loop of D1, which is involved in hormone binding in the GHR and PLR, is buried in the domain interface of the class 2 receptors. As a result, the AB-loop of the class 2 receptors is unavailable for ligand interactions. Additional differences between the receptor classes are found within each domain. Generally, the domains of the class 1 and class 2 receptors display a conserved β-sheet core with structurally distinct loops. Significant differences in loop size and structure are observed for the BC-loop of D2, which is 10 residues longer in TF and the IFN-γRα than in the GHR/PLR, and the AB-loop of D2 which is 7 residues shorter than in the GHR and PLR. As a result of the shorter AB-loop, D2 of IFN-γRα and TF are incapable of forming receptor-receptor interactions analogous to binding site 3 in the GH receptor complex. Another notable difference between the GH/PLR and IFN-γRα/TF are the interactions of the C′ β-strand in D1. In the GHR/PLR, the C′ β-strand forms hydro-gen bonds with β-strand C of one β-sheet as well as β-strand E from the other. In

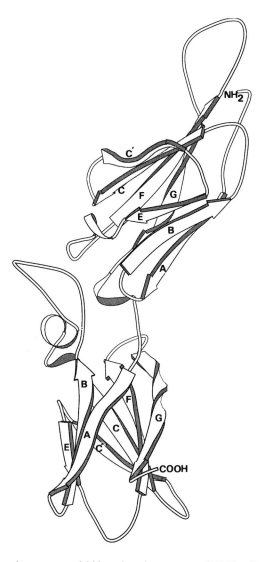

FIGURE 10   The cytokine receptor fold based on the structure of TF. The diagram was generated with Molscript.[78]

IFN-γRα/TF, β-strand C′ hydrogen bonds only to β-strand C. Conserved among class 1 and class 2 receptor structures is the presence of serine and threonine residues which form hydrogen bonds with the mainchain nitrogen atoms of the adjacent β-strands F and G.[57,58] The importance of this structural feature is currently unknown.

D. RECEPTOR LIGAND COMPLEXES
1. Growth Hormone and Prolactin Receptor Complexes

The crystal structure of the growth hormone receptor complex revealed two GH receptors bound to one molecule of GH (Figure 11).[57] Especially remarkable,

FIGURE 11 The growth hormone receptor complex. The diagram was generated with Molscript.[78]

the study showed that two identical receptor molecules using identical binding loops could interact with two chemically and structurally distinct sites on the hormone. The receptor binding site is formed from six loops located at the domain interface. Four of the loops are contributed from D1 (AB, CC', and EF loops, and the linker) and two (BC and FG loops) from D2. The receptor loops undergo small local conformational changes to adapt to the different hormone surfaces. The first hormone binding site is comprised of residues on helix A, D, and the AB-loop. Approximately 1300 Å^2 of molecular surface area is buried in this interaction site. The second receptor binding site (site 2) is formed from helices A and C which buries ~900 Å^2 of surface area. A third interaction site (site 3) is formed between the membrane proximal domains (AB-loop of D2) of the complex which buries an additional 500 Å^2 of surface area. The structural analysis of these binding sites supports a sequential mechanism for receptor dimerization where a second receptor may bind to GH only after the first receptor has bound.[59] Based on the complex, binding of the second receptor is thought to be stabilized by the receptor interactions formed at binding site 3. Studies in other systems suggest that this method of activation may be general to the entire cytokine-receptor family. Further mutagenesis work on the GH system has found that more than 85% of the total binding energy is contributed from only 8 of 30 residues in the interface.[60] Especially critical for high-affinity binding are Trp-104 and Trp-169, which bury the most solvent-accessible surface area in the complex. These two residues alone contribute nearly three fourths of the binding energy of the complex.

The structure of the 1:1 GH:PLR complex represents an inactive intermediate of the signaling process.[61] The domains of the PLR could be superimposed with an r.m.s. deviation of less than 1 Å onto the GHR from the GH:GHR complex. Although the individual domains are structurally equivalent, the relative orientation of the domains has undergone a significant shift (~5 Å translation, 8° rotation). Despite the conformational changes, the same receptor loops and many of the same residues used by the GHR are also used by the PLR to bind GH. In contrast to the local conformational changes of the binding loops used to interact with the two distinct surfaces of GH, the predominant structural changes in the PLR which lead to high-affinity binding are the rigid body movements of the domains.

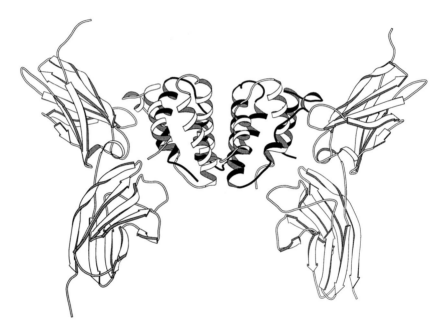

FIGURE 12   The IFN-γ receptor complex. The diagram was generated with Molscript.[78]

2. Interferon γ Receptor Complex

The crystal structure of the IFN-γ receptor complex provided the first view of a class 2 cytokine-receptor complex (Figure 12).[58] The active signaling complex of IFN-γ is comprised of a high-affinity receptor (IFN-γRα) which must bind to IFN-γ prior to the binding of a second receptor, the IFN-γRβ.[62] Like the GH:PLR complex, the IFN-γ receptor complex is an inactive intermediate complex consisting of one IFN-γ dimer and two IFN-γRαs. In the crystal, two IFN-γRα molecules bind the identical twofold-related surfaces of the IFN-γ dimer. The twofold axis of IFN-γ is oriented approximately perpendicular to the putative position of the cell membrane. In contrast to the GH receptor complex, the two IFN-γRαs do not interact with one another and are separated by 27 Å. The orientation of IFN-γ and GH on their receptors differs by approximately 60° with IFN-γ assuming the more upright position. The binding site on IFN-γ is comprised of helices A, the AB-loop, and helix F (equivalent to helix D in GH). Upon binding to its receptor, the flexible AB-loop of IFN-γ adopts a 3_{10} helical conformation. A total of 960 Å^2 of accessible surface is buried in the binding interface. As observed in the GH:GHR complex, the aromatic receptor residues Tyr-52, Trp-85, and Trp-210 contribute almost one third of the buried surface area. Based on the GHR mutagenesis work, these residues likely contribute significantly to the binding free energy.

ACKNOWLEDGMENTS

I would like to thank Dr. Ramaswamy Rahakrishnan and Alan Long for figure preparation, Maxine Rice for composing Tables 1 and 2 and the references, and Leigh J. Walter for reading the manuscript prior to submission.

APPENDIX

FURTHER READING

Blundell, T. L. and Johnson, L. N., *Protein Crystallography,* Academic Press, San Diego, 1976.

Colowick, S. P., Kaplan, N. O., Wyckoff, H. W., Hirs, C. H. W., and Timasheff, S. N., Eds., *Methods in Enzymology, Diffraction Methods for Biological Macromolecules,* Volume 115, Part B. Academic Press, Orlando, FL, 1985.

Colowick, S. P., Kaplan, N. O., Wyckoff, H. W., Hirs, C. H. W., and Timasheff, S. N., Eds., *Methods in Enzymology, Diffraction Methods for Biological Macromolecules,* Volume 114, Part A. Academic Press, Orlando, FL, 1985.

Helliwell, J. R., *Macromolecular Crystallography with Synchrotron Radiation,* Cambridge University Press, Cambridge, 1992.

McPherson, A., Jr., *The Preparation and Analysis of Protein Crystals.* John Wiley & Sons, New York, 1982.

McRee, D. E., *Practical Protein Crystallography,* Academic Press, San Diego, 1993.

Stout, G. H. and Jensen, L. H., *X-Ray Structure Determination: A Practical Guide,* 2nd ed., John Wiley & Sons, New York, 1989.

REFERENCES

1. Stahl, N., and Yancopoulos, G. D., The alphas, betas, and kinases of cytokine receptor complexes, *Cell,* 74, 587, 1993.

2. Heldin, C.-H., Dimerization of cell surface receptors in signal transduction, *Cell,* 80, 213, 1995.

3. McPherson, A., Malkin, A. J., and Kuznetsov, Y. G., The science of macromolecular crystallization, *Structure,* 3(8), 759, 1995.

4. Ultsch, M., de Vos, A. M., and Kossiakoff, A. A., Crystals of the complex between human growth hormone and the extracellular domain of its receptor, *J. Mol. Biol.,* 222, 865, 1991.

5. Ultsch, M., and de Vos, A.M., Crystals of human growth hormone-receptor complexes, *J. Mol. Biol.,* 231, 1133, 1993.

6. Windsor, W. T., Walter, L. J., Syto, R., Fossetta, J., Cook, W. J., Nagabhushan, T. L., and Walter, M. R., Purification and crystallization of a complex between human interferon-γ receptor (extracellular domain) and human interferon-γ, *PROTEINS: Struc. Func. Genet.,* 26, 108, 1996.

7. Dyda, F., Hickman, A. B., Jenkins, T. M., Engelman, A., Craigie, R., and Davies, D. R., Crystal structure of the catalytic domain of HIV-1 integrase: similarity to other polynucleotidyl transferases, *Science,* 266, 1981, 1994.

8. D'Arcy, A., Crystallizing proteins — a rational approach?, *Acta Crystallogr.,* D50, 469, 1994.
9. Veesler, S., Marcq, S., Lafont, S., Astier, J. P., and Boistelle, R., Influence of polydispersity on protein crystallization: a quasi-elastic light-scattering study applied to α-amylase, *Acta Crystallogr.,* D50, 355, 1994.
10. Baldwin, E. T., Crumley, K. W., and Carter, C. W., Jr., Practical, rapid screening of protein crystallization conditions by dynamic light scattering, *Biophys. J.,* 49, 47, 1986.
11. Ducruix, A., and Giegé, R., Methods of crystallization, in *Crystallization of Nucleic Acids and Proteins,* Ducruix, A., and Giegé, R., Eds., Oxford University Press, New York, 1992, chap. 4.
12. Trakhanov, S., and Quiocho, F. A., Influence of divalent cations in protein crystallization, *Protein Sci.,* 4, 1914, 1995.
13. Cudney, B., Patel, S., Weisgraber, K., Newhouse, Y., and McPherson, A., Screening and optimization strategies for macromolecular crystal growth, *Acta Crystallogr.,* D50, 414, 1994.
14. Carter, C. W., Jr., and Carter, C. W., Protein crystallization using incomplete factorial experiments, *J. Biol. Chem.,* 254, 12219, 1979.
15. Jancarik, J. J., and Kim, S.-H., Sparse matrix sampling: a screening method for crystallization of proteins, *J. Appl. Crystallogr.,* 24, 409, 1991.
16. Cook, W. J., Windsor, W. T., Murgolo, N. J., Tindall, S. H., Nagabhushan, T. L., and Walter, M. R., Crystallization and preliminary X-ray investigation of recombinant human interleukin 10, *PROTEINS: Struct., Funct. Genet.,* 22, 187, 1995.
17. Stura, E. A., and Wilson, I. A., Seeding techniques, in *Crystallization of Nucleic Acids and Proteins,* Ducruix, A., and Giegé, R., Eds., Oxford University Press, New York, 1992, chap. 5.
18. Matthews, B. W., Solvent content of protein crystals, *J. Mol. Biol.,* 33, 491, 1968.
19. Rodgers, D. W., Cryocrystallography, *Structure,* 2, 1135, 1994.
20. Bragg, W. L., The diffraction of short electromagnetic waves by a crystal, *Proc. Cambridge Philos. Soc.,* 17, 43, 1913.
21. Arndt, U. W., X-ray position-sensitive detectors, *J. Appl. Crystallogr.,* 19, 145, 1986.
22. Walter, R. L., Thiel, D. J., Barna, S. L., Tate, M. W., Wall, M. E., Eikenberry, E. F., Gruner, S. M., and Ealick, S. E., High-resolution macromolecular structure determination using CCD detectors and synchrotron radiation, *Structure,* 3, 835, 1995.
23. Otwinoski, Z., Oscillation data reduction program in *Data Collection and Processing,* Sawyer, L., Isaacs, N. and Bailey, S., Eds. SERC Daresbury Laboratory, Warrington, U.K., 1993, 56-62.
24. Rossmann, M. G., and Blow, D. M., The detection of sub-units within the crystallographic asymmetric unit, *Acta Crystallogr.,* 15, 24, 1962.
25. Tollin, P., and Rossmann, M. G., A description of various rotation function programs, *Acta Crystallogr.,* 21, 53, 1966.
26. Navaza, J., AMoRe: an automated package for molecular replacement, *Acta Crystallogr.,* A50, 157, 1994.
27. Fitzgerald, P. M. D., MERLOT, *J. Appl. Crystallogr.,* 21, 273, 1988.
28. Brünger, A. T., Extension of molecular replacement: a new search strategy based on Patterson correlation refinement, *Acta Crystallogr.,* A46, 46, 1990.
29. Crick, F. H. C., The theory of the method of isomorphous replacement for protein crystals. I, *Acta Crystallogr.,* 9, 901, 1956.
30. Harker, D., The determination of the phases of the structure factors of noncentrosymmetric crystals by the method of double isomorphous replacement, *Acta Crystallogr.,* 9, 1, 1956.
31. Herzenberg, A., and Lav, H. S. M., Anomolous scattering and the phase problem, *Acta Crystallogr.,* 22, 24, 1967.
32. Rossmann, M. G., Arnold, E., and Vriend, G., Comparison of vector search and feedback methods for finding heavy-atom sites in isomorphous derivatives, *Acta Crystallogr.,* A42, 325, 1986.
33. Anon., The CCP4 suite: programs for protein crystallography, *Acta Crystallogr.,* D50, 760, 1994.
34. Wang, B. C., Resolution of phase ambiguity in macromolecular crystallography, in *Methods in Enzymology,* Volume 115 B, Wyckoff, H.W., Hirs, C.H.W., and Timasheff, S.N., Eds., Academic Press, Orlando, FL, 1985, 90.
35. Zhang, K. Y. J., and Main, P., The use of sayre's equation with solvent flattening and histogram matching for phase extension and refinement of protein structures, *Acta Crystallogr.,* A46, 377, 1990.

36. Bricogne, G., Methods and programs for direct-space exploitation of geometric redundancies, *Acta Crystallogr.,* A32, 832, 1976.

37. Hendrickson, W. A., Determination of macromolecular structures from anomalous diffraction of synchrotron radiation, *Science,* 254, 51, 1991.

38. Hendrickson, W. A., Horton, J. R., and LeMaster, D. M., Selenomethionyl proteins produced for analysis by multiwavelength anomalous diffraction (MAD): a vehicle for direct determination of three-dimensional structure, *EMBO J.,* 9, 1665, 1990.

39. Jones, A. T., Bergdall, M., and Kjeldgaard, M., O: A Macromolecule Modeling Environment, in Crystallographic and Modeling Environment, in *Crystallographic and Modeling Methods in Molecular Design,* Bragg, C. E. and Ealick, S. E., Eds., Springer-Verlag, New York, 1990, pp. 189.

40. Sack, J. S., CHAIN — a crystallographic modeling program, *J. Mol. Graphics,* 6, 225, 1988.

41. Jones, T. A., Zou, J.-Y., and Cowan, S. W., Improved methods for building protein models in electron density maps and the location of errors in these models, *Acta Crystallogr.,* A47, 110, 1991.

42. Jones, T. A., and Thirup, S., Using known substructures in protein model building and crystallography, *EMBO J.,* 5, 819, 1986.

43. Ponder, J. W., and Richards, F. M., Tertiary templates for proteins. Use of packing criteria in the enumeration of allowed sequences for different structural classes, *J. Mol. Biol.,* 193, 775, 1987.

44. Brünger, A. T., Kuriyan, J., and Karplus, M., Crystallographic *R* factor refinement by molecular dynamics, *Science,* 235, 458, 1987.

45. Brünger, A. T., Free R value: a novel statistical quantity for assessing the accuracy of crystal structures, *Nature,* 355, 472, 1992.

46. Lakowski, R. J., Macarthur, M. W., Moss, D. S., and Thronton, J. M., Procheck: a program to check stereochemical quality of protein structures, *J. Appl. Crystallogr.,* 26, 283, 1993.

47. Kleywegt, G. J., and Jones, T. A., Where freedom is given, liberties are taken, *Structure,* 3, 535, 1995.

48. Abdel-Meguid, S. S., Shieh, H.-S., Smith, W. W., Dayringer, H. E., Violand, B. N., and Bentle, L. A., Three-dimensional structure of a genetically engineered variant of porcine growth hormone, *Proc. Natl. Acad. Sci. U.S.A.,* 84, 6434, 1987

49. Sprang, S. R., and Bazan, J. F., Cytokine structural taxonomy and mechanisms of receptor engagement, *Curr. Opinion Struct. Biol.,* 3, 815, 1993.

50. Bazan, J. F., and McKay, D. B., Structural design and molecular evolution of a cytokine receptor superfamily, *Proc. Natl. Acad. Sci. U.S.A.,* 87, 6934, 1990.

51. Metcalf, D., Hemopoietic regulators, *TIBS,* 17, 286, 1992.

52. Kossiakoff, A. A., Somers, W., Ultsch, M., Andow, K., Muller, Y. A., and De Vos, A. M., Comparison of the intermediate complexes of human growth hormone bound to the human growth hormone and prolactin receptors, *Protein Sci.,* 3, 1697, 1994.

53. Harlos, K., Martin, D. M. A., O'Brien, D. P., Jones, E. Y., Stuart, D. I., Polikarpov, I., Miller, A., Tuddenham, E. G. D., and Boys, C. W. G., Crystal structure of the extracellular region of human tissue factor, *Nature,* 370, 662, 1994.

54. Muller, Y. A., Ultsch, M. H., Kelley, R. F., and de Vos, A. M., Structure of the extracellular domain of human tissue factor: location of the factor VIIa binding site, *Biochemistry,* 33, 10864, 1994.

55. Leahy, D. J., Hendrickson, W. A., Aukhil, I., and Erickson, H. P., Structure of a fibronectin type III domain from tenascin phased by MAD analysis of the selenomethionyl protein, *Science,* 258, 987, 1992.

56. Jones, E. Y., The immunoglobulin superfamily, *Struct. Biol.,* 3, 846, 1993.

57. de Vos, A. M., Ultsch, M., Kossiakoff, A. A., Human growth hormone and extracellular domain of its receptor: crystal structure of the complex, *Science,* 255, 306, 1992.

58. Walter, M. R., Windsor, W. T., Nagabhushan, T. L., Lundell, D. J., Lunn, C. A., Zauodny, P. J., and Narula, S. K., Crystal structure of a complex between interferon-γ and its soluble high-affinity receptor, *Nature,* 376, 230, 1995.

59. Cunningham, B. C., Ultsch, M., de Vos, A. M., Mulkerrin, M. G., Clauser, K. R., and Wells, J. A., Dimerization of the extracellular domain of the human growth hormone receptor by a single hormone molecule, *Science,* 254, 821, 1991.

60. Clackson, T., and Wells, J. A., A hot spot in binding energy in a hormone-receptor interface, *Science,* 267, 383, 1995.

61. Somers, W., Ultsch, M., De Vos, A. M., and Kosslakoff, A. A., The X-ray structure of a growth hormone-prolactin receptor complex, *Nature,* 372, 478, 1994.

62. Soh, J., Donnelly, R. J., Kotenko, S., Mariano, T. M., Cook, J. R., Wang, N., Emanuel, S., Schwartz, B., Miki, T., and Pestka, S., Identification and sequence of an accessory factor required for activation of the human interferon γ receptor, *Cell,* 76, 793, 1994.

63. Ultsch, M. H., Somers, W., Kossiakoff, A. A., and de Vos, A. M., The crystal structure of affinity-matured human growth hormone at 2 Å resolution, *J. Mol. Biol.,* 236, 286, 1994.

64. Diederichs, K., Boone, T., and Karplus, P. A., Novel fold and putative receptor binding site of granulocyte-macrophage colony-stimulating factor, *Science,* 254, 1779, 1991.

65. Walter, M. R., Cook, W. J., Ealick, S. E., Nagabhushan, T. L., Trotta, P. P., and Bugg, C. E., Three-dimensional structure of recombinant human granulocyte-macrophage colony-stimulating factor, *J. Mol. Biol.,* 224, 1075, 1992.

66. Hill, C. P., Osslund, T. D., and Eisenberg, D., The structure of granulocyte-colony-stimulating factor and its relationship to other growth factors, *Proc. Natl. Acad. Sci. U.S.A.,* 90, 5167, 1993.

67. Pandit, J., Bohm, A., Jancarik, J., Halenbeck, R., Koths, K., and Kim, S.-H., Three-dimensional structure of dimeric human recombinant macrophage colony-stimulating factor, *Science,* 258, 1358, 1992.

68. Bazan, J. F. and McKay, D. B., Unraveling the structure of IL-2, *Science,* 257, 410, 1992.

69. Walter, M. R., Cook, W. J., Zhao, B. G., Cameron, R. P., Jr., Ealick, S. E., Walter, R. L., Jr., Reichert, P., Nagabhushan, T. L., Trotta, P. P., and Bugg, C. E., Crystal structure of recombinant human interleukin-4, *J. Biol. Chem.,* 267, 20371, 1992.

70. Wlodaver, A., Pavlovsky, A., and Gustchina, A., Crystal structure of human recombinant interleukin-4 at 2.25 Å resolution, *FEBS,* 309(1), 59, 1992.

71. Milburn, M. V., Hassell, A. M., Lamber, M. H., Jordan, S. R., Proudfoot, A. E. I., Graber, P., and Wells, T. N. C., A novel dimer configuration revealed by the crystal structure at 2.4 Å resolution of human interleukin-5, *Nature,* 363, 172, 1993.

72. Walter, M. R., and Nagabhushan, T. L., Crystal structure of interleukin 10 reveals an interferon γ-like fold, *Biochemistry,* 34, 12118, 1995.

73. Zdanov, A., Schalk-Hihi, C., Gustchina, A., Tsang, M., Weatherbee, J., and Wlodawer, A., Crystal structure of interleukin-10 reveals the functional dimer with an unexpected topological similarity to interferon γ, *Structure,* 3(6), 591, 1995.

74. Senda, T., Saitoh, S., and Mitsui, Y., Refined crystal structure of recombinant murine interferon-β at 2.15 Å resolution, *J. Mol. Biol.,* 253, 187, 1995.

75. Ealick, S. E., Cook, W. J., Vijay-Kumar, S., Carson, M., Nagabhushan, T. L., Trotta, P. P., and Bugg, C. E., Three-dimensional structure of recombinant human interferon-γ, *Science,* 252, 698, 1991.

76. Robinson, R. C., Grey, L. M., Staunton, D., Vankelecom, H., Vernallis, A. B., Moreau, J.-F., Staurt, D. I., Heath, J. K., and Jones, E. Y., The crystal structure and biological function of leukemia inhibitory factor: implications for receptor binding, *Cell,* 77, 1101, 1994.

77. McDonald, N. Q., Panayotatos, N., and Hendrickson, W. A., Crystal structure of dimeric human ciliary neurotrophic factor determined by MAD phasing, *EMBO J.,* 14, 2689, 1995.

78. Kraulis, P.J., MOLSCRIPT: a program to produce both detailed and schematic plots of protein structures, *J. Appl. Crystallogr.,* 24, 946, 1991.

INDEX